# Taxonomy of Fungi Imperfecti

Edited by BRYCE KENDRICK

Mycologists have been searching for a better system of classification of Fungi Imperfecti than that based on mature morphology. This volume documents an intensive phase of that search. It is largely an account of the proceedings of the First International Specialists' Workshop Conference on Criteria and Terminology in the Classification of Fungi Imperfecti held at the Environmental Sciences Centre of the University of Calgary, Kananaskis, Alberta. The invited contributors, all mycologists of international reputation, have had long experience with Fungi Imperfecti.

The first fifteen chapters follow the course of the conference: they reproduce the formal papers and the lively discussions which followed. Chapter 16 describes a new, experimental scheme of classification distilled from the conclusions reached at Kananaskis. Four chapters concerned with the application of this scheme and with a variety of techniques now being used to extend knowledge of the Fungi Imperfecti round out the volume. The text is illustrated throughout with numerous photographs and drawings.

In editing the volume, Professor Kendrick has given the text continuity by inserting short linking passages. The result is a readable and very informative account which conveys the unique atmosphere of this important conference.

BRYCE KENDRICK received a PhD from the University of Liverpool. After coming to Canada in 1958 he worked for six years as a Research Officer in the Mycology Unit of the Plant Research Institute in Ottawa. Since then he has been a member of the faculty of the University of Waterloo, where he is now Professor of Biology. He has initiated some unorthodox approaches to the taxonomy of Fungi Imperfecti, and has recently completed a comprehensive, illustrated treatment of the genera of Hyphomycetes for Volume IV of *The Fungi.*

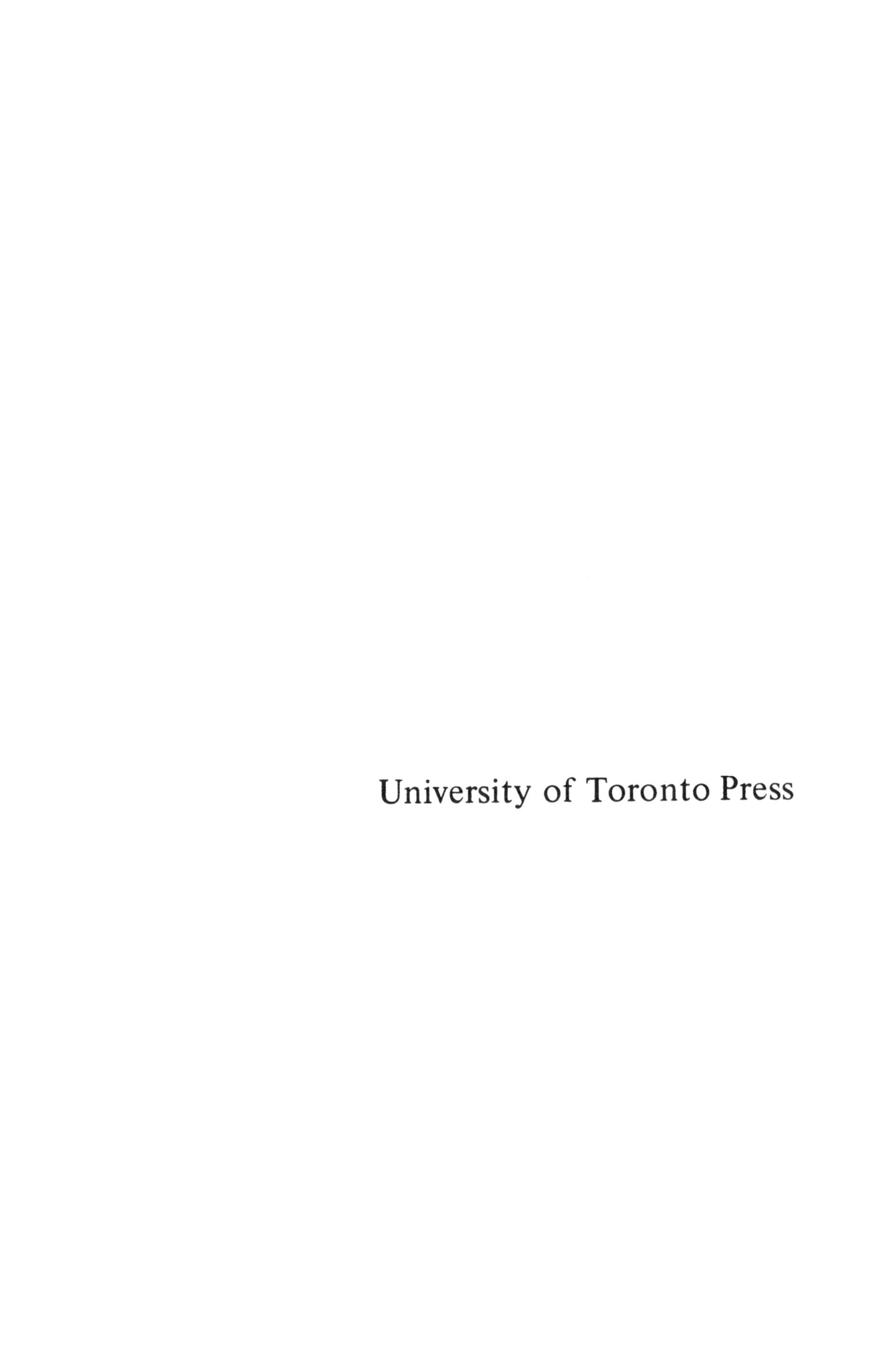

University of Toronto Press

# Taxonomy of Fungi Imperfecti

Proceedings of the
First International Specialists' Workshop-Conference
on Criteria and Terminology in the
Classification of Fungi Imperfecti
held at the Environmental Sciences Centre of the
University of Calgary,
Kananaskis, Alberta, Canada

Edited by BRYCE KENDRICK

Toronto and Buffalo
Reprinted in paperback 2017
ISBN 978-0-8020-1795-6 (cloth)
ISBN 978-1-4875-9223-3 (paper)
Microfiche ISBN 0-8020-0111-4
LC 79-163826

# Contents

# Preface

A friend once suggested that, from the purely scientific point of view, conferences were essentially a waste of time; that if new information was ready to be brought forth, it would be presented in the journals; that if ideas were to be worked out, they would be elaborated by individuals thinking in the privacy of their own laboratories; that if one scientist was wrong, another right, time would tell. My own experience of conference-going had done little to contradict these statements, but I had a few ideas of my own for improving the yield, and enthusiastically went ahead and planned the conference reported in this book. I am now convinced that something was achieved: nothing startling or apocalyptic, but a meeting of minds as well as men, a consensus and a new direction that might have taken years to achieve through the regular channels.

I have often been appalled at the rigid and repressive regime followed by most scientific meetings. A multitude of papers jostle for one's attention as several sessions run concurrently. It is customary for papers to be allotted fifteen or twenty minutes. It is equally customary for these monologues to spill over into the meagre five or ten minutes scheduled for discussion, thus effectively curtailing dialogue or suppressing it altogether. Anything more calculated to engender frustration and mental indigestion, I cannot conceive. Surely it is only through deference to everyone's wish (or administratively dictated need) to give a paper that such a system survives. It seems to be one of the more severe penalties that large conferences incur.

Even during the 1964 Botanical Congress in Edinburgh, when the idea for Kananaskis '69 first came to me, I promised myself that any mycological conference I organized would not follow this oppressive format, and I soon devised a more free-wheeling formula. First, there would not be a large number of participants - the meeting would be for specialists with common interests. Second, the subject matter of the conference would be discussed by a small, select panel before being thrown to the floor, and discussion would be allowed full development under the genial guidance of a strong chairman. With the right kind of participants, I reasoned, the discussions would be just as valuable an element of the conference as the papers.

What happened to this lofty vision? Needless to say it did not emerge unscathed, but the general principles - a small number of participants, invited keynote speakers, uninhibited discussion - were successfully put into practice at Kananaskis. Because the group was sufficiently compact, the initiation of each discussion by a small panel became unnecessary; but the panels, far from being redundant, soon assumed a new role: each had the responsibility of reviewing its keynote paper and the discussion following, and producing a committee report containing recommendations which would be placed before a closing session of the whole conference for acceptance or modification. These reports were to spark the final discussions during which so many of the novel ideas spawned at the meeting were articulated.

Many of us at Kananaskis realized that by some strange, intangible synergism, our thinking was being metamorphosed, subtly but surely. In these proceedings, I have tried to show something of this evolution in our thought.

I like to think that the success of the conference was in some small way due to its unusual modus operandi, so I wanted to record not only the academic flesh but also the organizational bones, and to suggest ways in which meetings of this type might be improved. All this because I believe that such small, specialized "think-tanks" are the wave of the future.

Many conference proceedings are dry and indigestible. They are often presented in the infamous "American telegraphic" style, which excludes personality, places no undue emphasis, advances no hypotheses, has little continuity, and is totally humourless. These attributes make them more than a little difficult to read. The development of concepts that took place during our conference was gradual; it will not be apparent to anyone who doesn't read the whole book. In editing the discussions, and in my narrative insertions (printed in italics), I have tried to give the book continuity, and to convey the (in my experience) unique flavour of the gathering. My comments have all the limitations inherent in their origin with a single observer. They are, inevitably, incomplete and biased, and I accept responsibility for any omissions and misinterpretations. Withal, I hope this book is at least an adequate record of Kananaskis '69.

B.K.

# Acknowledgments

It is a pleasure to record my indebtedness to the National Research Council of Canada, who were kind enough to consider my proposal for the conference worthy of financial support. Their grant made possible the attendance of four of the participants from overseas, and without it the meeting simply would not have taken place.

I am also grateful to a friend, Dr Dennis Parkinson, chairman of the Department of Biology, University of Calgary, for suggesting that the conference be held at Kananaskis, and for his encouragement and support throughout - moral, financial, and logistic. The venue was perfect, and the smoothness with which many of our arrangements were made can be traced to Dr Parkinson's beneficent influence. The conferees owe him special votes of thanks for sponsoring the reception with which the conference opened and, perhaps most of all, for making possible the magnificent bus trip up the Banff-Jasper highway to the Columbia Icefields, which was our mid-conference treat.

Dr J.B. Cragg, director of the Environmental Sciences Centre at Kananaskis, willingly gave his permission for us to use the excellent facilities of the Centre, which provided us with sleeping accommodation, excellent meals, conference room, and laboratories. Mr Tom Matheson took care of the many administrative chores efficiently and unobtrusively. Mr Girish Bhatt organized transportation to and from Calgary for participants. Thanks to him, and to Paul Widden and John Bissett, none missed his plane. Conferences, like armies, march on their stomachs, and Mrs Elsie Chaykowsky pampered us most delightfully in that quarter, even catering specially for our two vegetarian participants.

As editor of this book, I owe a special debt of gratitude to one of the members of the conference, Dr Bridge Cooke, who acted as recorder, making summaries of each day's proceedings to be read out at the beginning of the following day, and providing me with a brief account of conference activities that has proved invaluable in collating the pieces of the book.

Miss Marian Dawes and Mrs Audrey Nelson patiently typed and retyped the manuscript. Miss Dawes found and corrected some of the errors I had accidentally introduced; but readers, no doubt, will take great pleasure in pointing out other mistakes.

Finally, I must express my gratitude to all the mycologists whose work and thought is represented in this book, both those who attended Kananaskis '69 and also those who, though unable to come, contributed manuscripts or illustrations. Only their cooperation made this book possible.

ACKNOWLEDGMENTS
FOR PERMISSION TO REPRODUCE ILLUSTRATIONS

Figure 2.4 A and B are reproduced by permission of the Czechoslovak Academy of Sciences.

Figures 2.4 (C,D,F), 4.6, 4.7, 7.1, 7.7, 9.1, 11.2, 11.3, 11.6 are reproduced from the *Canadian Journal of Botany* by permission of the authors and of the National Research Council of Canada.

Figures 2.1, 2.2, 2.3, 2.4 (E,G,H), 7.4 are reproduced from *Mycologia* by permission of the authors and of the Mycological Society of America.

Figures 3.2 (a,b,f,g), 3.3 (a,b,c,e), 3.4e, 3.5g, 3.6 (b,h), 3.7 (b,c,e), 3.8 (a,b,d), 12.2 (d,j) are reproduced by permission of Dr G.L. Barron and the Williams and Wilkins Company.

Figures 7.2, 7.3 are reproduced from the *Journal of Bacteriology* by permission of the authors and of the American Society for Microbiology.

Figure 7.11 is reproduced from the *Annals of Botany* by permission of the authors and of the *Annals of Botany*.

Figure 2.6 is reproduced by permission of Georg Thieme Verlag, Stuttgart.

Figure 7.13 A is reproduced from the *Journal of General Microbiology* by permission of the authors and of the *Journal of General Microbiology*.

Figures 3.2 (c,d,e), 3.3f, 3.5 (a,e,h), 3.6 (c,d,e,f,g,i), 3.8f, and 12.2b are reproduced by permission of Dr S.J. Hughes.

Figures 3.4 (b,c,d), 3.5 (b,c,f), 3.7a, 12.2 (a,g), 12.3 are reproduced by permission of Dr M.B. Ellis.

Figure 7.8 is reproduced by permission of Dr K. Zachariah.

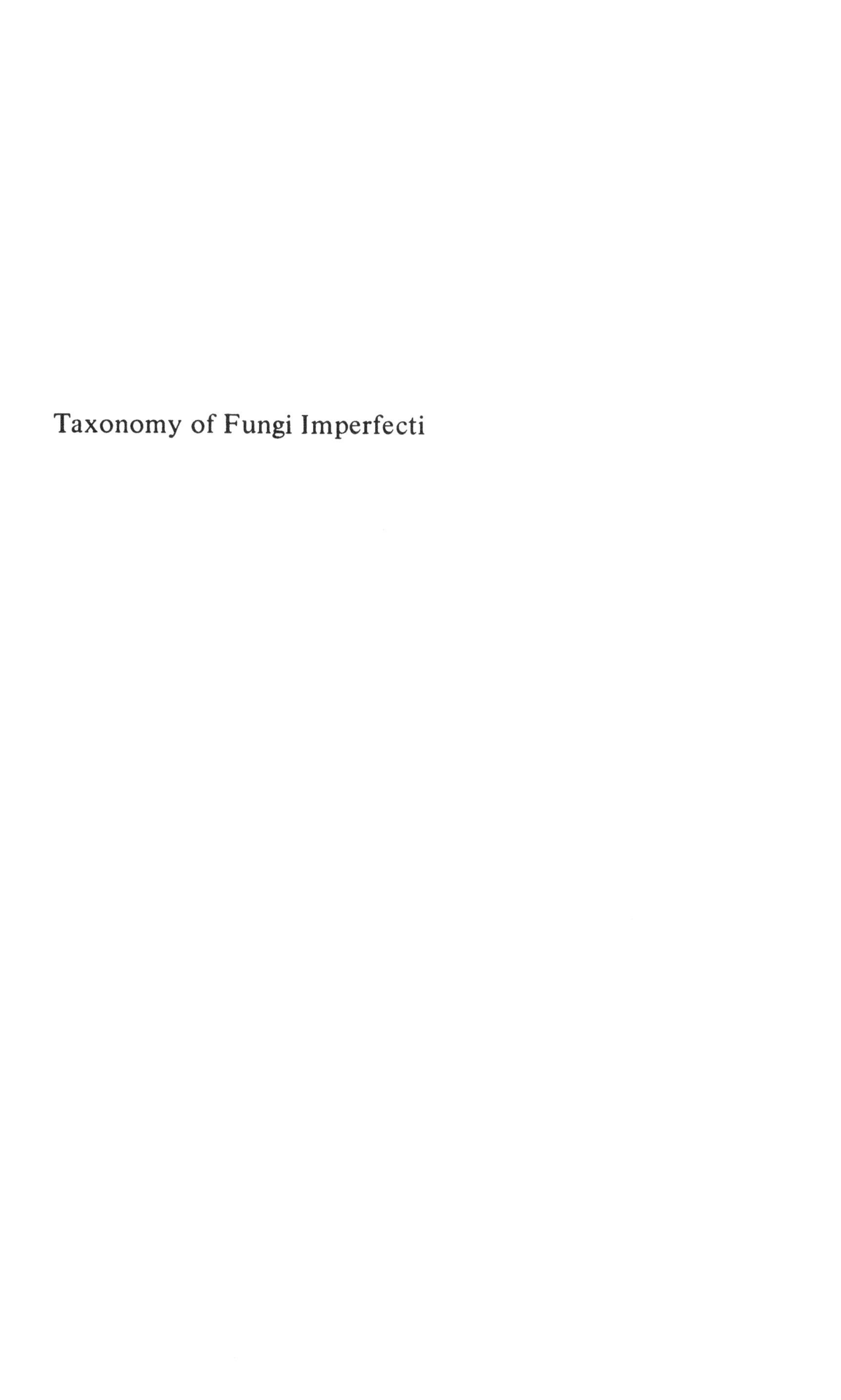

# Taxonomy of Fungi Imperfecti

It is a mistake to hand the delegates to a thinkers' conference seven pounds of thoughts. They take fright at the volume and hide from it in the bar, where their capacity to absorb thoughts - not equal to seven pounds' worth in the first place - deteriorates steadily.

Editorial, *Globe and Mail*, Toronto, 1969

Argument, which should play a vital role in science, is all but squeezed out of contemporary journals. The inevitable long delay in publication, coupled to the distaste of editors for polemics, has effectively eliminated argument as a public instrument of scientific progress.

D.E. Green, in *Science*, January 1964

Assigning terms is often much more problematical than ... conferring names ... There are practically no internationally established terminologies.

M. Scheele in *Library Science and Documentation*, vol. 2, 1961

A definition is the enclosing of a wilderness of ideas within a wall of words.

Samuel Butler, Note Books

# 1 Introductory Remarks

BRYCE KENDRICK

Critical accounts of the history of taxonomy in the Fungi Imperfecti have been given elsewhere by S.J. Hughes (*Can. J. Botany,* 31 [1953]: 577-80), G.L. Barron (*The Genera of Hyphomycetes from Soil* [Williams and Wilkins, Baltimore, 1968], pp. 1-8), and C.G.C. Chesters (in *The Fungi*, vol. III, edited by Ainsworth and Sussmann [Academic Press, New York, 1968], pp. 524-6). Those who wish to reap the maximum benefit from this conference should be familiar with these accounts.

The first phase of "Hyphomycetology" can be almost perfectly circumscribed by the nineteenth century. This phase was concerned with the initial collecting, describing, and classifying activities, which, perhaps inevitably, were almost entirely based on superficial morphological characters of fructification, conidiophore and conidium, or on apparent host preferences. That we have not advanced far enough beyond this stage of analytical or alpha taxonomy to be condescending towards it is truer than most of us like to admit. One has only to check works aimed at hyphomycete identification to see how much they rely on a straightforwardly morphological approach or, since this is often unsatisfactory in practice, descend to the level of "picture-book mycology."

The Hyphomycetes are still, I regret to say, at the stage where one either knows the beast or one does not. To my students I often liken a hyphomycete specialist to someone holding a large bunch of balloons (the better the specialist, the larger the bunch of balloons!). Each balloon is a hyphomycete and each has a separate, independent string leading to it.

At the beginning of the twentieth century, a few brave souls were still describing hyphomycetes, and the occasional original spirit like Vuillemin was attempting to dig beneath the morphology and unearth the developmental basis

Department of Biology, University of Waterloo, Waterloo, Ontario, Canada.

that would allow him to make taxonomically valuable generalizations. Vuillemin was ahead of his time - sixty years later we are just recognizing the validity of some of his concepts. In the 1920s and 1930s, a few workers like Mason kept the science alive by occasional injections of new and often controversial ideas, but it was not until the 1940s, during the Second World War, that interest in the Hyphomycetes became widespread. These fungi produced antibiotics and other useful biochemicals; they rotted clothing and equipment; they caused great losses of food, both while it was growing and while it was in storage. All of these aspects had positive or negative impacts on the war effort, and government agencies began to take these fungi seriously. This interest subsequently spread to the universities, where it sometimes took on a more rarefied aspect, perhaps culminating in some of the "ivory tower" research done at Waterloo!

The analytical process of discovery and description proceeded at an accelerated pace as the more senior workers introduced a new generation of students to this fascinating group of organisms. Attempts to synthesize new and better classifications became gradually more authoritative and convincing, and the thicket of terminology grew denser and more impenetrable by the month. By 1964 there was a large body of recent literature on the group, and the organizers of the 10th International Botanical Congress at Edinburgh thought it appropriate to arrange a brief symposium on the systematics of the Fungi Imperfecti.

Imagine a crowded lecture theatre, a smattering of hyphomycete specialists, a good sprinkling of mycologists of other persuasions, and a host of the botanical fraternity. Speakers came and went; each was eminently well qualified, but each gave a different interpretation, different terminology. The audience grew a little restive, more than a little confused. I was slated to speak last. Having no wish to confuse everyone further by offering yet another scheme, in all probability less adequate than those which had preceded it, I abandoned my text, and spoke on another topic altogether. Our symposium took place towards the end of the congress, and the participants had soon dispersed, literally to the ends of the earth, without having had much chance to resolve their differences. And there the matter rested for two years. Of course, I knew that the apparent disunity in our ranks concealed a seething ferment of fresh ideas; but it was equally obvious to me that, if the processes of communication between the principals in this debate were confined to the literature, then we were likely to see various different positions become more entrenched as the years went by. The only way to resolve this problem was to bring the principals together. In August 1966, I wrote to a number of mycologists, proposing an International Workshop-Conference, and solicited their comments. The response was enthusiastic, but pessimistic. Yes, it was a fine idea, but nobody would give financial backing to such a venture, seemed to be the consensus. I decided to go ahead anyway, to plan a conference, and to apply for support. That the conference - if it took place - should follow the 1969 Seattle Congress seemed logical, and the problem of locale was settled when Dr Parkinson suggested that, for a small group, the new Environmental Sciences Centre of the University of Calgary might be available, a suggestion which Dr Cragg, director of the Centre, looked upon kindly.

I proceeded with an application to the National Research Council of Canada; I made what I hoped was an eloquent plea for Canadian sponsorship of the conference, and in April 1967, I learned that we were in business.

And so we are all gathered here today. It is a proud moment for me, and I hope it will turn out to be a useful episode in the study of the Fungi Imperfecti. From my discussions with those who will be keynote speakers in our sessions, and from my examination of material I have solicited from other scientists who cannot be here in person, I know that many new ideas, and a few rather surprising phenomena, will be presented. We shall enjoy a most stimulating and productive experience. Dr Martin Ellis said to me the other day, "We need unequivocal definitions of our terms." I agreed, but on reflection, I thought, "It's no use having unequivocal definitions if people insist on redefining terms in their own way (also unequivocal), or if they refuse to use a term, however unequivocally defined." So it is my hope that a spirit of goodwill and international cooperation will prevail during our meetings, and that each of us will adhere to any agreements arrived at here, until such time as another meeting can re-evaluate the situation in the light of the rapidly growing stream of new information.

*After the constant turmoil of the International Botanical Congress at Seattle, which saw more than four thousand botanists milling around in a frequently vain attempt to separate the wood from the trees, the Environmental Sciences Centre of the University of Calgary at Kananaskis seemed to us a veritable Shangri-la. We were a small group - about one-half of one per cent of the total attendance at the congress. Not only this, we were in an idyllic setting. Forest closure, imposed after an exceptionally dry season, had nearly kept us out; but now we were in, we were almost grateful for the closure, because very few sounds broke the silence of our beautiful mountain valley.*

*We were quartered in an elegant lodge whose windows gave a superb panorama of forest sweeping up to earth's bared and rocky ribs all round the skyline. Here we slept and ate. Excellent meals kept our morale high and gave us the intestinal fortitude to work unflaggingly (well, more or less unflaggingly) through long academic sessions each day. With our physical well-being so well looked after, and amid such splendid surroundings, how could we avoid a certain amount of serendipity?*

*I had originally planned two academic sessions each day (from 9 to 12 in the morning, and 7 to 10 in the evening) during the five days of the conference; ten sessions in all; thirty hours of talking. I had hoped that the afternoons would be left free for private fungus forays, informal discussion groups, letter-writing, thinking - unscheduled activities, academic or otherwise. Had the conference lasted seven days, that might have been just the right formula; but I had not taken into account the enthusiasm with which the members waded into the discussions. It soon became apparent that two sessions a day would not bring us to the end of our agenda. It also transpired that collecting trips were severely curtailed by forest closure, so without too much hesitation I announced that there would be three sessions each day. Fortunately the news was not greeted by a general uprising, and this prodigious work-load only began to exact its toll towards the end of the meeting. Nevertheless, I would recommend my original scheme to conference-planners, with the rider that a conference scheduled to last three days should be extended to five, and a five-day conference to seven. Like the supposedly adequate supply of money which melts away when one is on vacation, conference time passes all too quickly when free discussion is allowed.*

# 2
# Phycomycetes, Basidiomycetes, and Ascomycetes as Fungi Imperfecti

S.J. HUGHES

"Wie in der Zoologie genaue Kenntniss der Lebensverrichtungen zur Bildung rationeller Gattungen nothwendig ist, so ist zur richtigen Begründung der Gattungen in der Familie der Faserschimmel, genaue Untersuchung der Entwickelung der Sporen, und ihrer normalen Lagerung primäre Bedingniss."

A.C.J. Corda, *Pracht-Flora europaeischer Schimmelbildungen*
(Gerhard Fleischer, Leipzig and Dresden, 1839)

In the *Dictionary of the Fungi* (Ainsworth 1961) it is stated that "most Fungi Imperfecti are imperfect states of Ascomycetes, but some Mycelia Sterilia and a very small number of other Fungi Imperfecti are, or are possibly, states of Basidiomycetes. Conidial forms of rusts, smuts, and Phycomycetes are generally put in these groups and not with the other Fungi Imperfecti." Generally speaking, students of Fungi Imperfecti have devoted little attention to the imperfect states of non-ascomycetous groups although all imperfect states belong to that vast assemblage "Fungi Imperfecti"; they produce spores to reproduce the species, to tide them over unfavourable conditions, or to fulfil a sexual function. All imperfect states should be accounted for in a consideration of spore ontogeny. Apart from the spore forms of the rusts, imperfect states of Basidiomycetes are poorly developed and are often known only in pure culture, but those of the higher phycomycetes have produced some elegant forms through the modifications of zoosporangia or sporangia and elaboration of the sporangiophores. It is in the Ascomycetes that the imperfect states have come into their own in their remarkable diversity.

The current tendency is to ascribe to conidium ontogeny a primary significance in the delimitation of form genera in ascomycetous Fungi Imperfecti. This account is a survey of the methods of producing reproductive structures, for the

**Plant Research Institute, Canada Department of Agriculture, Ottawa, Canada.**

most part in non-ascomycetous groups, especially with reference to the production of a plurality of such structures from a single cell or hypha; there is a tendency for vast numbers of these structures to be produced, and various fungi in different groups have gone about this in the same or in different ways.

The scope of this survey must be restricted, primarily because of my lack of knowledge of alien groups. However, many lucid accounts are available of imperfect states in Phycomycetes and Basidiomycetes and I have drawn freely from these; a few personal observations are also included. A brief summary is given of an account of spore ontogeny in the rusts (Hughes 1970) and some details included from one on percurrent proliferations in fungi, algae, and mosses (Hughes 1971a). No references are made to imperfect states of Trichomycetidae (Phycomycetes), or the Ustilaginales and some other orders in the Basidiomycetes.

The more comprehensive our knowledge of the development of imperfect states the easier it will be to recognize which ontogenetic methods are present or absent in different groups. Subsequently, the affinities of an imperfect state may become clearer and possible interrelationships of major groups may emerge. We find, for instance, that phialospores and meristem arthrospores are common to Ascomycetes and Uredinales but are not known with certainty in other groups. On the other hand, the methods of successive development of different reproductive structures by sympodial growth or percurrent proliferation of the subtending cell, or the more or less synchronous development of these structures, are represented in Phycomycetes and Basidiomycetes as well as the Ascomycetes. But, of course, such methods are by no means universal within these vast groups; for example, percurrent proliferation, as a method of producing a plurality of reproductive structures, is not found in the Zygomycetidae and occurs only sporadically in the Basidiomycetes.

In the ascomycetous Fungi Imperfecti in particular, at a lower taxonomic level, we find that certain kinds of form genera are genetically connected with certain kinds of perfect states: Dr Müller will be dealing with connections between perfect and imperfect states later (Chapter 13). Our experience enables us occasionally to forecast, sometimes with a considerable degree of success, the perfect state of some Fungi Imperfecti and what imperfect states are likely to occur in the same species. It is obvious that an extensive documentation is needed of the multifarious imperfect states occurring in the same species, with or without the perfect state. Such information, scattered in the literature, is being gathered together at Ottawa; so far as I am aware no such compilation has been published. Interesting results are already evident from a very partial compilation; but there are, of course, difficulties in the published records concerning the correct application of generic names, both perfect and imperfect, and often the necessary details of conidium development are insufficient. Nevertheless, we have a reasonable expectation that a very orderly pattern awaits us.

## PHYCOMYCETES

The cleavage of the contents of sporangia into motile or non-motile spores is characteristic of the majority of phycomycetous imperfect states; other reproductive structures may also be produced. In some species sporangia have acquired some or all of the characteristics of conidia. My intention now is briefly to consider imperfect states in some phycomycetes, especially the methods of producing a plurality of sporangia or their homologues. This covers some familiar ground, but in the context of an introductory survey it merits inclusion. 

*Oomycetidae*

It is well known that the higher members of the subclass Oomycetidae seem to have abandoned the aquatic habitat and have become to varying degrees terrestrial, as parasites of higher plants and animals, or as saprophytes. The colonization of the new environment is associated with modifications of the various species that are, for the most part, related to air dispersal. Thus, we find deciduous sporangia that may germinate directly by germ tube rather than by zoospore discharge; such sporangia are to all intents and purposes conidia. The change from sporangia which produce numerous zoospores in situ to dispersed conidia which produce germ tubes would result in a considerable reduction in the number of actual reproductive units. However, we find that various methods have evolved for increasing the number of "conidia" on each sporangiophore, which, in the higher members, has acquired the differentiation and dignity of a conidiophore.

First, there may be an increase in the fertile area of the conidiophore by the production of a terminal swelling. Secondly, there may be profuse branching, with each branch terminating in a solitary conidium. In these two methods conidium development is apparently synchronous, and there is a flush of conidia from a conidiophore, which then collapses; however, in a number of Peronosporaceae successive crops of conidiophores are produced from the same substomatal knot of hyphae, but this need not concern us here. Thirdly, successive conidia may develop sympodially on a single conidiophore, and fourthly, successive conidia may arise from percurrent proliferations through the conidium scars. In the last two methods conidium production by a single conidiophore is prolonged.

We have, therefore, in the Oomycetidae, methods of multiple conidium production which are well known in ascomycetous Fungi Imperfecti. But before proceeding directly to the conidium-bearing species of the Oomycetidae we should perhaps inquire into the methods of sporangium production in the lowlier members; after all, there seems little doubt that the conidia of the aerial Peronosporales are the homologues of the zoosporangia of the aquatic orders.

In the Chytridiales there are a number of examples of successive sporangia formed by the method of percurrent proliferation. This process is apparently common in *Cladochytrium* and has also been described in other genera (Sparrow 1960).

In the Blastocladiales, Sparrow (1960) recorded internal (percurrent) proliferations in species of *Blastocladia*, e.g. *B. prolifera*; in other species, e.g. *B. gracilis*, sporangia are produced successively and sympodially. In all species of *Allomyces* "secondarily formed sporangia may be found in basipetal succession" and it appears that this occurs through the involvement of successive cells of a hypha; this is in addition to the sympodial development of sporangia known in this genus.

In the Monoblepharidales sporangia develop either sympodially or by percurrent proliferation. In *Monoblepharis macrandra* the sporangia are "grouped in sympodial or fasciculate fashion, occasionally proliferating [percurrently]" (Sparrow 1960).

In the Saprolegniales both sympodial and percurrent proliferation methods of sporangium production are found, occasionally in the same species. Thus, *Saprolegnia* was described by Sparrow (1960) with sporangia "typically proliferating within the older ones in a 'nested' fashion" but often also as in *Achlya*, i.e. "renewed sympodially or by a basipetalous development and cymose branching." No percurrent proliferation is recorded in *Achlya* by Johnson (1956).

In some species of the Leptomitales there is a basipetal conversion of slightly modified segments into sporangia; in others the sporangia are "sympodially arranged along the hyphae," or formed (? successively) "in whorls or umbels at the tips of the segments," or they appear "lateral by sympodial branching of the supporting filaments." No percurrent proliferations are recorded in the Leptomitales by Sparrow (1960).

It is of particular interest to a hyphomycetologist, at least, to know that a succession of sporangia produced sympodially and by percurrent proliferation is well established in these entirely aquatic orders. The basipetal involvement of hyphal cells in sporangium production is also present, but so far as I am aware this has no counterpart in those genera of Oomycetidae in which sporangia have acquired the nature of conidia.

The Peronosporales includes the families Peronosporaceae, Pythiaceae, and Albuginaceae. In species of *Pythium* (Pythiaceae) sporangia may be intercalary or terminal, or a number may arise successively on a single undifferentiated branch by sympodial growth or by percurrent proliferation. Sympodially produced and percurrently proliferating sporangia may occur in the same species, e.g. *P. proliferum.* Butler (1907) showed that in this species the proliferous character may be lost in culture and thereafter sporangia are renewed sympodially. Perhaps this suggests that the sympodial method is more primitive than the percurrently proliferating method; however, it does indicate that the kind of sporangium production is not as genetically fixed as in other species. The sporangia of *Pythium* are apparently seldom deciduous.

In *Phytophthora* (Pythiaceae) both sympodial and percurrent proliferation methods of sporangium production are found, as shown by Blackwell (1949). Sparrow (1960) stated that both methods may be represented in the same species, e.g. *P. oryzae*. Blackwell illustrated sporangia developing in a lax or close sympodium on a more or less undifferentiated hypha in *P. cactorum*; on the other hand, Blackwell illustrated the percurrent proliferating method in *P. megasperma*. Sporangia of *Phytophthora* secede in some species and remain attached in others (Waterhouse 1963). The freeing of the sporangium from the sporangiophore was a tremendous step forward for these organisms, especially in their establishment in a partly aerial habitat. The secession involves a circumscissile rupture of the sporangiophore wall at a level below the septum delimiting the sporangium, because freed sporangia carry away with them a portion of the sporangiophore; furthermore, it presumably involves a method of sealing the broken end of the sporangiophore to prevent escape of its contents. Seceding sporangia are found only in those species which produce these structures sympodially; sporangia produced by percurrent proliferations do not secede. 

*Phytophthora infestans* is the well-known example of a *Phytophthora* with regularly seceding sporangia (conidia). These arise terminally, and it is reported that by subsequent subsporangial growth the original sporangium is pushed to one side. Each new extension of the sporangiophore is more or less flask-shaped; there is very little geniculation.

The Albuginaceae includes one genus, *Albugo*. It has been shown by Hughes (1971a) that the club-shaped conidiophores (sporangiophores) which produce basipetal chains of conidia (sporangia) do so by means of successive percurrent proliferations. It seems to me, therefore, that the Albuginaceae are far more closely related to the Pythiaceae than to the Peronosporaceae where proliferations are lacking. The conidia in *Albugo* secede readily and form the characteristic powdery fructifications.

The Peronosporaceae comprise genera whose species are obligate parasites which produce sporangiophores through the stomates of the host plant; the sporangia are deciduous and may now be referred to as conidia borne on conidiophores. In *Basidiophora* the conidiophores are unbranched and distally somewhat swollen, bearing stalked conidia which, following dispersal, germinate by zoospores or germ tubes. The conidiophores in other genera, on the other hand, are variously branched, as in *Plasmopara*, *Sclerospora*, and others, each branch or dichotomy bearing a single terminal conidium; in *Bremia* the conidia are produced marginally on a saucer-like inflation of each branch of the conidiophore. In these last-named genera the conidia germinate by zoospores or by germ tube. Finally, in *Peronospora*, the largest genus in the family, the non-papillate conidia are stated to germinate invariably by germ tubes.

According to the literature, one of the characteristic features of the Peronosporaceae is that the conidia develop more or less simultaneously as the blown-out ends of the variously elaborated conidiophores. For instance, Mangin (1891) stated, "les conidies ne se développent pas successivement, mais presque simultanément à l'éxtremité des stérigmates." The series of illustrations by Weston (1923) of various species of *Sclerospora* certainly suggests a synchronous development; it is of interest to note too that in at least one species of *Sclerospora*, Weston (1923) showed that "apparently all the conidia on one conidiophore customarily are ejected at once."

In the Peronosporales, therefore, the Peronosporaceae produce conidia more or less simultaneously on simple or branched conidiophores, and forcible discharge is known. In the Albuginaceae and Pythiaceae there is a succession of conidia or sporangia, by percurrent proliferation from a specialized sporogenous cell in the first-named family, and by sympodial and/or percurrent proliferations in the last-named.

CONIDIA AND CHLAMYDOSPORES IN OOMYCETIDAE Butler (1907) remarked that the "conidia of species of *Pythium* are potential sporangia, but their power of germination as such is gradually lost and the simpler process of direct germination substituted." Mathews (1931) considered conidia in *Pythium* as "indistinguishable from sporangia except in the method of germination," which is by germ tubes, and in some species such as *P. debaryanum* the conidia may fall off the hypha before germination. Such conidia are intercalary, terminal, or lateral, and in *P. catenulatum* three to eight may occur in a chain. Mathews (1931) remarked that conidia of *P. afertile* have been referred to as chlamydospores; he recorded terminal and intercalary, spherical, thick-walled chlamydospores in *P. undulatum*. In her key to *Pythium*, Waterhouse (1967) "decided to restrict the term sporangium to structures giving out zoospores. All other non-sexual enlargements of hyphae are termed vegetative hyphal bodies." Although hyphal bodies could be deciduous or thick-walled Waterhouse did not use the term conidium or chlamydospore.

Hesseltine (1952) recorded "thick-walled resting cells in the old mycelium" of members of the Saprolegniales; Sparrow (1960) referred to these as "'gemmae' (chlamydospores)."

Blackwell (1949) drew attention to the presence of chlamydospores in *Phytophthora*, and such thick-walled terminal or intercalary structures were illustrated in *P. parasitica* and *P. cactorum*. Chlamydospores were recorded for a number of other species by Waterhouse (1963) and they are apparently of some taxonomic value. Waterhouse (1963) stated that they are "always completely delimited by a smooth wall becoming up to 2$\mu$ thick by the deposition of an inner layer of cellulosic material." I found references to the production of chlamydospores only in Saprolegniales and Peronosporales (Pythiaceae) and the spores were always simple and devoid of ornamentation.

## *Zygomycetidae*

### Mucorales

Here, as in the Oomycetidae, there are "lines of evolution" leading to the production of conidia from sporangia. In the Peronosporales this seems to have been accomplished by a delay in, or complete suppression of the zoosporic state: in the Mucorales the production of sporangiospores is not delayed or suppressed but there is a reduction of the number produced so that in the highest forms sporangia which are much reduced in size are 1-spored. As a result of this the sporangial wall may sometimes be differentiated with great difficulty from the wall of the spore. For instance, it has only recently been established (Hawker 1966) that the conidium of *Cunninghamella elegans* is a 1-spored sporangium. In *Mycotypha africana*, in which zygospores are known, the conidia have been shown by Young (1969) to be 1-spored sporangia. The type species of the genus *Mycotypha, M. microspora,* has not been found in a zygosporic condition and has been classified variously as Moniliales and as Mucorales; in view of Young's work with *M. africana, M. microspora* is undoubtedly mucoraceous. *Sigmoideomyces* was included in the Mucorales by Zycha (1935) and Naumov (1939) but excluded by Hesseltine (1955); the name was included as "Mucorales or Phycomycete" by Ainsworth (1961). In the absence of zygospores, electron microscope and other studies may be required to decide the affinity of *Sigmoideomyces*. *Rhopalomyces elegans* is another imperfect state not connected to a perfect state; it has been regarded as a phycomycete by Boedijn (1927) and J.J. Ellis (1963) but excluded earlier by Hesseltine (1955). 

Benjamin (1959) published his beautifully illustrated account of merosporangiferous Mucorales and also discussed lines of evolution in the Mucorales as a whole. Of particular interest to us is the "*Piptocephalis-Kickxella*" line and a few points are included here.

In *Syncephalastrum*, for instance, the sporangiospores are produced uniseriately by the cleavage of the merosporangial contents into primordia around each of which a wall develops. (In this regard they are reminiscent of the first phialospores arising in the long neck of an unopened phialide.) The sporangiospores may be free within the merosporangium, and Benjamin stated that with gentle pressure they can be squeezed out; they are apparently as free as the sporangiospores in a *Mucor* sporangium and are freed by the fragmentation of the merosporangial wall. In *Piptocephalis* the merosporangial wall is delicate and evanescent whereas in *Syncephalis* it is evanescent or persistent; when persistent, it separates with the spores by circumscissile rupture so that a wrinkled part of the merosporangial wall remains on the spore. The secession of spores in this manner is reminiscent of the freeing of some arthrospores in ascomycetous Fungi Imperfecti.

In Dimargaritaceae the merosporangia are 2-spored: these are separated from each other by a gap left between the spores during their formation. In *Dispira cornuta* the merosporangial wall may be persistent and holds the spores together in pairs.

The Kickxellaceae shows the highest development in the Mucorales. The sporangia are 1-spored, and produced singly on "pseudophialides" which develop on specialized "sporocladia." The sporangial walls are persistent, but in *Dipsacomyces* (Benjamin 1961) the sporangial membrane separates readily from the spore and the upper part of the membrane is shed.

Decrease in the number of sporangiospores in the Mucorales, like the delay or suppression of the zoosporic stage in the Peronosporales (Oomycetidae), is accompanied by an increase in the production of sporangiola, merosporangia, or 1-spored sporangia. This has generally been accomplished by the profuse branching of the sporangiophore with each branch terminating in a sporangium, by the swelling of parts of the sporangiophore and the development of synchronous sporangia on these swellings, or by the development of successive sporangia. So far as I am aware there are no percurrent proliferations associated with sporangium production recorded in the Zygomycetidae; this provides another difference which serves to separate this group from the Oomycetidae, where this process is common (Hughes 1971a).

CHLAMYDOSPORES, ARTHROSPORES, AND YEAST-LIKE BUDDING Chlamydospores have been recorded in a large number of Mucorales; they are thick-walled, smooth, terminal, or intercalary and occur singly or in twos. Hesseltine (1954), for instance, stated that chlamydospores are present in the aerial mycelium in large numbers in *Mucor* section *Racemosus*; in some other sections they are "absent or in the substrate mycelium or, if present in the aerial mycelium, only in old cultures and uncommon."

Oidia (arthrospores) have been described in various species of *Mucor*, e.g. *M. genevensis* (Hesseltine 1954), in which they were described as "abundant in submerged mycelium, spherical to ovoid, with a heavy clear wall, ... single or in a series at the end of branches ... later breaking off and undergoing budding." Bartnicki-Garcia and Nickerson (1962) investigated "duality in morphogenesis, commonly termed mold-yeast dimorphism" in some species of *Mucor* and found that this is subject to environmental control. These authors differentiated rounded arthrosporic cells from yeast-like budding cells. Under certain conditions *Mucor rouxii* developed a yeast-like form with no trace of hyphal growth: the spherical cells showed characteristic multipolar spherical buds "formed by extrusion of protoplasm through a relatively small area of the surface of the swollen spore." Under different conditions the filamentous form of the *Mucor* developed together with spherical cells, but these "did not originate by budding but by arthrospore formation, a process of hyphal fragmentation which commenced with formation of septa at or near the tips of filaments. Segments of hyphae walled off by the nonperforate septa increased in volume and acquired a spherical shape. Individual arthrospores, or chains thereof were easily detached from the hyphae ..." This evidence certainly suggests that these mucoraceous arthrospores are equivalent to those found in ascomycetous Fungi Imperfecti.

## Entomophthorales

In the Entomophthoraceae and Basidiobolaceae the terminal, solitary conidium produced on the conidiophore is generally regarded as the homologue of a sporangium; with one exception these conidia are forcibly discharged. In *Empusa*, Thaxter (1888) has shown that the sporangial wall may swell in water and the spore has been illustrated as floating within it: the conidia germinate by a hypha or produce a conidiophore and a secondary conidium. In species of *Basidiobolus*, conidia can produce non-motile sporangiospores by cleavage of their contents, which may be taken as indicative of their sporangial nature. Drechsler (1955a) described two species of *Conidiobolus* whose conidia, following discharge, develop microconidia simultaneously and singly on short "sterigmata." He regarded these microconidia as "homologous with zoospores of Oomycetes and with sporangiospores of Mucorales." The conidia of species of *Basidiobolus* and some species of *Conidiobolus* frequently germinate to produce single secondary conidia which may then form tertiary conidia, each successive generation being accompanied by a reduction in size of the conidium; the synchronous production of many microconidia in the two species of *Conidiobolus* may be this sort of process and not at all related to the internal cleavage. However, following the description of abundant production of sporangiospores within conidia of *Basidiobolus meristosporus*, Drechsler (1955b) remarked that this confirmed the "correspondence between the endogenous and exogenous types of multiplicative sporulation found in the Entomophthorales." 

Drechsler (1946) described *Gonimochaete horridula* as a "primitive member of the Entomophthoraceae." Within the nematode, loose cells of the fungus enlarge and produce a neck-like extension outside the host's body. The protoplasm migrates into the cylindrical neck and the swollen basal part of the resulting flask-shaped cell becomes cleared of protoplasm. Within the neck the protoplasm is cleaved progressively from the tip backwards; the sporangiospores so formed are then liberated in groups through the opening at the apex of the neck.

Chlamydospores have been recorded for a number of species of *Conidiobolus*, e.g. *C. chlamydosporus* (Drechsler 1955a).

In the Zoopagaceae four kinds of spores can be distinguished; these are chlamydospores, sessile or stalked solitary spores, spores produced successively and sympodially on short or long conidiophores or on specialized branches, and arthrospores.

In Zoopagaceae chlamydospores do not occur, so far as I am aware, with any other kind of imperfect spore form in the same species, which is itself peculiar. Drechsler (1941a) wrote that as they develop "each receives granular protoplasm supplied through progressive evacuation ... of the parent hypha; successive steps in this evacuation being marked by deposition of consecutive retaining septa." In this regard chlamydospore formation is similar to that in a number of basidiomycetes (see below) where the retaining septa have been called "retraction septa."

Chlamydospores are intercalary or terminal on a short stalk, or lateral, ellipsoidal to obovoid or spherical, sometimes lobate, and moderately thick-walled. Drechsler (1941a) remarked that unlike aerial conidia generally, including those of the Zoopagaceae, these chlamydospores are not adapted for easy disarticulation.

Drechsler has illustrated and described a number of species with blastic conidia arising solitarily on a short or long conidiophore. Such conidia (Figure 2.1) vary considerably in shape in the different species and may bear one or more terminal appendages and sometimes a basal appendage as well. Such appendages are emptied during spore development through the complete or partial withdrawal of the protoplasm into the body of the spore and are finally delimited by septation.

In some other species the blastic conidia arise sympodially on a conidiophore of varying length (Figure 2.2) which may or may not show conspicuous geniculations. In *Euryancale* the conidia develop on specialized lateral sympodulae. Conidia vary in size and shape in different species and apparently secede readily at maturity.

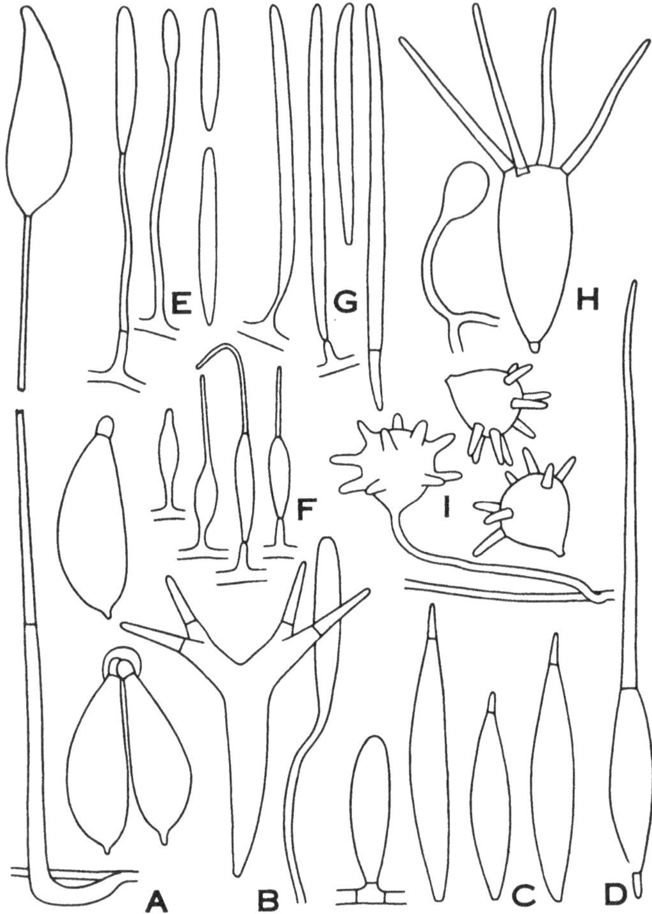


Figure 2.1 Conidia of various Zoopagaceae: A, *Stylopage rhyncospora* (after Drechsler 1939); B, *Acaulopage dichotoma* (after Drechsler 1945); C, *Endocochlus gigas* (after Drechsler 1936); D, *Acaulopage ceratospora* (after Drechsler 1935a); E, *Stylopage haploe* (after Drechsler 1935a); F, *Acaulopage cercospora* (after Drechsler 1936); G, *Acaulopage macrospora* (after Drechsler 1935a); H, *Acaulopage tetraceros* (after Drechsler 1935a); I. *Acaulopage acanthospora* (after Drechsler 1938).

Another conidial apparatus illustrated and described by Drechsler is shown in Figure 2.3. The erect hyphae extend by a terminal growing point without septum formation. Then, as Drechsler (1937) described in *Cochlonema megaspirema*, sporulation "takes place as in other catenulate members of the family, that is, through the withdrawal of contents from the short, slightly constricted isthmi perceptible at rather regular intervals in the aerial prolongations, followed by the laying down of septa at both ends of the protoplasts thus separated." Such conidia secede readily by a break across the isthmi.

Figure 2.2 Conidia of various Zoopagaceae: A, *Stylopage lepte* (after Drechsler 1935a); B, *Stylopage cephalote* (after Drechsler 1938); C, *Stylopage hadra* (after Drechsler 1935b); D, *Euryancale sacciospora* (after Drechsler 1939); E, *Stylopage scoliospora* (after Drechsler 1939).

Chlamydospores of Zoopagaceae can be assumed to be the homologues of those produced in other Entomophthorales but the zoopagaceous ones are only moderately thick-walled. Drechsler (1941a) drew attention to the similarity in development of chlamydospores of *Cystopage* and the conidia of *Endocochlus*; nevertheless, he added, "in view of their general characteristics they would seem to represent chlamydospores rather than conidia."

Conidia (and chlamydospores) of Zoopagaceae germinate by germ tube and not by the production of sporangiospores so that their homology is a matter of speculation. However, Bessey (1950) remarked that "in the Zoopagales the sporangia are reduced to indehisced sporangioles (or 'conidia')," an assertion which should be considered very seriously.

Drechsler (1938) considered the sporangial nature of catenulate conidia and stated, "though asexual reproductive apparatus in all species of *Cochlonema* and *Zoopage* was carefully examined, nothing has been observed that could be held

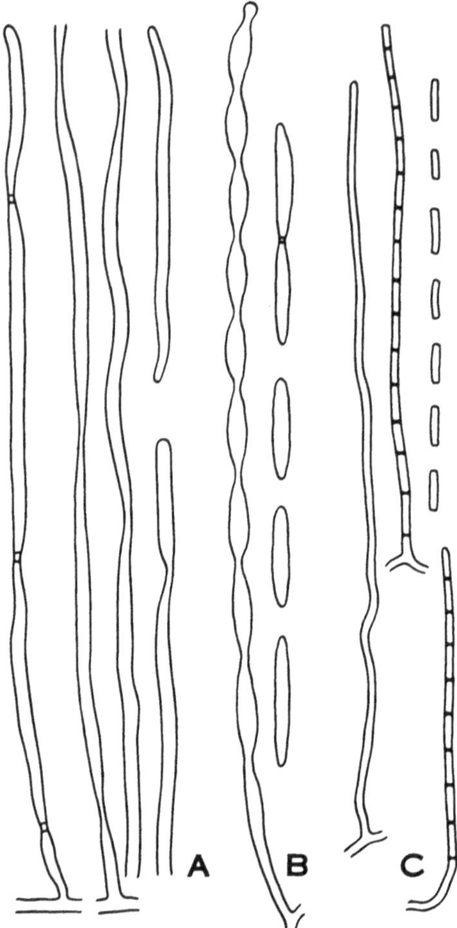


Figure 2.3 Conidia of various Zoopagaceae: A, *Zoopage atractospora* (after Drechsler 1936); B, *Zoopage thamnospira* (after Drechsler 1938); C, *Cochlonema pumilum* (after Drechsler 1939).

to argue in favor of endogenous development of the conidia in catenulate members of the family, or otherwise to sustain any supposition of homology between the conidial chains in these members and the rows of spores in the Piptocephalidaceae." Presumed arthrospores occur in the Mucoraceae and perhaps the catenulate conidia of the Zoopagaceae are of the same nature. However, the possible cleavage of the contents of a fertile hypha in Zoopagaceae into the homologues of sporangiospores should not be discounted because in highly advanced Mucorales, for instance, the sporangial nature of some conidia, although highly probable, is nevertheless difficult to prove. Convergent development of arthrospores of sporangial origin and of apparently indistinguishable structures of hyphal origin may have occurred. 

With regard to the nature of solitary and sympodially developed conidia there is perhaps a little evidence in favour of their being regarded as the homologues of sporangia. In *Basidiobolus* and *Conidiobolus* some species display a repetitional development of conidia (sporangia) (Drechsler 1955a, b); Drechsler (1939) showed that this occurs also in species of the zoopagaceous *Stylopage*, e.g. *S. rhyncospora*, in which secondary and even tertiary conidia can develop on "germ sporangiophores [*sic*]" and "each derived spore is appreciably smaller than its parent." However, Drechsler (1935b) had drawn attention to the occurrence of repetitional development "in many of the predaceous hyphomycetous forms referable to *Monacrosporium* and *Dactylaria*" so he considered the importance of repetitional development as an indication of affinity between Zoopagaceae and other Entomophthorales hardly to merit emphasis. Nevertheless, he concluded that "the suggestive correspondencies with the insectivorous Entomophthorales ... are at least deserving of mention."

Another, but very tenuous piece of evidence, favouring the acceptance of the sporangial nature of these conidia of Zoopagaceae, is the fact that the "adhesive conidia" (sporangia) of *Basidiobolus meristosporus* (Drechsler 1955b) have a globose mass of golden yellow, glutinous material at the tip. The conidia of a number of species of *Stylopage*, at least one species of *Acaulopage*, and the "chlamydospores" of a *Cystopage* also produce a similar yellow or golden yellow glutinous material. But of course this character could be adaptive because it is concerned with the capture of prey; indeed Drechsler (1935a) remarked that essentially the same method of capture is known among the predaceous hyphomycetes.

If the solitary and the sympodially produced conidia of the Zoopagaceae are not the homologues of sporangia, then they have no counterpart in the Phycomycetes mentioned above and must have arisen independently.

SUMMARY OF IMPERFECT STATES OF SOME PHYCOMYCETES Cleavage of sporangial contents into zoospores or non-motile spores occurs almost throughout the group. A multiplicity of sporangia may be produced by percurrent proliferation, by sympodial growth, or by the basipetal involvement of hyphal cells; in some of the higher members a synchronous or sympodial development occurs of conidium-like sporangia on a much branched or apically inflated conidiophore. *Albugo* alone produces a succession of conidia (sporangia) by percurrent proliferation. Arthrospores are produced in some Mucoraceae and some Zoopagaceae. Also in the Zoopagaceae, solitary and sympodially produced conidia are found; these may or may not be the homologues of sporangia. The chlamydospore has shown a considerable degree of conservatism and has remained a relatively simple structure. Phialides are lacking in Phycomycetes as a whole. Percurrent proliferations are not found in the Leptomitales or the Peronosporaceae (Peronosporales) and have not been recorded in the Zygomycetidae.

*Homobasidiomycetes - Aphyllophorales and Agaricales*

Lyman (1907) reviewed "the nature of polymorphism among the Hymenomycetes"; he studied some species in pure culture and recognized three kinds of secondary spores: conidia, oidia, and chlamydospores. From the work of Lyman and others it is apparent that most of these secondary spores are only recognized in pure cultures. Nobles (1948) illustrated and described cultural characteristics of 126 species of wood-rotting homobasidiomycetes, and she also recognized these three kinds of secondary spores.

Apart from a few examples, the known imperfect states of these basidiomycetes are seldom robust and lack the considerable elegant variations found in the ascomycetous Fungi Imperfecti.

Through the work of J. Eriksson (1958), Pouzar and Jechová (1967), Pouzar and Holubová-Jechová (1969), and others, at least eight species of the genus *Oidium* have been connected to species of *Botryobasidium*. The conspicuous *Oidium* imperfect states, with broad hyphae, produce blastospores (Hughes 1953) which may or may not occur in acropetal chains (Figure 2.4 A, E). It is of interest to note that in the Homobasidiomycetes the production of chains of blastospores is exceedingly uncommon.

In a number of species, e.g. *Fomes annosus* (Nobles 1948), *Stereum sulcatum*, *Vararia granulosa* (Figure 2.4 B), and *Corticium furfuraceum*, conidia are produced more or less simultaneously on tapering denticles on the terminal swelling of a conidiophore: these are the "botryose blastospores" of Hughes (1953). Nuclear behaviour during conidium production in the last three species was followed by Maxwell (1954) and in *Corticium effuscatum* by Nobles (1942). Similar fructifications were recorded by McKeen (1952) in three species of *Peniophora*.

Nobles (1937) described the production of conidia singly on tapering, sterigma-like outgrowths in *Corticium incrustans*: "on the haploid mycelium a cell may produce one to many conidia, which appear in clumps on small elevations or in whorls." A somewhat similar production of conidia is recorded in *Polyporus rutilans* by Nobles (1948). I believe that these conidia are also botryose blastospores but more scattered than aggregated.

In some other species of Homobasidiomycetes (Aphyllophorales) conidia are produced singly and successively as sympodioconidia (Figure 2.4 C): such conidia are illustrated and described by Maxwell (1954), who also included details of nuclear behaviour during conidium production. Eriksson and Hjortstam (1969) traced hyphae bearing conidiophores of *Costantinella micheneri* to hyphae of *Botryobasidium botryosum* and were convinced of the identity of the two states. *Costantinella* produces sympodioconidia on very characteristic sympodulae. Eriksson and Hjortstam suggested that *C. terrestris* might also have a *Botryobasidium* state, possibly *B. pruinatum*.

So far as I am aware from available accounts, conidia formed sympodially, and blastospores or botryose-blastospores, do not occur together in the same species; Dr Mildred K. Nobles (pers. comm.) concurs.

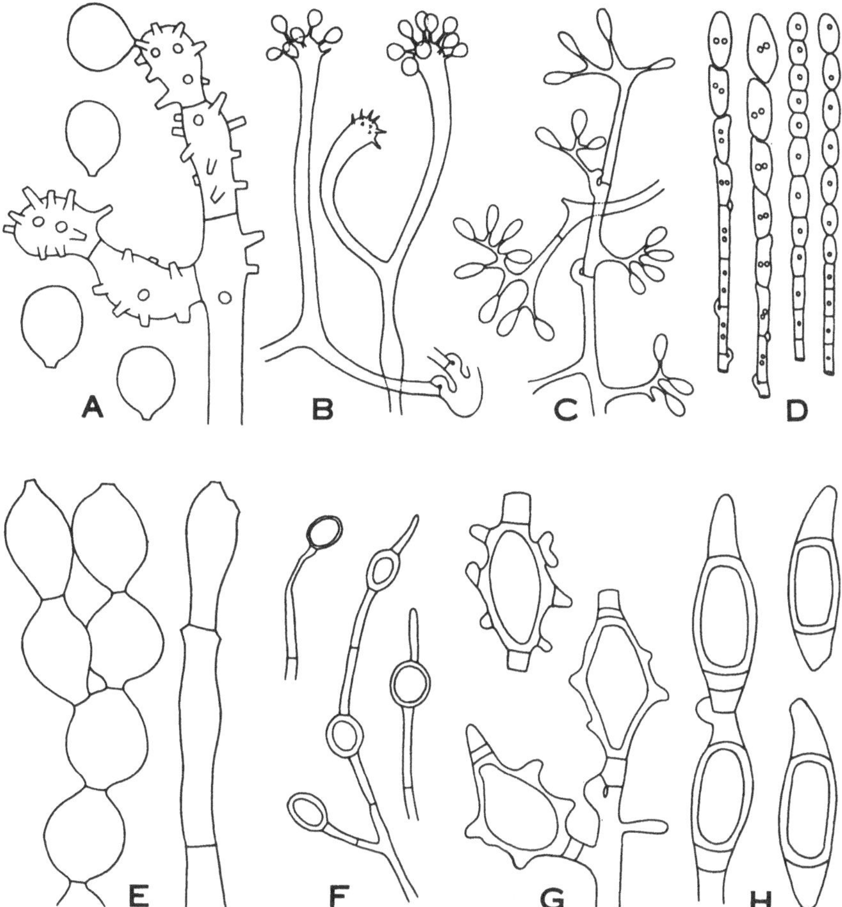


Figure 2.4 "Conidia, oidia, and chlamydospores" of homobasidiomycetes: A, *Botryobasidium robustior* (after Pouzar and Jechová 1967); B, *Vararia granulosa* (after Maxwell 1954); C, *Trechispora raduloides* (after Maxwell 1954); D, *Pleurotus corticatus* (after Kaufert 1935); E, *Botryobasidium simile* (after Pouzar and Holubová-Jechová 1969); F, *Corticium vellereum* (after Nobles and Nordin 1955); G, *Nyctalis asterophora* (after Thompson 1936); H, *Nyctalis parasitica* (after Thompson 1936).

The above examples include the "conidia" of Lyman's (1907) account. These conidia may be solitary, or in acropetal chains of blastospores, or they may be botryose blastospores or sympodioconidia.

Another imperfect state common to a large number of homobasidiomycetes is represented by "oidia" which are "produced by the fragmentation of ordinary vegetative hyphae" (Nobles 1948). These are arthrospores. Lyman (1907) discussed the presence of oidia in various groups of Basidiomycetes and concluded the oidia-formation is present in a large proportion of species in the higher families of the Hymenomycetes, but is not known in the lower families.

Kaufert (1935) described the production of oidia in black, glistening drops on tall, white coremia on both haploid (Figure 2.4 D, right) and dikaryotic mycelia (Figure 2.4 D, left) of *Pleurotus corticatus* (Agaricales). The oidia are formed by the basipetal fragmentation of the tips of the distal hyphae of the coremia. Kaufert explained that in coremia on dikaryotic mycelium there are prominent clamp connections at each cross-wall; disarticulation takes place so that the clamp connection becomes a part of the spore formed beneath the septum. Such arthrospores at first possess a terminal, beak-like projection which is the former clamp, but at maturity the projection is lost and the spore is finally ellipsoidal.

The third spore form found in the Homobasidiomycetes is the chlamydospore. Lyman (1907) considered the method of formation to agree "with the process known in other groups of fungi, as in *Mucor racemosus* ...; the condensing protoplasm draws away from the ends of the cell and concentrates in the middle region where the side-walls bulge to receive it. Here a resistant wall (endospore) forms about the encysting cell within and adnate to the hypha walls at each end. Continued contraction of the protoplasm may cause the abandonment of these end walls, and the formation of new walls farther in ... The mature chlamydospore is thick-walled, with dense, granular, refractive contents" and is "freed only by the decay of the empty portion of the parent hypha." Dr J.H. Ginns has used the term "retraction septa" for the walls laid down as the protoplasm migrates into the developing chlamydospore (pers. comm.); this term is derived from "les cloisons de retrait" of Boidin (1954). Chlamydospores of some phycomycetes develop in apparently the same manner and Drechsler (1941a) used the term "retaining septa" for essentially the same structures as "retraction septa." Nobles (1948) recorded terminal and intercalary chlamydospores in a number of wood-rotting homobasidiomycetes in culture. In *Corticium vellereum* (Figure 2.4 F), for instance, Nobles and Nordin (1955) found large numbers of chlamydospores produced on aerial and submerged hyphae, and also "in the tuft-like or diffuse fructifications formed in culture. Similar chlamydospores are usually very numerous in sporophores collected in nature." No other imperfect spore forms are produced in this species. The development of chlamydospores described by Nobles and Nordin is essentially similar to that given by Lyman (1907); in *C. vellereum* the subtending cells of the chlamydospores "may collapse and disintegrate, thus freeing the spores ... They have not been seen to germinate."

In the Polyporaceae chlamydospores appear to have no value in taxonomy above the species level (Nobles 1958), but this author does group a number of species in which chlamydospores are produced in abundance in cultures, and also in nature in association with the sporophores.

In nature, perhaps the chlamydospores of most species are normally produced in hyphae occurring in wood, soil, or other substrates, and are, therefore, seldom if ever seen. In these habitats they are possibly liberated from the subtending cells by the action of microorganisms; there seems to be little or no evidence of a specialized dehiscence mechanism to free the chlamydospores in pure culture. However, Dr J.H. Ginns (pers. comm.) has observed that chlamydospores of *Plicatura nivea* secede in pure culture by an apparent lysis of the subtending cells.

The chlamydosporic states of *Nyctalis asterophora* (Figure 2.4 G) and *N. parasitica* (Figure 2.4 H) have been discussed by Buller (1924); it appears that these two species reproduce themselves primarily by chlamydospores, and basidiospore-bearing fruit bodies are uncommon or rare. In *N. asterophora* the upper stratum of the pileus breaks up into a pulverulent, fawn-coloured mass of stellate chlamydospores. From Buller (1924), Thompson (1936), and an examination of a collection of *N. asterophora* (DAOM 110803) I gather that the terminal and intercalary cells, destined to develop into chlamydospores, produce irregular, tapering, finger-like and often branched processes: condensation of the cell contents and the laying down of a thick wall (ca. 3.5$\mu$) proceed so that the processes of the original cell wall are emptied of their contents. Perhaps the chlamydospores of *N. asterophora* secede from subtending cells by a natural dehiscence through a basal septum of terminal spores and through septa at each end of an intercalary spore, presumably with a circumscissile rupture of the outer wall. Each free chlamydospore is found with one or two, more or less cylindrical, broad processes, each of which represents the ends of the hyphal cell which gave rise to it. The ends of these processes are flat or slightly convex and are devoid of remnants of a contiguous cell, except perhaps for a slight indication of an outer circular frill. If the chlamydospores do secede schizolytically, then this is an innovation; in any case they have taken on the function of conidia. The presentation of a head of loose, dry spores would certainly favour their dispersal by wind; the multipronged nature of the spore possibly assists take-off and buoyancy.



In *N. parasitica* ellipsoidal chlamydospores are formed in the gills, which are ultimately composed almost entirely of these spores. Buller (1924) remarked that in this species "the hyphae which bear the chlamydospores become very pale and largely disappear, thus setting the chlamydospores free. Hence ripe chlamydospores readily fall from the gills in the form of a brown powder."

Burdsall (1969) illustrated and redescribed "stephanocysts" as unique structures in the Basidiomycetes, apparently limited to some species of *Hyphoderma* (Aphyllophorales); stephanocysts "do germinate and could act as asexual spores," and this author considered worthy of investigation the possibility of the forcible discharge of the upper cell of these two-celled structures.

Although the imperfect states of many homobasidiomycetes and ascomycetes share some common methods of conidium development, the form-generic names applied to these are very seldom mutually applicable. There are of course examples of form genera which are, or probably are, restricted to homobasidiomycetous Fungi Imperfecti, but the particular examples I have in mind are those conidial states produced by so many Aphyllophorales in pure culture. For instance, such a conidial state as that of *Fomes annosus*, and of other examples given above, has been referred to *Oedocephalum*; the species of this form genus apparently have a discomycetous affinity. The homobasidiomycetous counterparts can, I believe, be differentiated from species of *Oedocephalum* by their tapering and longer denticles. So far as I am aware no generic name has been proposed for this homobasidiomycetous botryose-blastosporic state, but here and elsewhere new names may be warranted, especially if the affinity of the state can be determined without reference to clamp connections (if these are produced).

SUMMARY OF IMPERFECT STATES OF SOME HOMOBASIDIOMYCETES With few exceptions the imperfect states of Homobasidiomycetes display little variation, regardless of the method of spore production involved. The paucity of generic names applied to such imperfect states is obvious from Donk's (1962) list of generic names of deuteromycetes with hymenomycetous affinity. Nevertheless, there are a few elegant forms such as *Riessia* (Goos 1967), *Costantinella* (the "handsome hyphomycete" of Grove 1936), *Oidium*, and *"Oedocephalum"*-like conidial states. Drechsler (1941b, 1960) has described some species which attack nematodes and amoebae, producing conidia or chlamydospores on mycelium bearing clamp connections. Also, Ingold (1959, 1961) illustrated elaborately branched spores of aquatic Fungi Imperfecti with clamp connections. But the Homobasidiomycetes have certainly not produced the wealth of form genera as have the Phycomycetes and Ascomycetes, or the Uredinales in the Heterobasidiomycetes.

Cleavage of cell contents is lacking in Homobasidiomycetes, as is the phialide and meristem arthrospore development. A multiplicity of conidia is produced either by sympodial development of the sporogenous cell or by the synchronous development of conidia on apical swellings of conidiophores. Acropetal chains of conidia are uncommon, and a succession of conidia produced by percurrent proliferations is not known although basidia have been described as developing in this way (Hughes 1971a). Chlamydospores are not uncommon in the Homobasidiomycetes but apart from those of *Nyctalis* they have remained more or less spherical thick-walled structures.

### *Heterobasidiomycetes - Uredinales and Septobasidiales*

An account of spore ontogeny in the Uredinales, considered in the light of those methods known to occur in ascomycetous Fungi Imperfecti, is published elsewhere (Hughes 1970) because it is too lengthy to include here. It is concluded in that article that spermatia are phialospores, aeciospores are meristem arthrospores, and urediniospores are sympodioconidia. Teliospores are, of course, held to represent the perfect state, at least so far as they are "cells of the kind giving rise to basidia" (Article 59); nevertheless, being spore-like in development, form, and structure, with a distributive function achieved by some, they are accordingly included in that account. It is further concluded that teliospores are produced either as terminal, chlamydospore-like cells of a hypha of determinate growth, as sympodioconidia, or as meristem arthrospores. The Uredinales alone in the Basidiomycetes display a series of distinctive imperfect states, and these are of a highly developed character: nevertheless, the spermatia, aeciospores, and urediniospores, generally speaking, show a remarkable uniformity. It is, for the most part, the teliospores which have given rise to diverse forms, and these are currently assigned to about one hundred generic names.

Nine or more species of *Septobasidium* described by Couch (1938) produce conidia of one kind or another. Some appear to be arthrosporic in nature, but the details available do not permit an assessment of their ontogeny. In about ten species of *Septobasidium* successive probasidia develop in a nested fashion as a result of percurrent proliferations of the subtending cell. 

## ASCOMYCETES

The majority of form genera of Fungi Imperfecti are undoubtedly based upon states of ascomycetes; they present us with an astonishing variety of form and it seems that every conceivable method of producing reproductive cells from other cells has been tried. The sporogenous cells themselves may be solitary, variously aggregated, or enclosed; this arrangement and disposition resulted in the distribution of imperfect states into Moniliales, Melanconiales, and Sphaeropsidales. Furthermore, the hyaline or coloured conidia themselves exploded into an astounding variety of sizes, shapes, and configurations with or without septation; these give rise to the Saccardoan spore groups.

Currently, conidium ontogeny is playing an increasing role in the taxonomic and classificatory processes in Fungi Imperfecti. This is particularly true in ascomycetous forms; these have received most attention because of their immense numbers and extraordinary variety. However, from the preceding fragmentary excursion into some phycomycetous and basidiomycetous groups, we find that the methods of producing solitary or a plurality of reproductive structures are essentially the same as those in the Ascomycetes. By this I mean that the sympodial, synchronous, and percurrent proliferating methods of producing a plurality of structures are common to the three groups although in some there may be only a sporadic occurrence of these methods. The arthrosporic disarticulation of hyphae is likewise widespread, and "chlamydospores" are commonly produced in all groups; acropetal chains of reproductive structures which secede, however, are apparently absent from the Phycomycetes.

Porospores and basauxic conidiophores are restricted to some ascomycetous Fungi Imperfecti. It has been pointed out that porospores are found only on some thick-walled, dematiaceous conidiophores; nevertheless, they have much in common and their perfect states have a peculiar affinity. Basauxic conidiophores are restricted to a few form genera; these and porospores will be the subjects of presentations by Dr K. Tubaki and Dr M.B. Ellis, respectively.

Of particular interest are: (1) the restriction of the method of cleavage of the contents of certain cells into spores to phycomycetes and some ascomycetes, and (2) the restriction of phialides and meristem arthrospores to Uredinales (and possibly Ustilaginales) and ascomycetes. It is about the phialide that I wish to make some comments.

## ON THE ORIGIN OF PHIALIDES

From the work of Olive (1944), on the formation of spermatia in *Gymnosporangium clavipes*, it is evident that these have a phialidic development similar to that found in a considerable number of ascomycetes Fungi Imperfecti. Olive described the sporogenous cells with an open collar (collarette) at the apex.

In ascomycetous phialides with an extended neck there is a basipetal succession of endogenous phialospores cleaved out of the distal contents of the unopened phialide without reference to the wall of the phialide itself. This can be observed in such phialides as those of *Sporoschisma* and *Chalara.* Somewhat similar observations had already been made by Lehman (1918) in a study of conidium (phialospore) formation in the (*Chalara*) sporogenous cells of "*Sphaeronema* [*Ceratocystis*] *fimbriatum.*" Lehman remarked thus on the first two conidia formed in an unopened sporogenous cell: "The wall of the conidium is nearly if not fully, as thick as that of the sheath surrounding it. This sheath does not appear thinner than the wall of the apical portion of the conidiophore before conidia had formed within it, nor yet less thick than the wall of the basal portion of the conidiophore ... the sheath takes no part in the formation of the conidial wall. The septum cutting off each conidium must arise from the new wall with which the protoplast invests itself and not directly from the wall of the conidiophore."

Some of Lehman's observations have been substantiated, in part, by Delvecchio, Corbaz, and Turian (1969), who investigated the *Chalara* phialide and phialospores of *Thielaviopsis basicola* with the electron microscope (see Figure 7.13 A). These authors found that the walls of phialides* were identical with those of vegetative hyphae, that the space between phialospores and phialide wall was devoid of cytoplasmic contents, and that the wall of the phialospore was never attached to the wall of the phialide; no evidence was found of an attachment between the wall of the meristematic protoplast in the lower part of the phialide and the wall of the phialide itself, and the wall of the meristematic protoplast appeared to be essentially the same as that of the phialospore.

It seems, therefore, that we have a kind of partial cleavage of the protoplasmic contents of a phialide to form the first one or more phialospores and these are formed independently of the wall of the phialide itself. The rupture of the apex of the phialide permits the release of the first-formed phialospores, the collarette thus becomes established, and a basipetal succession of other phialospores is produced from the remaining part of the protoplast, which behaves in a meristematic manner.

Whereas the cleavage of the entire contents of certain cells into spores is almost universal in the Phycomycetes, in the Ascomycetes the known examples of a complete cleavage of cell contents are few; nevertheless they are of considerable interest. Dring (1961a) is a recent author to demonstrate this cleavage in Ascomycetes; in the spermogonia of *Mycosphaerella brassicicola* he showed that the protoplast of the inner cells divided into four, sometimes a smaller number, and that these "protoplasts are extruded through a sterigma of which there is usually one but sometimes two." The protoplasts, which acquire their bacillar

*Delvecchio, Corbaz, and Turian used a different terminology from that which I use here, for phialospores and parts of the phialide:

shape as they make their exit, apparently remain "attached to their sterigmata until pushed from them by the next emerging single 'spore', or until the empty mother cell collapses." 

This process has been illustrated and described in a series of papers by Higgins (1920, 1929, 1936) and by Jenkins (1930, 1938, 1939) in other species of *Mycosphaerella*, and by Wolf and Barbour (1941) in *Systremma acicola*. Killian (1928), too, seemed to indicate a similar type of development. Dring (1961b) also described endogenous spermatia within spermogonia of *Ramularia armoraceae*. In some of these accounts it is recorded that the mother cell has one nucleus which divides twice and the cell contents then divide into four parts, but no walls are formed. The spermatium initials pass out one at a time. Occasionally the divided protoplasts lie end to end within the mother cell (Higgins 1936; Jenkins 1938, 1939), but usually they are in tetrads. In *Mycosphaerella bolleana*, Higgins (1920) found that spermatia may begin to develop even before the wall of the spermogonium is formed. Spermatium formation proceeds until the wall of spermogonia is very thin. Sutton (1964) listed this method of spermatium formation as one of "at least three logical explanations for the copious production of minute conidia" in pycnidial fungi.

Such mother cells as described above are not phialides because they possess no meristem, the entire contents of the cells being converted into spore initials. In this regard the mother cells are sporangial in nature. The observations of Higgins and others are of particular interest because they provide clues to the precursor of a kind of phialide; and it is, after all, in relation to the Phycomycetes that we should look for the derivation of some structures in ascomycetes.

At this juncture it is relevant to mention the spermatiophore and the development of spermatia in *Corynelia uberata* (Ascomycetes) as illustrated and described by Huguenin (1969). The spermatiophores, enclosed within spermogonia, are fusiform, simple, in tufts, thin-walled, and contain a large [basal] vacuole "dont le volume varie en fonction du développement de la spore." These spermatiophores, he continues, are simple phialides which produce their spores by successive apical budding. The illustration and text suggest that the increase in the size of the vacuole results in a reduction in the amount of protoplasm towards the apex of the phialide and at the same time the extruded protoplasm develops into a succession of spermatia at the apex of the phialide. There is a meristem involved so the spermatiophores are phialides but they differ from ordinary ones because of the presumed finite number of spores formed as the vacuole enlarges and the protoplasm is depleted. Also, they differ from the spermatium mother cells of some of the *Mycosphaerella* spp. mentioned above because there is no complete cleavage of the protoplasm within the mother cell.

Chadefaud (1960) derived the phialide from ancestral "sporocystes conidiogènes, ou 'conidiocystes'" and he referred to the spermatium mother cells of *M. tulipiferae* in regard to these. What I wish to do is to indicate the possible steps required for the initiation of a phialide from a cell of a sporangial nature, with reference to present-day fungi and some variations which have been observed in them.

Higgins (1936) and Jenkins (1938) recorded linearly arranged spermatial initials within the mother cells. If, in these, the basal initial acquired the ability to grow and divide repeatedly, then a succession of spermatia would result, each spermatium escaping through the "sterigma" or open end of the mother cell: a collarette would thus be established and the structure would possess all the basic attributes of a phialide. Of course, for this derivation we have to assume that the basal initial within the mother cell made that significant advance and kept growing and dividing; but this does not seem too fanciful because of certain features recorded in the sporangia of some phycomycetes.

In *Gonimochaete*, Drechsler (1946) described ellipsoidal cells, within the host, which develop a cylindrical neck into which the protoplasm migrates, the basal part of the cell being occupied by a large vacuole: the protoplasm in the neck cleaves, apparently basipetally, into spores which are liberated in groups through an apical aperture. There is complete cleavage of the protoplasm and no meristem, but it is of interest to note the linear arrangement of spores within the neck of this flask-shaped structure. A similar basal vacuole occurs in the spermatiophores of the ascomycete *Corynelia uberata*.

In *Basidiobolus meristosporus* (Drechsler 1955b) the contents of both globose and "adhesive conidia" are capable of cleaving into sporangiospores; occasionally only a part of the contents divides into spores, the part which remains undivided retaining its meristematic power because it is at least capable of producing a germ tube.

The conjecture is, therefore, that a phialidic process could well have arisen from a partially cleaved mother cell, with the undivided protoplast meristematically producing a basipetal succession of spores through the collarette or remains of the distal wall of the mother cell. This derivation would fulfil the principle "ontogeny recapitulates phylogeny," but this is not a denial that the phialide itself evolved further. After all, many phialides do not display such a clear delimitation of conidia with a long collarette; in *Penicillium corylophilum*, according to the reconstruction of Cole and Kendrick (1969) based partly on the electron microscope studies of Zachariah and Fitz-James (1967), the tip of the phialide is already open whilst the first phialospore is completely delimited only distally, being still in continuity with the phialide apex within the collarette. Furthermore, the Cole and Kendrick illustrations indicate that in this *Penicillium* the first phialospore increases in length and width after the distal, outer wall of the phialide is opened: in phialides with phialospores formed and maturing entirely within the collarette there is seldom further increase in width after their release.

It appears to me that one line of development in the phialide has led to a shortening of the neck and collarette, to an early rupture or dissolution of the wall enclosing the first phialospore, and finally to a lowering of the point of rupture so that it is close to the location of the future meristematic apex of the phialide itself with the result that the collarette is ill-defined and difficult to observe. A diagrammatic representation of the conjectured derivation of the phialide and of speculations on its further modification is shown in Figure 2.5.

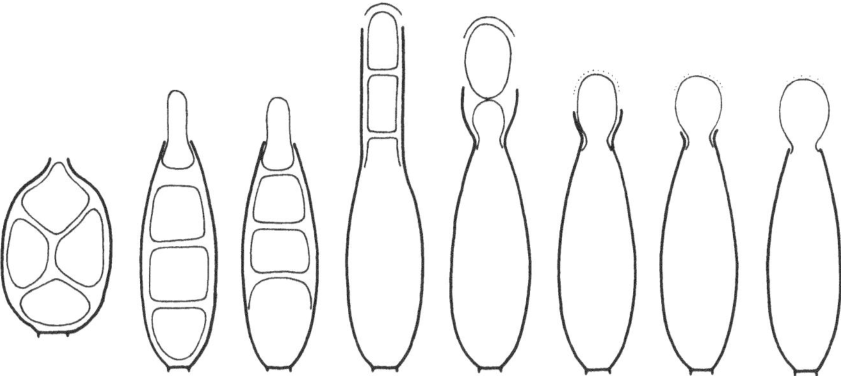

Figure 2.5 Stages in conjectured derivation of phialidic method of conidium development.

This certainly stresses the need for electron microscope studies of the formation and liberation of the first phialospore, in phialides which display a conspicuous collarette and in those in which the collarette is difficult to observe.

The basipetal succession of phialospores resulting from the meristematic activity of the protoplast within the opened phialide is equivalent to the development of meristem arthrospores from a mother cell. In this regard I draw attention to Dangeard's (1896) illustration and description, including details of nuclear behaviour, of the sequence of events in the "*Oidium*" conidiophore of *Sphaerotheca castagnei*. It is curious that no phialides (used in a restricted sense) have been recorded in the Erysiphaceae (Dr E. Müller, pers. comm.): Langeron (1945) and Chadefaud (1960), however, included, among other examples, such *Oidium* states as phialides, but they were using this term in a broad sense.

In Fritsch's (1945) treatment of the Chamaesiphonales (and, indeed, of many other orders of algae) we find some tantalizing illustrations and accounts concerning the production of reproductive structures. At this time I will mention only one: in his account of "exospores" Fritsch wrote that "when a *Chamaesiphon*-individual has reached a certain size, the membrane (probably the cell sheath) ruptures apically and the exposed protoplast abstricts spherical spores successively from its tip, much in the same way as conidia are produced in many Fungi." The protoplast "is not used up in the formation of the spores." These statements suggest that this process of spore formation may be compared with the phialidic or meristem-arthrosporic kind of conidium ontogeny found in fungi. But this is not the place to attempt an interpretation of illustrations and accounts of developing reproductive structures in algae; nevertheless, there is evidence (Hughes 1971a) that algae and fungi show some similar methods of producing a plurality of reproductive structures. Further similarities in the basic methods of producing conidia in fungi, and gemmae, not only in algae but also in bryophytes and pteridophytes, are briefly documented elsewhere (Hughes 1971b).

The preparation of this and other related articles has taught me, belatedly, that studies restricted to ascomycetous Fungi Imperfecti are not the only way, perhaps not even the best way of attempting to solve secrets of conidium ontogeny in this group.* I started with a relevant quotation from Corda's *Pracht-Flora* published 130 years ago and I cannot but think that he would have delighted in a gathering of mycologists intending to devote a week to a consideration of conidium ontogeny.

ACKNOWLEDGMENTS

I have benefitted considerably from frequent discussions with colleagues at the Plant Research Institute, and I wish to thank them. Dr C.W. Hesseltine kindly supplied useful references to oidia in Mucorales.

REFERENCES

Ainsworth, G.C. 1961. Ainsworth and Bisby's Dictionary of the Fungi. 5th ed. Commonwealth Mycological Institute, Kew

Bartnicki-Garcia, S., and Nickerson, W.J. 1962. Induction of yeastlike development in *Mucor* by carbon dioxide. J. Bacteriol. 84: 829-40

Benjamin, R.K. 1959. The merosporangiferous Mucorales. Aliso 4: 321-433

- 1961. Addenda to "The merosporangiferous Mucorales." Aliso 5: 11-19

Bessey, E.A. 1950. Morphology and taxonomy of fungi. The Blakiston Company, Philadelphia

Blackwell, E. 1949. Terminology in *Phytophthora*. C.M.I. Mycol. Pap. 30

Boedijn, K.B. 1927. Über *Rhopalomyces elegans* Corda. Ann. Mycol. Berl. 25: 161-6

Boidin, J. 1954. Essai biotaxonomique sur les Hydnés résupinés et les Corticiés. Etude spéciale du comportement nucléaire et des mycéliums. Thèse, Fac. Sci. Lyon, No. 202

Buller, A.H.R. 1924. Researches on fungi. Vol. 3. Longmans, Green and Co., London

Burdsall, H.H. 1969. Stephanocysts: unique structures in the Basidiomycetes. Mycologia 61: 915-23

Butler, E.J. 1907. An account of the genus *Pythium* and some Chytridiaceae. Dept. of Agric., India, Mem., Bot. Ser. 1 (5): 1-161

Chadefaud, M. 1960. Les végétaux non vasculaires (Cryptogamie). *In* Chadefaud, M., and Emberger, L. Traité de botanique systématique. Vol. 1. Masson et Cie, Paris

Cole, G.T., and Kendrick, W.B. 1969. Conidium ontogeny in hyphomycetes. The phialides of *Phialophora, Penicillium*, and *Ceratocystis*. Can. J. Botany 47: 779-89

Couch, J.N. 1938. The genus *Septobasidium*. University of North Carolina Press, Chapel Hill

Dangeard, P.-A. 1896. Second mémoire sur la reproduction sexuelle des Ascomycètes. Le Botaniste 5: 245-84

Delvecchio, V.G., Corbaz, R., and Turian, G. 1969. An ultrastructural study of the hyphae, endoconidia and chlamydospores of *Thielaviopsis basicola*. J. Gen. Microbiol. 58: 23-7

*Somewhat modified from the last paragraph in the Epilogue in W.B. Grove's *British Stem- and Leaf-Fungi (Coelomycetes)*, vol. 2 (Cambridge University Press, London, 1937).

Donk, M.A. 1962. The generic names proposed for hymenomycetes, XII. Deuteromycetes. Taxon 11: 75-104
Drechsler, C. 1935a. Some non-catenulate conidial phycomycetes preying on terricolous amoebae. Mycologia 27: 176-205
- 1935b. A new species of conidial phycomycete preying on nematodes. Mycologia 27: 206-15
- 1936. New conidial phycomycetes destructive to terricolous amoebae. Mycologia 28: 363-89
- 1937. New Zoopagaceae destructive to soil rhizopods. Mycologia 29: 229-49
- 1938. New Zoopagaceae capturing and consuming soil amoebae. Mycologia 30: 137-57
- 1939. Five new Zoopagaceae destructive to rhizopods and nematodes. Mycologia 31: 388-415
- 1941a. Four phycomycetes destructive to nematodes and rhizopods. Mycologia 33: 248-69
- 1941b. Some hyphomycetes parasitic on free-living terricolous nematodes. Phytopathology 31: 773-802
- 1945. Several additional phycomycetes subsisting on nematodes and amoebae. Mycologia 37: 1-31
- 1946. A nematode-destroying phycomycete forming immotile spores in aerial evacuation tubes. Bull. Torrey Bot. Club 73: 1-17
- 1955a. Two new species of *Conidiobolus* that produce microconidia. Am. J. Botany 42: 793-802
- 1955b. A southern *Basidiobolus* forming many sporangia from globose and from elongated adhesive conidia. J. Wash. Acad. Sci. 44: 49-56
- 1960. A clamp-bearing fungus using stalked young chlamydospores in capturing amoebae. Sydowia 14: 246-57
Dring, D.M. 1961a. Studies on *Mycosphaerella brassicicola* (Duby) Oudem. Trans. Brit. Mycol. Soc. 44: 253-64
- 1961b. *Ramularia armoraciae* Fuckel. Trans. Brit. Mycol. Soc. 44: 333-6
Ellis, J.J. 1963. A study of *Rhopalomyces elegans* in pure culture. Mycologia 55: 183-98
Eriksson, J. 1958. Studies in the Heterobasidiomycetes and Homobasidiomycetes - Aphyllophorales of Muddus National Park in North Sweden. Symb. Bot. Upsal. 16 (1): 1-172
Eriksson, J., and Hjortstam, K. 1969. Studies in the *Botryobasidium vagum* complex (Corticiaceae). Friesia 9: 10-17
Fritsch, F.E. 1945. The structure and reproduction of the Algae. Vol. II. Cambridge University Press, London
Goos, R.D. 1967. Observations on *Riessia semiophora*. Mycologia 59: 718-22
Grove, W.B. 1936. A handsome hyphomycete. J. Botany 74: 50-2
Hawker, L. 1966. Germination: morphological and anatomical changes. *In* The fungus spore. Edited by M.F. Madelin. Butterworths, London, England. pp. 151-61
Hesseltine, C.W. 1952. A survey of the Mucorales. Trans. New York Acad. Sci., Ser. 2, 14: 210-14
- 1954. The section *Genevensis* of the genus *Mucor*. Mycologia 46: 358-66
- 1955. Genera of Mucorales with notes on their synonymy. Mycologia 47: 344-63
Higgins, B.B. 1920. Morphology and life history of some ascomycetes with special reference to the presence and function of spermatia. Am. J. Botany 7: 435-44
- 1929. Morphology and life history of some ascomycetes ... II. Am. J. Botany 16: 287-96

- 1936. Morphology and life history of some ascomycetes ... III. Am. J. Botany 23: 598-602

Hughes, S.J. 1953. Conidiophores, conidia, and classification. Can. J. Botany 31: 577-659

- 1970 [1971]. Ontogeny of spore forms in Uredinales. Can. J. Botany 48: 2147-57

- 1971a. Percurrent proliferations in fungi, algae, and mosses. Can. J. Botany 49: 215-31

- 1971b. On conidia of fungi and gemmae of algae, Bryophytes, and Pteridophytes. Can. J. Botany (in press)

Huguenin, B. 1969. Micromycètes du Pacifique sud (Huitième contribution). Ascomycètes du Nouvelle-Calédonie (II). Cahiers du Pacifique 13: 295-308

Ingold, C.T. 1959. Aquatic spora of Omo Forest, Nigeria. Trans. Brit. Mycol. Soc. 42: 479-85

- 1961. Another aquatic spore-type with clamp connexions. Trans. Brit. Mycol. Soc. 44: 27-30

Jenkins, W.A. 1930. The cherry leaf-spot fungus, *Mycosphaerella cerasella* Aderh., its morphology and life history. Phytopathology 20: 329-37

- 1938. Two fungi causing leaf spot of peanut. J. Agric. Res. 56: 317-32

- 1939. The development of *Mycosphaerella berkeleyii*. J. Agric. Res. 58: 617-20

Johnson, T.W. 1956. The genus *Achlya:* morphology and taxonomy. University of Michigan Press, Ann Arbor

Kaufert, F. 1935. The production of asexual spores by *Pleurotus corticatus*. Mycologia 27: 333-41

Killian, C. 1928. Études comparatives des caractères culturaux et biologiques chez les Deuteromycètes et les Ascomycètes parasites. Ann. Sci. Nat. Bot. Sér. 10, 10: 101-292

Langeron, M. 1945. Précis de Mycologie. Masson et Cie, Paris

Lehman, S.G. 1918. Conidial formation in *Sphaeronema fimbriatum*. Mycologia 10: 155-63

Lyman, G.R. 1907. Culture studies on polymorphism of Hymenomycetes. Proc. Boston Soc. Nat. Hist. 33: 125-209

Mangin, L. 1891. Sur la désarticulation des conidies chez les Péronosporées. Bull. Soc. Bot. France 38: 232-6

Mathews, V.D. 1931. Studies on the genus *Pythium*. University of North Carolina Press, Chapel Hill

Maxwell, M.B. 1954. Studies of Canadian Thelephoraceae. XI. Conidium production in the Thelephoraceae. Can. J. Botany 32: 259-80

McKeen, C.G. 1952. Studies of Canadian Thelephoraceae. IX. A cultural and taxonomic study of three species of *Peniophora*. Can. J. Botany 30: 764-87

Naumov, N.A. 1939. Clés des Mucorinées. Encyclopédie mycologique. Vol. IX. Lechevalier, Paris

Nobles, M.K. 1937. Production of conidia by *Corticium incrustans*. Mycologia 29: 557-66

- 1942. Secondary spores in *Corticium effuscatum*. Can. J. Res. 20: 347-57

- 1948. Studies in forest pathology. VI. Identification of cultures of wood-rotting fungi. Can. J. Res. C, 26: 281-431

- 1958. Cultural characters as a guide to the taxonomy and phylogeny of the Polyporaceae. Can. J. Botany 36: 883-926

Nobles, M.K., and Nordin, V.J. 1955. Studies on wood-inhabiting hymenomycetes. II. *Corticium vellereum* Ellis and Cragin. Can. J. Botany 33: 105-12

Olive, L.S. 1944. Spermatial formation in *Gymnosporangium clavipes*. Mycologia 36: 211-14

Pouzar, Z., and Holubová-Jechová, V. 1969. *Botryobasidium simile* spec. nov., a perfect state of *Oidium simile* Berk. Česká Mykologie 23: 97-101

Pouzar, Z., and Jechová, V. 1967. *Botryobasidium robustior* spec. nov., a perfect state of *Oidium rubiginosum*. Česká Mykologie 21: 69-73
Sparrow, F.K. 1960. Aquatic phycomycetes. 2nd ed. University of Michigan Press, Ann Arbor
Sutton, B.C. 1964. *Phoma* and related genera. Trans. Brit. Mycol. Soc. 47: 497-509
Thaxter, R. 1888. The Entomophthoreae of the United States. Mem. Boston Soc. Nat. Hist. 4: 133-201
Thompson, G.E. 1936. *Nyctalis parasitica* and *N. asterophora* in culture. Mycologia 28: 222-7
Waterhouse, G.M. 1963. Key to the species of *Phytophthora*. C.M.I. Mycol. Pap. 92
- 1967. Key to *Pythium* Pringsheim. C.M.I. Mycol. Pap. 109
Weston, W.H. 1923. Production and dispersal of conidia in the Philippine *Sclerosporas* of Maize. J. Agric. Res. 23: 239-78
Wolf, F.A., and Barbour, W.J. 1941. Brown-spot needle disease of pines. Phytopathology 31: 61-74
Young, T.W.K. 1969. Electron and phase-contrast microscopy of spores in two species of the genus *Mycotypha* (Mucorales). J. Gen. Microbiol. 55: 243-9
Zachariah, K., and Fitz-James, P.C. 1967. The structure of phialides in *Penicillium claviforme*. Can. J. Microbiol. 13: 249-56
Zycha, H. 1935. Pilze II, Mucorineae. *In* Kryptogamenflora der Mark Brandenburg. Vol. VIa. pp. 1-264



## DISCUSSION

DR MÜLLER

It is interesting that there are so many similarities in the development of two completely different groups of fungi, the Oomycetes and the rest; this shows that we are confronted with a general problem which is not dependent on the origin of the organisms. Oomycete conidia probably do not have the same origin as those of other groups. Oomycete conidia are converted sporangia. We aren't sure whether the conidia of Ascomycetes have the same origin. Conidia of the Zygomycetes *may* have the same origin, because they show developmental stages from real sporangia with cleavage to unisporic sporangia which may be like those of the Oomycetes, but they are different because spores form within the sporangia, whereas in the Oomycetes a new spore is never formed within the sporangium, as far as we know. So there are differences, but it is astonishing that in such different groups there appear to be rules that produce similar forms.

DR KENDRICK

I think Dr Hughes has given us ample evidence that parallel or convergent evolution has taken place in these groups of organisms. This is one of the problems that face us when we attempt to set up a classification for the Fungi Imperfecti. We never know how many times a particular method of conidium production has evolved. Has Dr Hughes any idea how many times the phialide mechanism has evolved?

DR S.J. HUGHES

The phialide as currently understood is being recognized in Ascomycetes and Uredinales. But this is not a denial that the phialidic process occurs elsewhere.

DR HENNEBERT

I studied the "conidial" state (the uredinial state) of *Melampsora*, and found sympodial proliferation of the conidiophore during spore production. This kind of perspective - grouping all imperfect states - will necessitate a number of revisions in nomenclature. For instance, *Mucor*. This has been considered as the generic name of a perfect fungus, even though the type specimen of the type species did not bear zygospores. According to this point of view, *Mucor* should remain an imperfect genus, and we should have a name for the perfect state. We should consider Mucorales along with the pleomorphic fungi which are not yet satisfactorily dealt with in the Code of Botanical Nomenclature.

DR S.J. HUGHES

It was not my intention to give the impression that certain states of rusts and so on should be included *in* Fungi Imperfecti. I merely wanted to draw attention to the essential identity of some methods of producing reproductive structures in diverse groups.

DR ELLIS

The idea of bringing all these groups together for consideration is a good one - it is of fundamental importance - but unless some hitherto unknown method of spore production is found in any of these forms (and apparently it has not been), perhaps we should concentrate on the Hyphomycetes where the different methods of spore production are much more easily seen and recognized. The study of conidial states in the Uredinales and Phycomycetes is, perhaps, more difficult and could be left until the Hyphomycetes have been properly worked out.

DR KENDRICK

Certainly you can't grow some of these other groups in culture, which doesn't help. But it's interesting to note that the cytology of many of these other groups has been explored very fully, particularly in the rusts, whereas that of the Hyphomycetes has been almost totally neglected. Dr Hughes's paper has emphatically confirmed his original contention that there can only be a limited number of ways in which a spore can be formed, and that these ways will have been exploited by all sorts of different groups. It is interesting to note the direction or directions apparently taken by each group, and the limitations consequently imposed on it.

DR CARROLL

Dr Streiblová's work on yeasts is another example of how parallelism exists in the ways in which different fungal groups form cells. Figure 2.6 m,n represents the formation of bud cells in *Saccharomyces cerevisiae*. As I interpret them, they are essentially blastospores. The interesting thing is that, although one yeast cell can form a number of bud cells, only a single daughter cell can arise from any given place on the parent cell. The figures show a number of scattered bud scars, each of which represents only one new cell. The practical effect of this is that the reproductive potential of a cell of *S. cerevisiae* is limited by the amount of available cell surface.

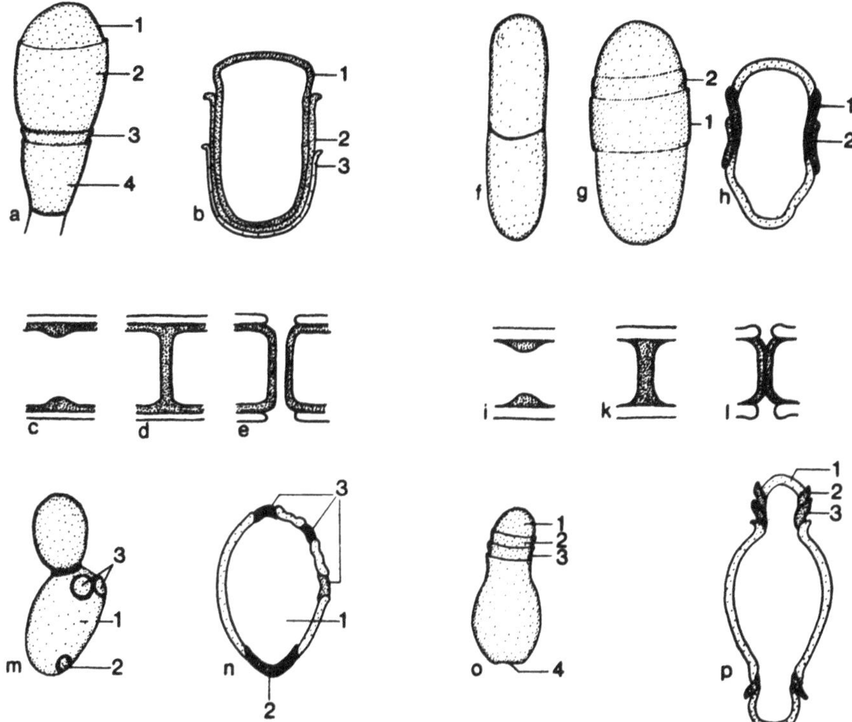


Figure 2.6 Conidium formation in yeasts; see text (after Streiblová; from Müller and Loeffler, *Mykologie* [Thieme-Verlag, Stuttgart, 1968]).

Fluorescence microscopy enables us to find out what happens to the old wall when new wall is formed, and because of the way in which fluorescent dyes are differentially absorbed by old and new wall material, we can resolve things we just couldn't see with ordinary light microscopy. I am most familiar with the case of *Saccharomycodes ludwigii* (Figure 2.6 o,p). Using fluorescence microscopy, Dr Streiblová has shown that bud cell formation here involves repeated percurrent proliferation, resembling annelloconidium production rather than blastoconidium production.

DR MÜLLER

In *Endomyces magnusii*, the inner wall of the mother cell lays down a transverse septum (Figure 2.6 c,d). The two layers of this septum eventually separate, and a concurrent circumscissile split in the outer wall releases the daughter cell (Figure 2.6 e). As the mother cell elongates, its new portion is initially clad only in the original inner wall, but a complete new inner wall is soon laid down, which then produces another septum to cut off a second daughter cell. After this whole process has been repeated several times, the mother cell is surrounded by the cup-like remnants of the successive wall layers (Figure 2.6 a,b).

Cell division in *Schizosaccharomyces octosporus* is somewhat similar. The septum is formed only by a more localized inner wall layer (Figure 2.6 i,k), and the separation occurs as in *Endomyces* (Figure 2.6 l). The cell elongates after the separation, and the ring-like remnant of the outer wall surrounds its middle portion. As cell division is repeated, the cells may become surrounded by systems of ring-like remnants (Figure 2.6 g,h).

# 3
# Characters of Conidiophores as Taxonomic Criteria

K.A. PIROZYNSKI

Replying to M.C. Cooke's criticism of the newly published first volume of *Sylloge Fungorum*, as a work advocating an artificial system of classification of Pyrenomycetes, Saccardo (1882) compared the carpological approach of Fries and his followers with that proposed, in the seventeenth century, for the higher plants by de Tournefort, who divided phanerogams into trees, shrubs, and plants without, as Saccardo put it, "taking any account of the characteristics which are much more important but much more difficult to preserve." He referred, of course, to the reproductive organs (notably flowers and structures derived from them) with which he considered fungal spores analogous. When devising his system of classification of Hyphomycetes, Saccardo (1886) also attached great significance to the morphology of conidia by making this character the basis for the major divisions of his ill-defined families - Mucedinaceae, Dematiaceae, Tuberculariaceae, and Stilbaceae - which, in turn, were based on colour and configuration of conidiophores. Saccardo's concept of the conidiophore was very broad and, consequently, vague: a spore-bearing hypha. In fact, by emphasizing the taxonomic importance of different configurations of these hyphae, Saccardo classified Hyphomycetes into "herbs, shrubs, trees, and forests."

This concept of the conidiophore lives. Snell and Dick (1957) defined the conidiophore as "a specialized hypha or sporophore bearing conidia." To define a diagnostic structure in terms as broad as these leaves much room for diverse interpretation and misinterpretation. For example, the conidiophore of an aleuriosporic fungus, represented diagrammatically in Figure 3.1, can be described as: (1) well-differentiated conidiophores lacking; conidia more or less sessile on vegetative hyphae; (2) conidiophores reduced, borne laterally on vegetative

Mycology Unit, Plant Research Institute, Ottawa, Canada.

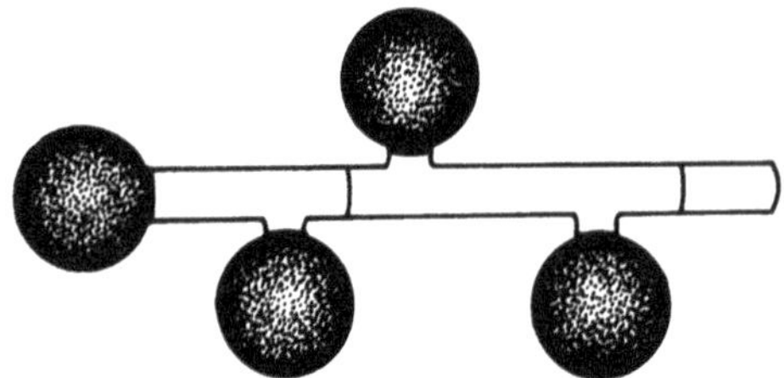

Figure 3.1 Conidiophore of a hypothetical "aleuriosporic" fungus.

hypha; or (3) (should the hypha happen to be more or less ascending) conidiophores well differentiated, producing conidia on prominent lateral denticles.

Such diverse interpretations of what is essentially the same type of simple fertile hypha are, nevertheless, within the accepted definition of the conidiophore. With the increasing complexity of conidium-bearing structures the terminology becomes even more chaotic as hordes of auxiliary names are introduced: conidiophore (or main) axis, stalk, trunk, or stipe bearing sporogenous or fertile branches, sporogenous, conidiogenous, or formative cells, sterigmata, branchlets, pedicels, prophialides, ampullae, metulae, collar hyphae, and a whole family of hyphae: parent, mother, daughter, etc., most of which can be used and are used in different senses. Yet, paradoxically, despite this proliferation of terms and the freedom of expression granted to authors, or perhaps because of it, many modern descriptions have not advanced, as far as clarity is concerned, beyond those written in the "flocci sporidiferi" era and, unless accompanied by an illustration, fail to create in the reader's mind a true picture of the organism. There are many conidiophores which defy the terminology and do not yield themselves to a "standardized" description; and, if such a description is provided, it may not only be meaningless, but also quite misleading. For example, the conidiophore of *Beltraniella porosa* (Figure 3.3 d) begins life as a minute conical cell closely appressed to the host's cuticle, and may produce conidia on successive terminal denticles. It then proliferates to form a vertical, sterile hypha bearing several lateral denticulate cells, and eventually becomes a complex structure composed of branched chains of denticulate conidiogenous cells intermixed with vegetative cells with some of the branches terminating in setae. Depending on the stage of development in which the specimen is found, the conidiophores can be described as one-celled, 5-10 μ high; simple, setiform, 100-120 μ high; or branched, setiferous, 300-350 μ high. Thus it is possible to emphasize, in the diagnosis, characters which are least diagnostic: the developmental stages comparable to seedlings, saplings, and trees. In considering examples such as this the question arises: is the terminology still inadequate or is it unsuitable; and if the configuration and dimensions of some conidiophores are of little diagnostic value, are there, within a system of spore-bearing hyphae, cells or groups of cells which should be singled out and emphasized?

As I have mentioned earlier, Saccardo stressed the importance of spore morphology in the classification of fungi. However, we now know that the spore is not, in taxonomic and phylogenetic considerations, equivalent to flowers of higher plants. In the Ascomycetes it is the ascus and particularly the apical apparatus specifically adapted for ascospore release that is far more significant than the ascospore itself. Likewise in the Fungi Imperfecti it is the conidiogenous cell, and particularly the area which participates in conidium production and release, that overshadows the conidium structure as a taxonomic criterion. Consequently, mycologists are beginning to realize that the term conidiophore, as applied to those hyphae which elevate conidia above the substrate, is ambiguous because the hyphae are not comparable. In recent literature one finds "conidiophores" cited in parentheses. Attempts are made to redefine the term. Ainsworth (1961) defined the conidiophore as "a simple or branched hypha (fertile hypha), with or without phialides," and Barron (1968) as a "main axis or branch bearing sporogenous cells." 

Already in 1953 Hughes had pointed out that "whereas a large number of well or poorly differentiated terms are available for various spore forms the reverse seems to be true of 'conidiophores'" for which "the need for more precise terms is evident." He was also well aware of the need for differentiating between different kinds of "actual structures bearing conidia" as well as between conidiophores of different origins as those of, for example, *Annellophora* and *Stysanus*. More recently Chesters (1969) called for a clear definition of "conidiophore" and suggested that one solution "might be to use the term fertile hyphae for the assemblage of hyphae which bear formative cells, to describe the system of branching of these hyphae by descriptive terms and to restrict the term conidiophore to the formative cell." Although the idea is highly commendable, the proposal seems rather drastic for the following reasons: the term conidiophore would replace conidiogenous cell and other more specific terms (e.g. phialide, annellophore) which are rapidly becoming established in mycological literature, and, at the same time, need would be created for changing the century-old concept of the conidiophore and replacing it with another term. The term conidiophore should be retained in its current sense, i.e. to describe the entire system of fertile hyphae. The structure must, however, be analysed into separate functional units which can then be evaluated in a more precise manner and in true perspective.

In the preliminary programme Dr Kendrick dedicated this session to "conidiophores (sporogenous cells and supporting structures)." The entry contains all the ingredients for a definition which, in my opinion, will differentiate "flowers" from "stems, branches, and trunks": the conidiophore is a conidiogenous cell or a system of conidiogenous cells with or without differentiated supporting structures.

In Table 3.1 I have attempted to classify diverse types of conidiophores into groups based on characters of conidiogenous cells and supporting structures. In the top column the recognized types of conidiogenous cells are arranged in a sequence which incorporates the schemes put forward by Barron (1968), Subramanian (1965), and Tubaki (1963), as well as that originally proposed by Hughes (1953). This arrangement is not designed to compare or to contrast

TABLE 3.1. Classification of conidiophores

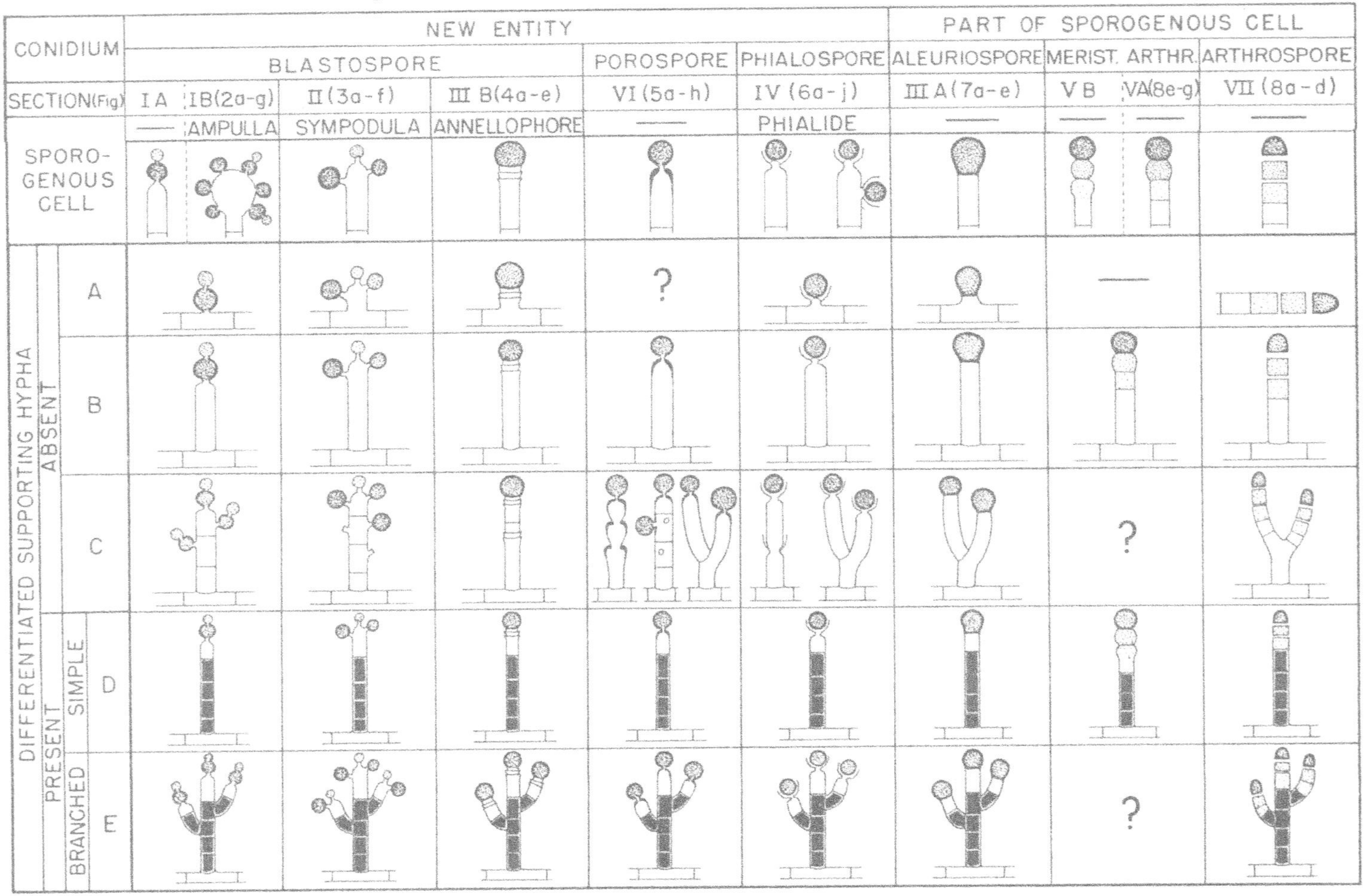

| CONIDIUM | | | NEW ENTITY | | | | | | PART OF SPOROGENOUS CELL | | |
|---|---|---|---|---|---|---|---|---|---|---|---|
| | | | BLASTOSPORE | | | POROSPORE | PHIALOSPORE | ALEURIOSPORE | MERIST. ARTHR. | | ARTHROSPORE |
| SECTION(Fig) | | | I A | IB(2a-g) | II (3a-f) | III B(4a-e) | VI (5a-h) | IV (6a-j) | III A(7a-e) | V B | VA(8e-g) | VII (8a-d) |
| | | | — | AMPULLA | SYMPODULA | ANNELLOPHORE | — | PHIALIDE | — | — | — | — |
| SPORO-GENOUS CELL | | | | | | | | | | | | |
| DIFFERENTIATED SUPPORTING HYPHA | ABSENT | A | | | | | ? | | | — | | |
| | | B | | | | | | | | | | |
| | | C | | | | | | | | ? | | |
| | PRESENT | SIMPLE D | | | | | | | | | | |
| | | BRANCHED E | | | | | | | | ? | | |

systems of classification based on conidium ontogeny alone with those based partly on conidium ontogeny and partly or wholly on conidiogenous cell types. However, I would like to call for an agreement on what we are going to call a conidiogenous cell: any cell which is directly involved in conidiogenesis or "a spore-bearing cell which is morphologically distinguishable from the ordinary sporiferous cells or conidiophores or hyphae," as suggested by Subramanian (1965). 

Illustrated in Table 3.1 (in rows A-E) are five types of conidiophores, i.e. conidiogenous cells and supporting hyphae. In groups A, B, and C the conidiophore is a simple conidiogenous cell or complex of conidiogenous cells; differentiated supporting hyphae are lacking.

In group A, the conidiophore is represented by a single conidiogenous cell, which can be discrete, as in some yeasts, incorporated in a vegetative thallus (e.g. stroma) or reproductive thallus (e.g. conidium or ascospore), or which, as in most cases, occupies an intercalary or terminal position in a hypha composed of similar vegetative or conidiogenous cells. In this group conidiogenous cells belonging to Hughes's sections I, II, IIIB, IV, IIIA, and VII are represented. The respective examples are: *Beniowskia sphaeroidea* (Figure 3.2 a), *Dactylaria purpurella* (Figure 3.3 a), *Saccharomyces ludwigii* (Figure 3.4 a) or *Annellophorella faureae* (see Ellis 1963), *Phialophora heteromorpha* (Figure 3.6 a), *Pithomyces chartarum* (Figure 3.7 a), and *Trichosporon cutaneum* (Figure 3.8 a). In all of these except *Saccharomyces* the conidiogenous cells are intercalary and as such cannot abstrict meristem arthroconidia (Hughes's section V) but can, at least theoretically, bear poroconidia (Hughes's section VI). When in a terminal position the conidiogenous cells are not restricted to the spore types illustrated. Groups B and C differ only in having more or less ascending conidiophores and conidiogenous cells which are more or less distinctly morphologically differentiated and cut off from vegetative hyphae by a septum.

In the fungi classified in group B the conidiophore is also a single conidiogenous cell, but it is borne laterally on a vegetative hypha, a stroma, or a reproductive body (as noted above). In this group each type of conidiogenous cell is represented: IA, *Monilia cinerea* (Figure 3.2 b) or *Septonema hormiscium* (Figure 3.2 c); IB, *Cephaliophora tropica*; II, *Rhinocladiella* sp. (Figure 3.3 b); IIIB, *Stigmina pedunculata* (Figure 3.4 b); VI, *Exosporium tiliae* (Figure 3.5 a); IV, *Gliomastix murorum* (Figure 3.6 b) or *Capnophialophora* states of *Strigopodia* (Figure 3.6 c, ascospore of *S. resinae*; Figure 3.6 d, *Hormisciella* conidium of *S. batistae*); IIIA, *Monodictys* sp. (Figure 3.7 b); VA, *Phragmotrichum chailletii* (Figure 3.8 f); VB, *Basipetospora rubra* (Figure 3.8 e); and VII, *Geotrichum candidum* (Figure 3.8 b).

In the fungi classified in group C the conidiophore is composed largely or entirely of conidiogenous cells. The vegetative cells which are often incorporated result, at least in some cases, from the division of spent conidiogenous cells. The conidiophores are built up of conidiogenous cells which proliferate by synchronous or sympodial branching, by apical meristematic growth and progressive acropetal, basipetal, or random septation with each newly formed cell assuming a

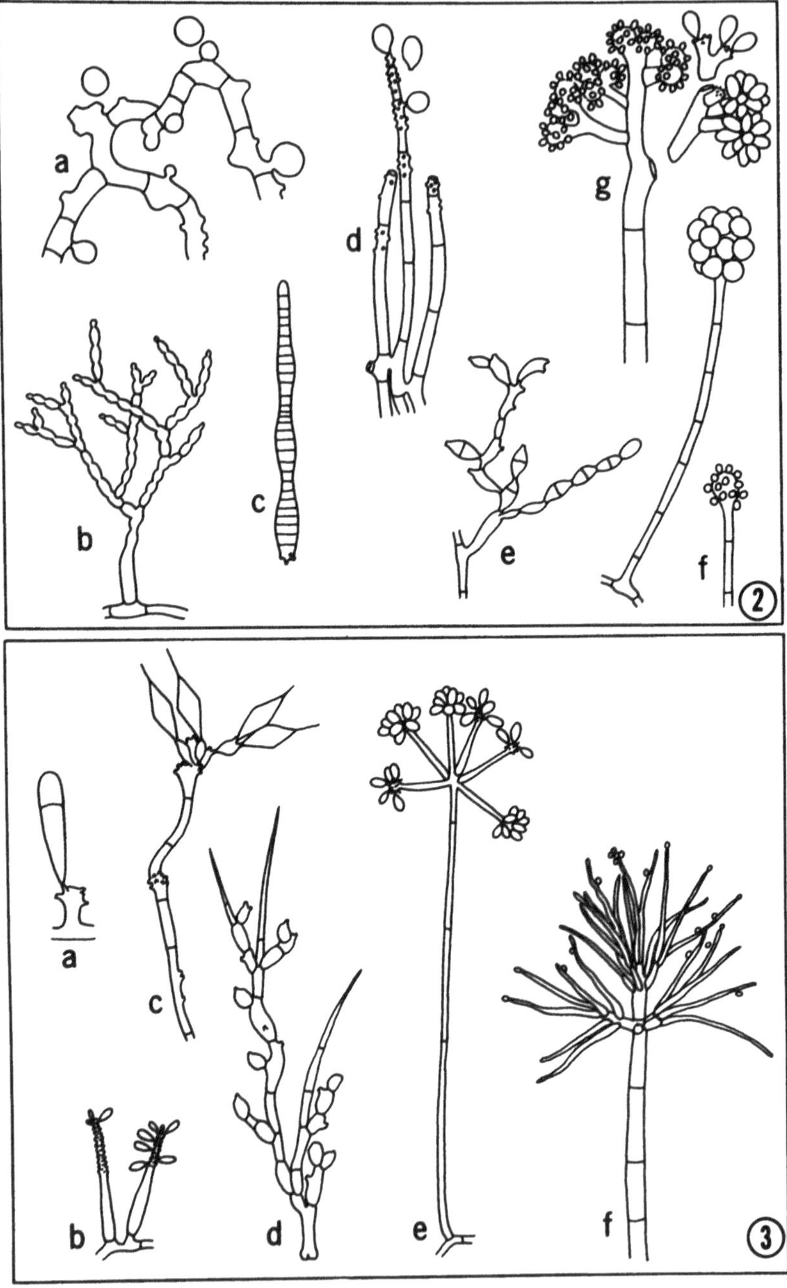


Figure 3.2 a, *Beniowskia sphaeroidea*; b, *Monilia cinerea*; c, *Septonema hormiscium*; d, *Oidium conspersum*; e, *Diploospora rosae*; f, *Gonatobotryum* sp.; g, *Botrytis cinerea.*

Figure 3.3 a, *Dactylaria purpurella*; b, *Rhinocladiella* sp.; c, *Beltrania rhombica*; d, *Beltraniella porosa;* e, *Pseudobotrytis bisbyi*; f, *Verticicladium* state of *Desmazierella acicola.*

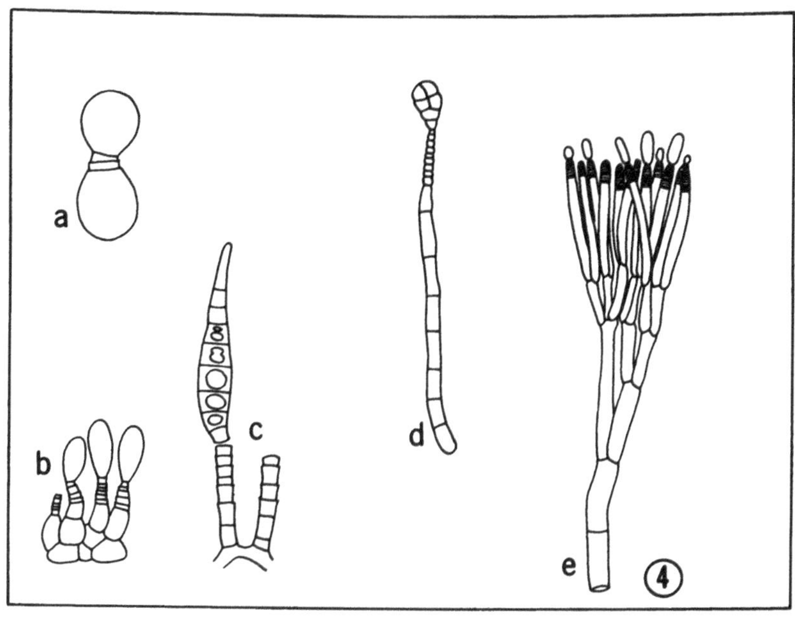


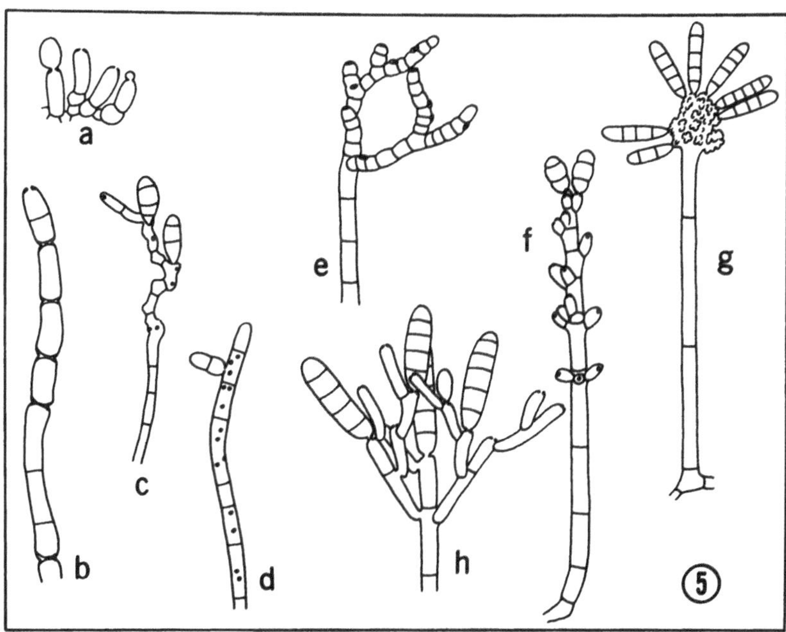


Figure 3.4 a, *Saccharomyces ludwigii* (redrawn from Streiblová et al. 1964); b, *Stigmina pedunculata*; c, *Sporidesmium subulatum*; d, *Acrodictys elaeidicola*; e, *Leptographium lundbergii*.

Figure 3.5 a, *Exosporium tiliae*; b, *Corynespora polyphragmia*; c, *Curvularia fallax*; d, *Spadicoides bina*; e, *Dendryphion laxum*; f, *Spondylocladiella botryoides*; g, *Dichotomophthora* sp.; h, *Dendryphiopsis atra*.

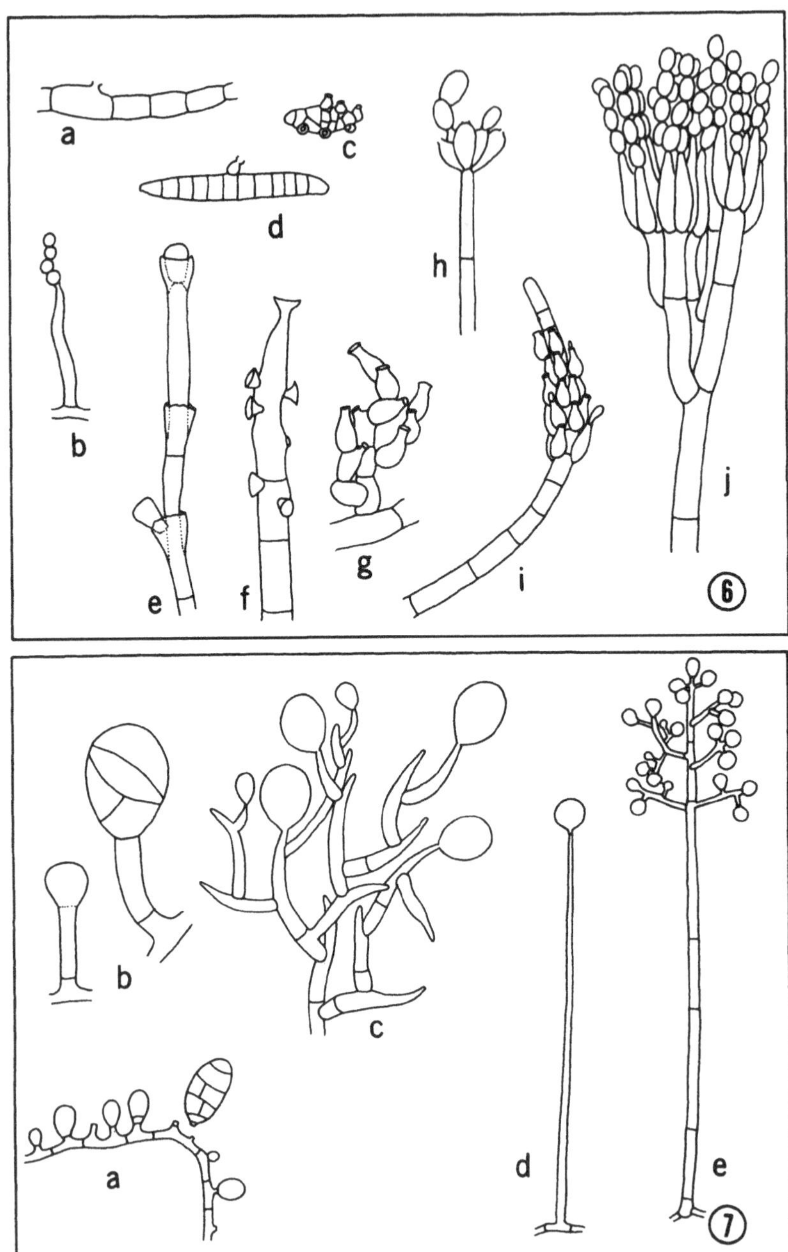


Figure 3.6 a, *Phialophora heteromorpha*; b, *Gliomastix murorum*; c, ascospore of *Strigopodia resinae*; d, *Hormisciella* conidium of *Strigopodia batistae*; e, *Catenularia cuneiformis*; f, *Codinaea fertilis*; g, *Capnophialophora* state of *Strigopodia batistae*; h, *Stachybotrys atra*; i, *Zanclospora brevispora*; j, *Penicillium expansum* (redrawn from Raper and Thom 1949).

Figure 3.7 a, *Pithomyces chartarum*; b, *Monodictys* sp.; c, *Acremoniella atra;* d, *Domingoella asterinearum*; e, *Staphylotrichum coccosporum*.

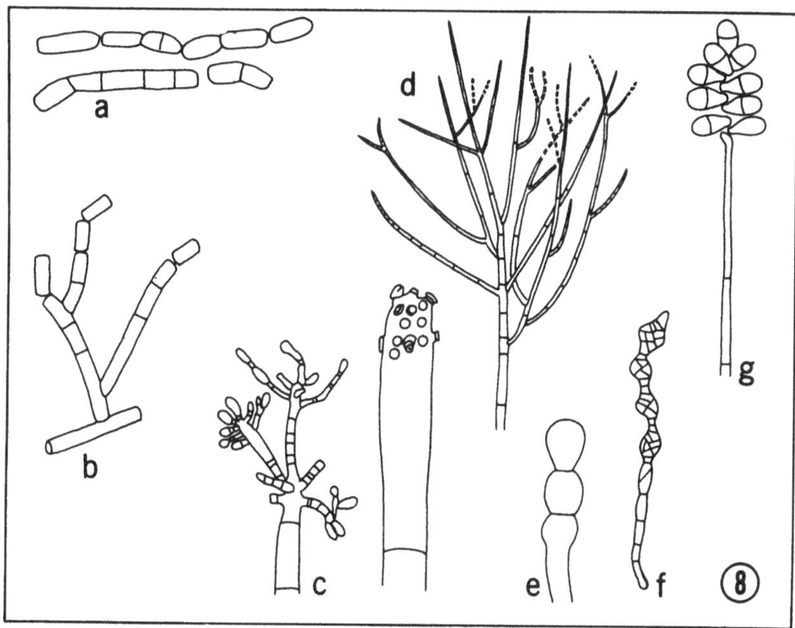


Figure 3.8 a, *Trichosporon cutaneum*; b, *Geotrichum candidum*; c, *Amblyosporium spongiosum* (redrawn from Pirozynski 1969); d, *Polyscytalum* sp. (redrawn from Pirozynski and Patil 1970); e, *Basipetospora rubra* (redrawn from Cole and Kendrick 1968); f, *Phragmotrichum chailletii*; g, *Trichothecium roseum.*

sporogenous function, or percurrently through previously formed functional or spent conidiogenous cells. Examples are known for each section except V: I, *Diploospora rosae* (Figure 3.2 e) or *Oidium conspersum* (Figure 3.2 d); II, *Beltrania rhombica* (Figure 3.3 c) or *Beltraniella porosa* (Figure 3.3 d); IIIB, *Sporidesmium subulatum* (Figure 3.4 c) or *Doratomyces* spp.; VI, *Spadicoides bina* (Figure 3.5 d), *Curvularia fallax* (Figure 3.5 c), or *Corynespora polyphragmia* (Figure 3.5 b); IV, *Catenularia cuneiformis* (Figure 3.6 e), *Codinaea fertilis* (Figure 3.6 f), or *Capnophialophora* state of *Strigopodia batistae* (Figure 3.6 g); IIIA, *Acremoniella atra* (Figure 3.7 c); and VII, *Geotrichum candidum* (Figure 3.8 b).

As already pointed out, there is considerable intergradation between the fungi classified in groups A, B, and C. Although in some species the conidiophores are always either of type A or of type B, in most taxa both forms commonly occur. In *Annellophorella faureae* (section IIIB), for instance, the conidiogenous cells are in both the intercalary and lateral position. The same is true of many species of *Phialophora* (IV), *Pithomyces* (IIIA), *Dactylaria* (II), *Mammaria* (I), and *Geotrichum* (VII).

The fungi classified in group C invariably begin life as members of group A and/or B, and the conidiogenous cells proliferate to produce a conidiophore which, as a rule, is of variable configuration, and as such has little diagnostic value.

Very different are the conidiophores illustrated in groups D and E. The sterile supporting hyphae are, especially when branched, of definite structure and configuration. In most cases, they are diagnostic for a species, group of species, or a genus as, for example, in *Penicillium* where they serve to separate Monoverticillata from Biverticillata and, within the latter, Symmetrica from Asymmetrica. Numerous examples of fungi with simple differentiated supporting structures can be cited. Those bearing conidiogenous cells representing each of Hughes's sections are: I, *Gonatobotryum* sp. (Figure 3.2 f); II, *Pseudobotrytis bisbyi* (Figure 3.3 e); IIIB, *Acrodictys elaeidicola* (Figure 3.4 d); VI, *Dendryphion laxum* (Figure 3.5 e), *Spondylocladiella botryoides* (Figure 3.5 f), or *Dichotomophthora* sp. (Figure 3.5 g); IV, *Stachybotrys atra* (Figure 3.6 h) or *Zanclospora brevispora* (Figure 3.6 i); IIIA, *Domingoella asterinearum* (Figure 3.7 d); V, *Trichothecium roseum* (Figure 3.8 g); and VII, *Amblyosporium spongiosum* (Figure 3.8 c). Examples of fungi which bear sporogenous cells on branched, differentiated supporting structures can also be found in each section of Hughes's scheme with the exception, perhaps, of section V: I, *Botrytis cinerea* (Figure 3.2 g); II, *Desmazierella acicola* (Figure 3.3 f); IIIB, *Leptographium lundbergii* (Figure 3.4 e); VI, *Dendryphiopsis atra* (Figure 3.5 h); IV, *Penicillium expansum* (Figure 3.6 j); IIIA, *Staphylotrichum coccosporum* (Figure 3.7 e); and VII, *Polyscytalum* sp. (Figure 3.8 d).

*Dr Pirozynski postponed consideration of Hughes's section VIII ("Basauxic conidiophores") until after Dr Tubaki's keynote address on that group; see Chapter 12.*

## REFERENCES


Ainsworth, G.C. 1961. Ainsworth and Bisby's Dictionary of the Fungi. 5th ed. Commonwealth Mycological Institute, Kew

Barron, G.L. 1968. The genera of Hyphomycetes from soil. Williams and Wilkins, Baltimore, Md.

Chesters, C.G.C. 1969. Morphology as a taxonomic criterion. In The Fungi. Edited by G.C. Ainsworth and A.S. Sussman. Academic Press, New York and London. Vol. 3, pp. 517-42

Cole, G.T., and Kendrick, W.B. 1968. Conidium ontogeny in hyphomycetes. The imperfect state of *Monascus ruber* and its meristem arthrospores. Can. J. Botany 46: 987-92

Ellis, M.B. 1963. Dematiaceous hyphomycetes. IV. C.M.I. Mycol. Pap. 87

Hughes, S.J. 1953. Conidiophores, conidia, and classification. Can. J. Botany 31: 577-659

Pirozynski, K.A. 1969. Reassessment of the genus *Amblyosporium*. Can. J. Botany 47: 325-34

Pirozynski, K.A., and Patil, S.D. 1970. Some setose hyphomycetes of leaf litter in south India. Can. J. Botany 48: 567-81

Raper, K.B., and Thom, C. 1949. A manual of the penicillia. Williams and Wilkins, Baltimore, Md.

Saccardo, P.A. 1882. Saccardo's "Sylloge." Grevillea 11: 66-7

- 1886. Sylloge Fungorum. Vol. 4. Patavia

Snell, W.H., and Dick, E.A. 1957. A glossary of mycology. Harvard University Press, Cambridge, Mass.

Streiblová, E., Beran, K., and Pokorný, V. 1964. Multiple scars, a new type of yeast scar in apiculate yeasts. J. Bacteriol. 88: 1104-11
Subramanian, C.V. 1965. Spore types in the classification of the Hyphomycetes. Mycopathologia 26: 373-84
Tubaki, K. 1963. Taxonomic study of hyphomycetes. Ann. Rep. Inst. Fermentation, Osaka 1: 25-54 

## DISCUSSION

DR ELLIS

At this stage, we ought to hammer out a firm description of the conidiophore. I would suggest: "A simple or branched hypha which bears conidia on either integrated or discrete conidiogenous cells." This should cover *every* type of conidiophore, both micronematous and macronematous. The term supporting structure is useful for macronematous conidiophores, but would be misleading if applied to micronematous conidiophores.

DR SUBRAMANIAN

Our definition should include the term conidiogenous cell. The complete picture includes conidiophore, conidiogenous cell, and conidium. We can add prefixes to the word conidium to indicate the different kinds of conidium.

DR KENDRICK

I agree wholeheartedly! In our demonstration at the recent Botanical Congress, Dr Cole and I deliberately used the term conidium rather than spore. We used arthroconidium, blastoconidium, etc., because we think that the conidium is a definite enough concept to allow its segregation from the very general concept of the spore. A conidium is a specialized, non-motile, asexual propagule, not formed by cleavage (as are sporangiospores). Such spores are produced by higher phycomycetes, and by ascomycetes and basidiomycetes, as well as by many fungi whose perfect state (if any exists) is unknown.

DR CARMICHAEL

I think Dr Ellis's definition of conidiophore doesn't make the necessary clear distinction between conidiophore and conidiogenous cell. I would like a functional definition: "Any structure which serves the purpose of holding the spores up and away from the assimilative mycelium, stroma, or subiculum." Conidiophores may be clearly distinct from the assimilative hyphae, scarcely distinct, or there may be *no* conidiophores at all. A fungus may have conidiogenous cells without any conidiophores to hold them up or away from the mycelium. Such fungi simply don't have conidiophores. In *Annellophora* the conidiogenous cell is also a conidiophore because it is very long. In many fungi the conidiophore is separate from, or in addition to, the conidiogenous cell. My definition could refer to either the supporting structure and the conidiogenous cell, or to the supporting structure alone.

DR ELLIS

I must disagree with Dr Carmichael. From its original definition, a conidiophore is a structure which itself bears conidia. You cannot alter it to mean a structure bearing conidiogenous cells, just like that. We already have a term, stipe, for any structure supporting conidiogenous cells. "Micronematous" applies to conidiophores which morphologically resemble ordinary hyphae in almost every way, but differ functionally in that they produce conidia. The conidiophore is the whole of the supporting structure plus the conidiogenous cells. If you don't incorporate the conidiogenous cells, the term conidiophore has lost its etymological meaning. The word *means* conidium bearer.

DR KENDRICK

May I ask Dr Ellis about the case of *Geotrichum*: I know it is regarded as being micronematous, but does it have a conidiophore at all, or does it merely have assimilative hyphae that *become* conidia rather than *bear* conidia?

DR ELLIS

It has micronematous conidiophores.

DR COLE

It is difficult, in the case of arthroconidia, to define structures which are clearly predestined to become conidiophores, without knowing when conversion will begin, or how far it will go.

DR KENDRICK

Dr Ellis thinks that all our fungi have "conidiophores"; Dr Carmichael thinks that some of them don't.

DR PIROZYNSKI

I think a conidiophore is a conidiogenous cell or cells plus any supporting structures. The conidiogenous cell is the "flower" and should be singled out as the most important part. Any hypha bearing conidiogenous cells is, in a sense, a supporting structure, but only those which are differentiated and of definite configuration have any diagnostic value and should be regarded as part of the conidiophore. Where conidiogenous cells are intercalary or lateral on differentiated vegetative hyphae, singly or as systems of proliferated cells, it is more logical to regard these single cells, or systems of cells, as conidiophores.

DR KENDRICK

Everybody has assumed that there is a thing called the conidiogenous cell, and that we can define it clearly; but I know of a number of cases where this isn't really possible. I think there are many fungi in which what we call the conidiogenous cell just happens to be delimited by a rather haphazard laying down of septa, which bear little if any relationship to the conidogenous mechanism. Figure 3.9 shows a conidiogenous cell. The part that is functioning is marked X. There is a sessile collarette - a reduced phialide. I don't think the location of the septum indicated by the arrow has anything to do with the production of conidia at the collarette. Polyphialides can go on proliferating sympodially and the septa can be laid down later. How do you delimit the conidiogenous cell here? In many of these fungi I don't think we have a cell concept comparable with that in animals or higher plants. I think that in a hypha we are dealing with a tube that is divided up into compartments by bulkheads. Cytoplasm and nuclei can pass through the central hole in each bulkhead. There may be many nuclei in each compartment. I have recently tended to use "compartments" rather than "cells" in my teaching, to emphasize these characteristics. We have, therefore, a problem in defining "conidiogenous cell."

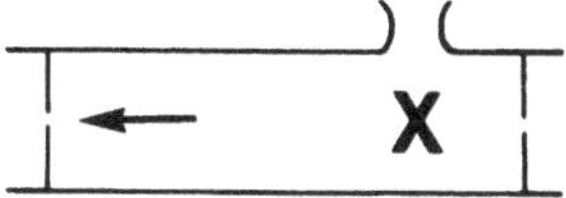


Figure 3.9 Diagram of a conidiogenous "cell"; see text.

MRS POLLACK
We could clarify matters if we didn't limit ourselves to what we see, but considered what the compartment does. Our definition should include not only the morphology, but also the meristematic activity.
DR KENDRICK
We know more or less where this activity goes on in terms of wall deposition, but we don't know the limits of activity in cytoplasmic or nuclear terms.
DR COLE
I'd like to re-emphasize one of Dr Pirozynski's points, that we should give taxonomic importance to the region of the conidiogenous cell involved in conidium formation. In Dr Kendrick's diagram (Figure 3.9), this is a very small region of the cell.
DR ELLIS
Nevertheless, I think that if we reject the term conidiogenous cell, we shall run into far more difficulty than if we retain it. The fact that most of the activity takes place in a part of the cell does not really alter the fact that the cell itself is conidiogenous. Furthermore, if we do away with the term cell for these things, we can scarcely use the term at all in botany, because there is protoplasmic continuity between all cells by means of pores of one sort or another.
DR KENDRICK
Yes, but not *nuclear* migration. Nevertheless, you're probably right. From the pragmatic angle, we may need a working definition of a cell, despite the valid theoretical objections that can be raised.
DR CARROLL
Dr Kendrick pointed out that the conidiogenous cell can be a scarcely differentiated compartment of a vegetative hypha. I'd like to mention that the conidiogenous cell can also be a previously formed spore, as in *Alternaria* and certain ascospores.
EDITOR
*At this point we agreed to postpone our final decision on the definition of "conidiophore" until our closing session on terminology. The discussion may be picked up early in Chapter 15. The definition reached and almost unanimously accepted during that final discussion was that given by Dr Pirozynski in this chapter.*

# 4
# Blastospores, Aleuriospores, Chlamydospores

J.W. CARMICHAEL

Because of the nature of the discussion arising from Dr Pirozynski's paper, I shall begin with some definitions culled from a manuscript I wrote in 1962 but never published.

*Cell* (fungus) Any unit of a fungus thallus or spore which is morphologically separated from neighbouring units by a wall or septum. Contiguous cells do not necessarily contain individual protoplasts. Cytoplasm and nuclei may pass from one cell to another.

*Sporogenous cell* Any cell from which, or within which, spores are directly produced.

*Sporogenous hypha* A hypha which produces spores by fragmentation or fission.

*Conidium initial* A cell or part of a cell which will become a conidium by differentiation.

*Meristem* Any place on a hypha, sporogenous cell, or conidium where growth occurs which results in an increase in volume.

*Conidiophore* Any structure which serves the purpose of holding the spores up or away from the assimilative mycelium, stroma, or subiculum. Conidiophores may be clearly distinct from the assimilative hyphae, scarcely distinct, or completely absent. They may consist of a single hypha, branched or not, or of hyphae united in a coremium or synnema. It is convenient to use the term conidiophore to mean either the structure supporting the sporogenous cells or both the support and the sporogenous cells. Rarely a single sporogenous cell functions as a conidiophore.

When Mason published his 1933 paper on spore terminology a correspondent wrote to him: "I shy immediately at your introduction of additional termino-

Mold Herbarium and Culture Collection, University of Alberta, Edmonton, Alberta, Canada.

logy and attempts at defining various types of conidia, such as chlamydospore, meristem spores, radula spores, etc. There are such a host of conidial types and so many transitional and aberrant forms, that any attempt to force them into a terminology seems to me to lead to a straining of terms or conceptions to fit preconceived ideas rather than a broad attitude towards relationships."* I have been forced by the fungi to have considerable sympathy with this view. It is easy to construct a clear-cut classification of spore types based on a small sample of fungi. However, the more fungi you examine, the more kinds of spore production you find and the less distinct the kinds become. I have concluded that the various methods of spore production described by previous workers are not separate and distinct kinds, but outstanding parts of a continuously intergrading and overlapping spectrum of methods for releasing propagative elements. Nonetheless, I feel that a better classification of the Fungi Imperfecti depends on a careful, though arbitrary, categorizing of the methods of spore production, and the establishment of a suitable nomenclature for spore types. In the part that follows I have made every effort to fit the terms and conceptions to the fungi, rather than to preconceived ideas. I have also endeavoured to maintain the general sense of usage of older terms and to keep the introduction of new terms to a minimum.



There are, I believe, only five basic processes by which fungi produce spores:

1 *Free cell formation* - where spores are formed within the protoplast by a concentration of material in localized areas which are then each surrounded by a wall. Examples: ascospores, oospores (and bacterial endospores).

2 *Cleavage* - where the protoplast splits into fragments, each of which becomes surrounded by a wall. Example: sporangiospores.

3 *Fragmentation* - where the cytoplasm becomes concentrated in certain cells of the thallus and the remaining cells are exhausted. Example: arthroaleuries and aleuries.

4 *Fission* - where the cells of the filament break apart at the septa, which are double. Example: fission arthrospores.

5 *Extrusion* - where the spores are produced as extrusions from the ends or the sides of the hyphae. Examples: blastospores and phialospores.

I would restrict the use of the term endospores to the spores produced by the first two methods. Spores produced by the other three methods can be covered by the term conidia, which then includes all asexual fungus spores except sporangiospores. This classification, based on the relation of the spores to a filamentous thallus, is useful in that it clearly separates endospores and conidia. However, even the broad classification of conidia into three groups runs afoul of intermediate types. In *Trichosporon cutaneum*, for example, we find all gradations between fission and extrusion. In *Chrysosporium merdarium* we find all gradations between fragmentation and extrusion. It also suffers from the defect that some spores which are similar in other features are placed in separate groups.

*E.W. Mason, "Annotated account of fungi received at the Imperial Mycological Institute. List II (Fascicle 3 - General Part)," *C.M.I. Mycol. Pap.* 4, p. 81.

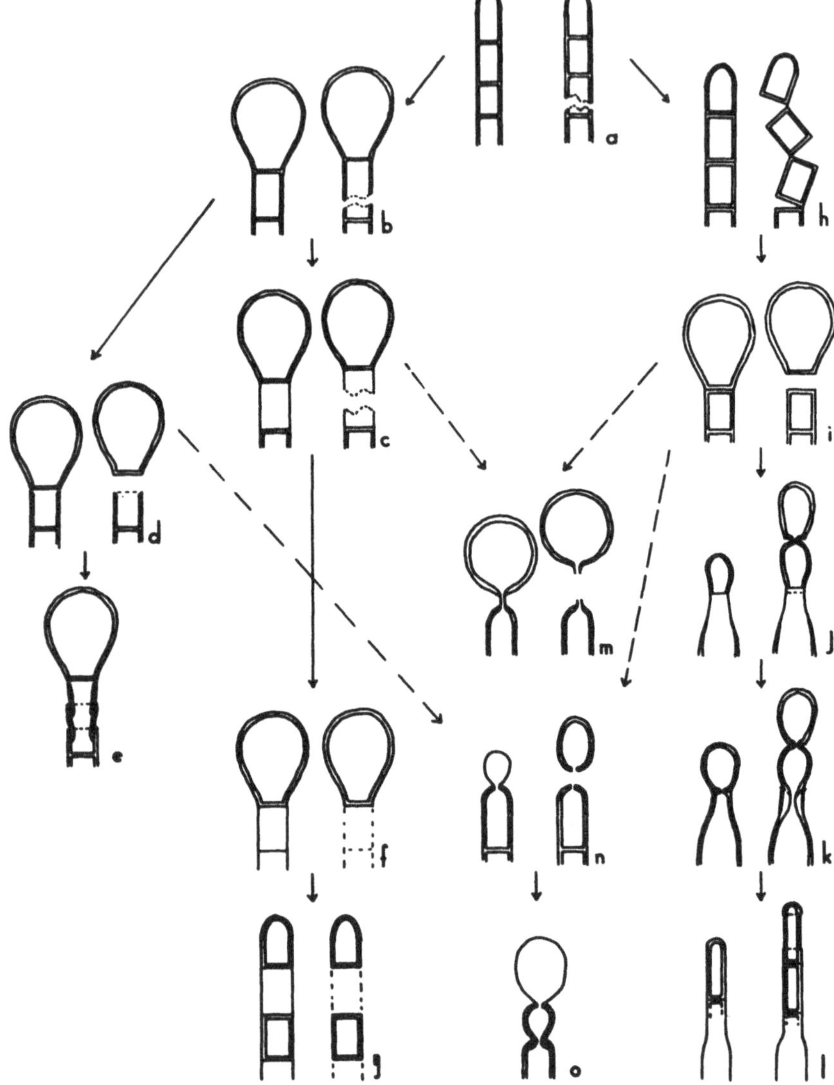


Figure 4.1 Conidia classified by dehiscence: a, an undifferentiated hypha releasing a propagule by fracture; b-g, conidia released by sacrifice of a supporting cell; h-1, conidia released by fission of a double septum; m-o, conidia released by fracture of a fine connective.

In most of the categorizing of spores that has gone on until now, the emphasis has been on the manner in which the spores are produced. In distinguishing chlamydospores, aleuriospores, and blastospores, I would like to focus attention on the manner in which the fungus releases the spores once it has produced them. In other words, on the manner of their dehiscence. If we think of a simple septate hypha, the least modification required before some of it can be released is simply that it breaks (Figure 4.1 a). This is the kind of dehiscence, or indehiscence if you like, that I think of as characterizing chlamydospores. They are simply resistant cells, either terminal or intercalary, which are released by mechanical fracture of a non-differentiated cell wall. I would restrict the use of the term chlamydospore to this kind of spore. Thinking of more specialized ways of releasing spores, we have three general kinds of mechanism. The first is by having a fragile intervening cell which is sacrificed to effect the release of the spore. This can be a simple, thin-walled cell (Figure 4.1 c), a cell with a thin abscission ring (Figure 4.1 d), or a cell whose wall is dissolved by enzymatic action (Figure 4.1 f,g). The second method is to have a double septum at the base of the spore. Release can be accomplished, without the loss of a cell, by a splitting of the double septum (Figure 4.1 h,l). The third method is to produce the spore on such a fine connection that no further mechanism is needed either to release the spore or to prevent the escape of cytoplasm from the severed connection (Figure 4.1 m,n). Although these three types of mechanism seem quite distinct, any attempt to sort the fungi into groups based on them will soon disclose numerous doubtful cases, especially in the thin-walled, hyaline forms. However, the fungi I have chosen to illustrate chlamydospores, blastospores, and aleuriospores mostly provide clear-cut examples.

Figure 4.2 a shows what I would call chlamydospores produced by *Helicoma isiola*. They are highly differentiated and multicellular, but what makes them chlamydospores is the lack of any mechanism for liberation, and they are firmly attached. I presume that in nature the supporting hyphae will eventually be broken down by bacterial action, by other fungi, or by weathering, or they may be eaten by something - I don't know. There doesn't seem to be any differentiated mechanism to weaken the wall of the supporting cells, and this I would take as the hallmark of the chlamydospore. In Figure 4.2 b the spores appear at first glance to be chlamydospores, but on close examination you can see that there is a thin ring around the cell immediately beneath the basal cell of the spore, where the spores always break off (arrows). I have no term for spores released by this mechanism. Figure 4.2 c shows chains of dry blastospores, where each spore starts out as a small bud, or blown-out end, which eventually is separated from its neighbour by a double septum. These spores are released by fission of the double septum. Further back on the same sporulating hypha (Figure 4.2 d), it can be seen that arthrospores are formed and that it is impossible to draw a sharp line between blastospores and fission arthrospores. If there are definite constrictions as the sporogenous hypha grows, we call them blastospores; if there are no constrictions, they are arthrospores. In this fungus all gradations between a slightly moniliform hypha which later breaks up into arthrospores, and a hypha which is constricted enough so that we think of each spore initial as being a blown-out end, are found. The *Monilia* state of *Neurospora sitophila* shows similar intermediate forms.

The distinction between blastospores and porospores in the less differentiated

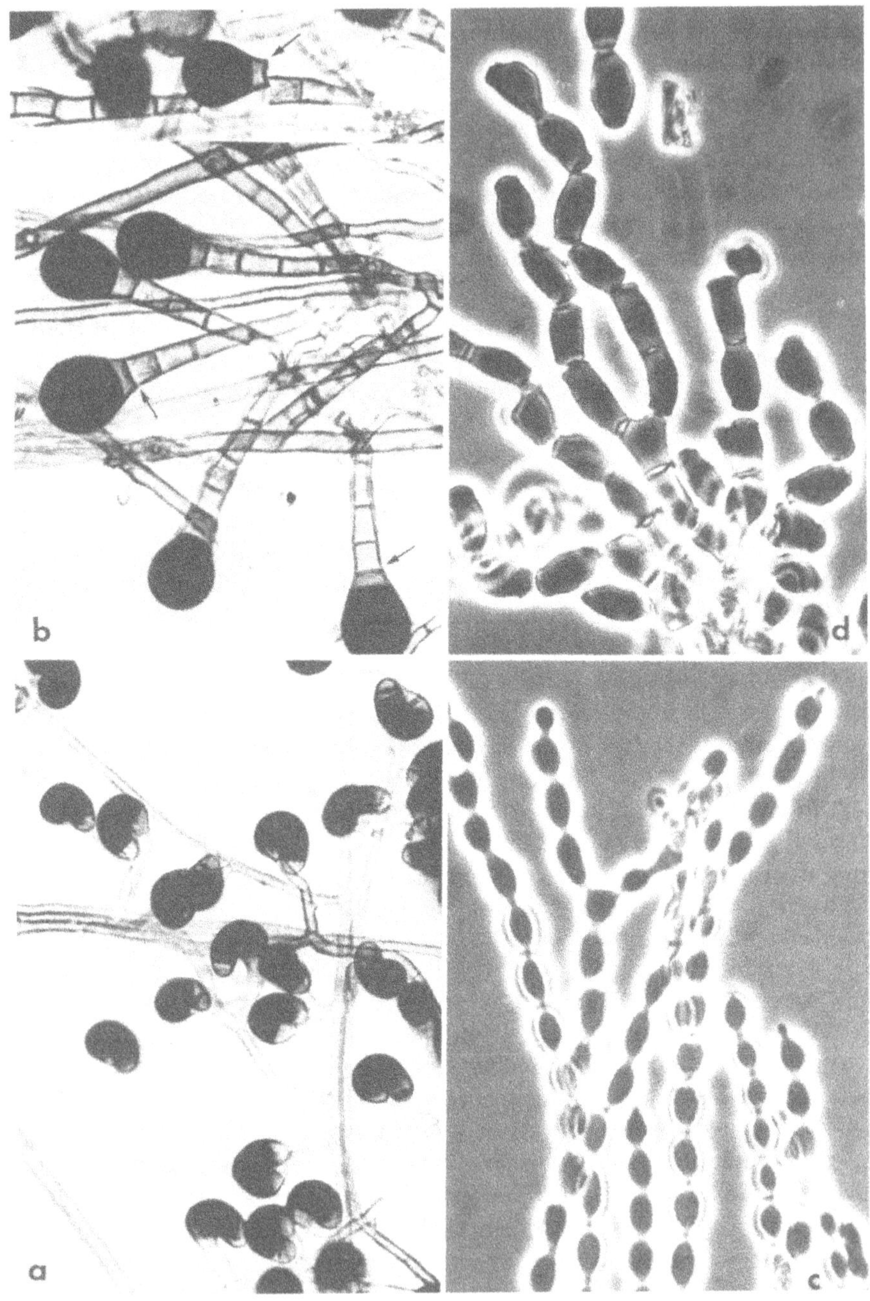


Figure 4.2 a, *Helicoma isiola* (UAMH 1359); b, *Endophragmia biconstituta* (UAMH 1112); c,d, *Monilia* state of *Sclerotinia laxa* (UAMH 1781).

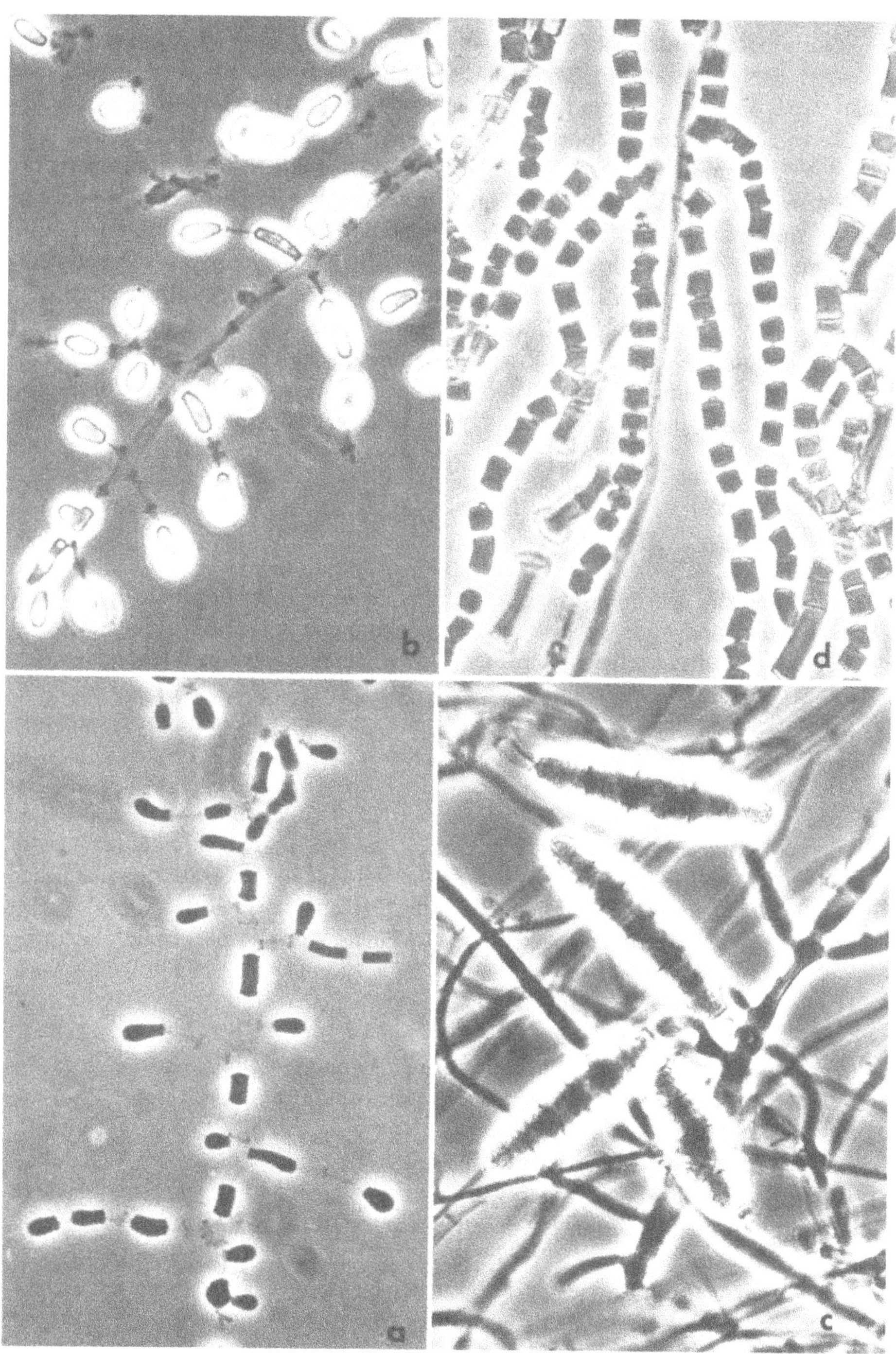


Figure 4.3 a, (?) *Trichophyton* sp. (UAMH 1733); b, *Chrysosporium tropicum* (UAMH 1534); c, a *Microsporum* of the *Gypseum* group (UAMH 1615); d, *Coremiella ulmariae* (UAMH 1513).

kind of porospores is also not very clear. Figure 4.3 a shows a good example of aleuriospore formation in a *Trichophyton* species. The spores are in situ, but the hyphae that bore them have vanished. Remnants of walls can be seen here and there, but in many places they have been completely lysed, and this I take to be the hallmark of the aleuriospore - one or more cells are dissolved to release the spore. Figure 4.3 b shows *Chrysosporium keratinophilum*, where only parts of the hyphae have dissolved. In *Microsporum gypseum* (Figure 4.3 c) we have a specialized separating cell at the base of the spore, which dissolves to release the spore. In *Coremiella ulmariae* (Figure 4.3 d) the cells between the alternate arthrospores or arthro-aleuries have completely dissolved. How these fungi can so completely dissolve the cell wall after most of the contents appear to be gone, I am not quite sure. Enzymes from the maturing spores may assist in the process. Some of the things other people have called aleuriospores I would call blastospores because they have a fission type of dehiscence involving a double septum.

## DISCUSSION

DR KENDRICK

Regarding the sporogenous, or as I prefer to call it conidiogenous, cell, what about the annellophore which has become secondarily septate? (Figure 4.4). Is each of the cells an annellophore? Or is the whole thing an annellophore?

DR S.J. HUGHES

The septa do not really matter because they are probably laid down for support; such septa are also found in long polyphialides.

DR KENDRICK

But we're talking about conidiogenous *cells*. Dr Carmichael has given us a definition of a sporogenous cell which, despite the fact that it only defines a morphological unit, is still widely used. So the fact that the lower cells have annellations doesn't mean *they* are annellophores. Perhaps only the fruiting cell at the tip should be called an annellophore. In Figure 4.5, the poroconidia are produced at random from several cells of the conidiophore. Are all of those cells conidiogenous cells? Yes, they are.

DR ELLIS

Annellophore is a term comparable with phialophore - a special kind of conidiophore. It does not, or should not, refer to the conidiogenous cell alone.

DR HENNEBERT

There are two parallel series of names, one for conidiogenous cells (e.g., phialide), and one for conidiophores (e.g., phialophore). I think the term annellide will fit into the first series, and annellophore into the second.

DR PIROZYNSKI

Dr Carmichael, I'm interested in your comments on blastoconidia and arthroconidia in the *Monilia* state of *Sclerotinia fructigena.* Did you say they are basically not very different?

DR CARMICHAEL

It is the *same* branch of the fertile hypha in both illustrations. It is growing at

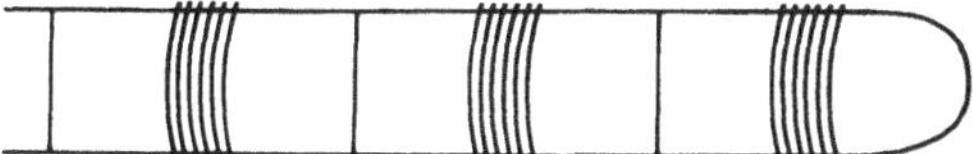

Figure 4.4. Diagram of annellophore with secondary septa; see text.

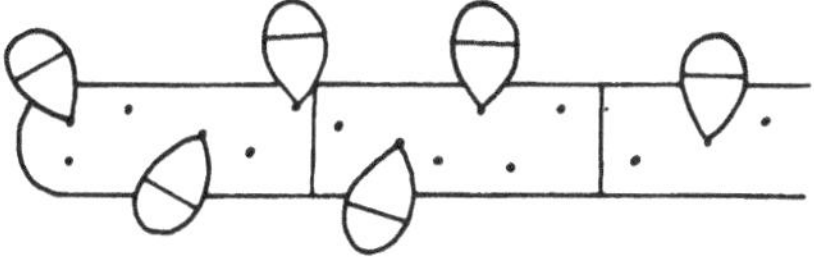

Figure 4.5. Diagram of conidiophore producing "poroconidia" from several cells; see text.

the tip by budding, and further back it is breaking up into arthroconidia.

DR PIROZYNSKI

I have always thought that a blastic chain with the youngest spore at the apex must be considered as a unit - "a multiple conidium."

DR CARMICHAEL

Not necessarily. These things can differentiate to the point where they are only held together perhaps by a minute strand of cytoplasm through the pore, and by the thin outer wall; but, nonetheless, the food material is still going through and they're still growing out at the apex, even while they're differentiating backward at the base.

DR ELLIS

One can only consider them as conidia if each part is capable of germination. This condition is met in the *Monilia* state of *Sclerotinia*: both the terminal, rounded-off portions and the fragments of the basal part of the conidiophore germinate readily.

DR KENDRICK

Are the terminal spores blastoconidia, or will they become arthroconidia later on?

DR ELLIS

Oh, no!

DR CARMICHAEL

My point is that it is impossible to make fundamental distinctions between blastoconidia and fission arthroconidia. They intergrade.

DR CARROLL

I was bothered by the statement that the only difference between the two is the presence or absence of a constriction. It seems to me that there is a fundamental developmental difference between these two types of conidia: whereas the precursor of arthroconidia is a hyphal form of growth, which is later converted into conidia, blastoconidia are produced by a budding process that is quite different.

DR CARMICHAEL

I'm sure you'll agree that Figure 4.2 c represents budding, but the development of septa at the constriction is delayed. So what we have in essence is simply a moniliform hypha which later lays down double septa and can disarticulate.

DR CARROLL AND DR COLE

Have you recorded this development process by time-lapse photomicrography?

DR CARMICHAEL

I haven't in this species, but I have for *Monilia sitophila* and *Trichosporon*, where exactly the same things happen (cf. Figure 4.2 c,d).

DR KENDRICK

The only difference I can see between the blastic phenomenon and the formation of arthroconidia, as most of us have envisaged them in the past, is that arthroconidia are not formed until extension growth of the conidiophore has ceased.

DR CARMICHAEL

But if the conidiophore keeps extending, you call them meristem arthroconidia.

DR KENDRICK

Only if it's extending at the base of the chain, not at the tip. If you want to call these particular conidia arthroconidia, they are of another kind. Unfortunately, although I recognize six kinds already, as we'll see later, that is not one of them!

DR CARMICHAEL

This is my whole point. Each fungus you look at produces its conidia in a slightly different way, sometimes in several ways; and when we distinguish and define spore types, we are picking out the high spots. The various methods actually intergrade.

MRS POLLACK

It seems to me that the first question that needs to be asked in understanding whether *Monilia* forms blastoconidia or arthroconidia is: how are the conidia initiated? The subsequent development and abstriction involve other processes. I regard the method of initiation as most significant, but spore separation must also be considered.

DR CARMICHAEL

They are both initiated by the elongation of the hyphal tip. The difference lies in how constricted it is as it grows.

DR HENNEBERT

I have observed that all monilioid hyphae elongate first and become septate later. Septation occurs, sometimes from the base, sometimes from the apex, sometimes irregularly, when budding has almost stopped.

Mrs Pollack raised an interesting point in speaking of the initial and ripening phases of conidium formation. Should we identify the type of conidium on the basis of the initial phase? Personally, I think a conidium is only a conidium when ripe and liberated. Otherwise, it is just a potential conidium. We should really consider both initial and ripening phases to know what a conidium is. In the *Monilia*, the conidium is only a conidium when septation has occurred, and I would consider it an arthroconidium. But there might be two kinds of arthroconidia, one kind initiated by budding of the hypha, the other by an involvement of the full width of the hypha.

DR CARROLL

I don't think *Monilia* has been studied biochemically, but in *Mucor*, for example, budding can be induced by putting it in a nitrogen atmosphere and the processes of hyphal growth and budding can be shown to be quite different biochemically - in terms of what is laid down in the wall, and in terms of the fine structure, how it is laid down; this has been studied in yeasts and mycelial systems. Therefore, I find it difficult to accept that there can be a gradation between these two processes along the same fertile hypha.

DR CARMICHAEL

The conditions that are necessary to induce budding in *Mucor* are quite different from those under which it grows as hyphae, and one might expect that wall composition and physiology would be somewhat different. I would suspect that there could also be intermediate conditions which would produce partly contorted hyphae, partly budding. I think the biochemists have picked two extremes, and I'm sure intergrades could be found.

DR CARROLL

I didn't see intergradations in your two pictures. Do you have them?

DR CARMICHAEL

I can see all degrees of intergradation in *Trichosporon* between a hypha which becomes wider and narrower in a very irregular and indeterminate way, and a yeast-like budding.

DR S.J. HUGHES

In the chain of blastic potential conidia in *Monilia*, the septa appear only at predestined positions, at the constrictions; but this is not true of the arthrospores of, say, *Geotrichum*. Granted, the conidia at the base of the chain in *Monilia* fall apart, but this is not unusual in fungi. Conidiophores and mycelium can fall apart in an arthroconidial way after they have produced other conidia and perhaps a perfect state, so arthroconidia are rather widespread. This is characteristic of the sooty moulds, where I regard it as a facultative or supplementary measure. Contrast this with *Geotrichum*, where arthroconidia are the only propagules employed - obligate arthroconidium development.

DR ELLIS

*Monilia* forms a much more definite conidium than is the case in *Geotrichum* and many other arthroconidial genera where septum formation is irregular, often subdividing recently delimited segments until a minimum size is reached. I think Dr Hughes's idea of a definite, predetermined place for septation can be used to separate these *Monilia* conidia from arthroconidial forms, and allow us to designate them as true blastoconidia.

DR CARMICHAEL

In *Geotrichum*, under unfavourable conditions and in slide cultures, there *is* irregular breaking up. But in young, vigorous cultures, the striking feature is the extreme regularity. The typical pattern is for the hyphae to break up into very regular lengths. We can't see in advance where the septum is going to be, but maybe the fungus has it all worked out.

DR ELLIS

I can't agree with Dr Carmichael, because Dr Cole and Dr Kendrick's film shows that after a segment has been cut off additional septa are often laid down within it. It's all very irregular.

DR KENDRICK

Do you think there is any connection between the sphere of influence of a nucleus and the volume or length of a conidium? If a *Geotrichum* conidium is short, it can be thick, but if long, it may be thin; the relative volumes might be quite comparable.

MRS POLLACK

We need to determine whether the meristematic activity in *Monilia* is continually moving forward as the nuclei divide and progress. In the *Geotrichum* type, meristematic activity ceases before basipetal segmentation begins.

DR KENDRICK

I agree entirely with Mrs Pollack. There's lots of room for developmental and cytological work on many of these fungi. Some of my graduate students are beginning to attack the problems from these angles. [*See Chapter 18*]

DR SUBRAMANIAN

It looks to me as if there are two types of propagules that are of general occurrence. One is the chlamydospore; the other is the arthroconidium. Sometimes what looks like one conidium is subsequently subdivided: for example, what looks like a chlamydospore in *Thielaviopsis basicola* is later transformed into endoarthroconidia. Even septate phialoconidia may split up into smaller units. The original product can be called a phialoconidium, but the final product is an arthroconidium, especially since each of the final bits is capable of germinating.

DR KENDRICK

Perhaps we should define an arthroconidium as a propagule arising from a preformed element, whether this is a hypha or another kind of conidium.

DR S.J. HUGHES

Are there arthroconidia which show a moniliform configuration of the hyphal walls? Perhaps Dr Pirozynski could say something about *Amblyosporium*. Are the septa laid down at predetermined locations?

DR PIROZYNSKI

In *Amblyosporium spongiosum* the moniliform configuration is secondary: a result of enlargement and rounding off of conidia from cells delimited previously in a preformed conidiogenous cell. I do not know how the septa are laid down, but I would expect to find greater uniformity in the size of conidia and the separating cells if the septa were laid down at predetermined locations.

DR HENNEBERT

May I raise the question of the distinction between aleuriospores and chlamydospores? Chlamydospores can be interpreted as thick-walled intercalary spores which need dissolution of the hypha before dispersal is possible. Or they can be interpreted in a morphological sense only, as spores with thick walls. Some blastospores, aleuriospores, and arthrospores may have thick walls and are in that sense chlamydospores.

DR KENDRICK 

You are perhaps aiming at a functional rather than a developmental classification here, and that is certainly more in the spirit of the original definition of chlamydospores.

DR HENNEBERT

Actually, I think we should avoid the word chlamydospore because it is too general. An aleuriospore is also a chlamydospore because it is thick-walled.

DR KENDRICK

Not according to Dr Carmichael, who is concerned chiefly with the method of dehiscence.

DR GOOS

"Chlamydospore" has a meaning to many people that has nothing to do with its development, and the term is so entrenched in the literature that it is now completely unsuitable to apply to a spore distinguished by its method of formation *or* dehiscence.

DR CARMICHAEL

I don't like to distinguish spores because one is terminal and another is intercalary. This is a very superficial distinction in some types, and we can find all intermediate conditions. Vuillemin gave a very useful concept of the aleuriospore - a spore intermediate between the chlamydospore, which was indehiscent, and the conidiospore, which was dehiscent. Although he understood and clearly described the unique dissolution mechanism by which aleuriospores are freed, he then confused, perhaps himself, and certainly subsequent workers, by calling them indehiscent.

DR GOOS

Yes. When Ingold used the term in the aquatic hyphomycetes, he changed the concept, and it has subsequently become increasingly confused.

DR S.J. HUGHES

Fungi grow in substrates with a mixture of microorganisms which may contribute to the dissolution of the walls of the presumably dead separating cells, thus freeing the spores.

DR KENDRICK

In cultures of *Sporendonema*, the outer wall of empty cells between conidia persists. In *Coremiella*, it doesn't. How fundamental is the difference between the two, particularly since the wall, which in one case persists in culture, disappears so rapidly in the soil?

DR CARMICHAEL

I can only repeat that these different spore types intergrade, but I still think these concepts are useful. The chlamydospore, which is liberated by outside agencies (freezing, bacteria, etc.), and the aleuriospore, where there is a clear, built-in dissolution mechanism. I think *Sporendonema purpurascens* is variable in this respect. Sometimes the outer wall dissolves, sometimes it does not.

DR KENDRICK

Many people call the conidia of *Sporendonema* endoarthrospores; you apparently want to call them something else.

DR CARMICHAEL

I would call them alternate arthroconidia - separated from each other by an intervening dead cell - and whether they are aleuriospores or chlamydospores depends on whether the wall breaks down or not.

DR S.J. HUGHES

I find that in *Coremiella* two neighbouring conidia will separate by a split across the septum, except for a connection or isthmus maintained in the middle of the common septum, and the circumferential outer wall still remains intact. The space thus formed mimics a separating cell.

DR CARMICHAEL

I'm sure this happens in *Oidiodendron*, too. There isn't a whole cell lost. It is a space that has developed.

DR S.J. HUGHES

Drechsler described the same thing in some of his Zoopagaceae.

DR COLE

In our time-lapse studies of *Geotrichum candidum* and the *Geotrichum* state of *Endomyces magnusii*, we found that the extension growth ceases when septation begins. The fertile hypha becomes septate in a rather random way, and finally breaks up. Concerning wall relations in blastoconidia, I agree with Mrs Pollack: we should examine the conidium initial. In *Cladosporium* the outer wall of the parent cell is continuous around the new conidium initial. In *Gonatobotryum apiculatum* small conical swellings appear on the expanded apex of the conidiophore. Soon the wall breaks at the apex of each bump, and the wall of the conidium which emerges is derived from the inner wall layer of the conidiophore. Perhaps there are two kinds of blastoconidia which can be differentiated on the basis of wall relations.

DR ELLIS

Or perhaps, in the case of *Gonatobotryum*, as modifications of the poroconidium. In poroconidia the same thing happens - the inner wall of the conidiophore comes out and forms the outer wall of the conidium. One difference is that in *Gonatobotryum* the outer wall ruptures, whereas in poroconidium formation it dissolves by enzymic action.

DR KENDRICK

I'm not sure that this distinction can be made as readily as Dr Ellis suggests.

DR ELLIS

I saw this phenomenon in Dr Cole and Dr Kendrick's demonstration at the Botanical Congress in Seattle, and was struck by this new method that hasn't been described before - the break in the outer wall of a double-walled conidiogenous cell. The denticles themselves show this well.

DR SUBRAMANIAN

*Drechslera* typically produces poroconidia, but when it is incubated at higher temperatures, as Luttrell reported for *D. sorokiniana*, it may produce blastoconidia and also aleuriospores or gangliospores. All three types of conidium can be found in the same culture, sometimes on the same conidiophore. What I think happens is as follows. In the case of the blastospore, a very small zone of the conidiophore becomes elastic and blows out, so that ultimately there is something that looks like a constriction. But where an aleuriospore or gangliospore is formed, a larger zone of the conidiophore becomes elastic and blows out, so no constriction is seen. In fact, when one looks through a large number of fungi, intergradations from one extreme to the other will be found, depending on how much of the tip of the conidiophore becomes elastic and blows out. Electron microscopy of *Oedocephalum*, carried out at Bristol, shows that all layers of the conidiophore wall are involved in the blowing out.

DR G.C. HUGHES

In conjunction with this discussion of blastospore ontogeny, I would like to show you a series of electron micrographs that my assistant and I have taken of blastospore development in the *Chromelosporium* state of *Peziza ostracoderma*. The general morphology of the spore-bearing apparatus is shown in Figure 4.6 A (X425) by the single ampullar head as viewed with the light microscope. Figure 4.6 B shows an electron micrograph of an early stage in conidium formation in which the completely formed denticle is beginning to "balloon" at the tip (X16,000). Note especially the location of the nucleus (N) just below the young conidium. We surmise that nuclei move from this position into the denticle and thence into the conidium. The nucleus in Figure 4.6 C (X27,000) was fixed as it entered the conidium; the limits of the nuclear envelope (NE) are indicated by arrows. In Figure 4.6 D (X24,000) we see lomasomes appressed to the walls of a conidium (arrows) and in the denticle, whereas Figure 4.7 A (X24,100) is an oblique section of a developing conidium which shows the incorporation of lomasomes into the wall (arrow) and free multivesicular bodies (X) in the cytoplasm. Vesicles (V) are also evident in the periphery of this conidium. The occurrence of lomasomes in such close proximity to actively developing walls (arrow, Figure 4.7 A) leads us to accept the view that they function in wall synthesis. However, we don't have any more real evidence of their function than other workers who ascribe quite different functions to them. Two stages of septum formation in denticles of this *Chromelosporium* are shown in Figure 4.7 B (X27,000) and C (X24,000). In B we see the incomplete septum, whereas in C, in the near-median section of a serial series, we see the complete septum. The septa do not form in denticles until the conidia are nearly mature. This is indicated by the condition of the cytoplasm of the ampulla in Figure 4.7 B and C, as compared with Figure 4.6 C and D. Figure 4.7 B and C also serve to show, along with Figure 4.6 B and D, that the ampulla wall does not rupture when denticles form in this *Chromelosporium*.

Figure 4.7 D (X17,500) is a view of the ampulla following septum formation and maturation of the conidia. Note that it is virtually empty, a condition that can also be seen in the mature ampulla shown in Figure 4.6 A. The final figure (Figure 4.7 E, X16,200) shows a section through a mature conidium of *Chromelosporium*. We have been unable to determine whether two nuclei move into the conidium from the ampulla or whether only one enters and then divides by the time the conidium is mature. At any rate, most of the mature conidia seem to have two nuclei.

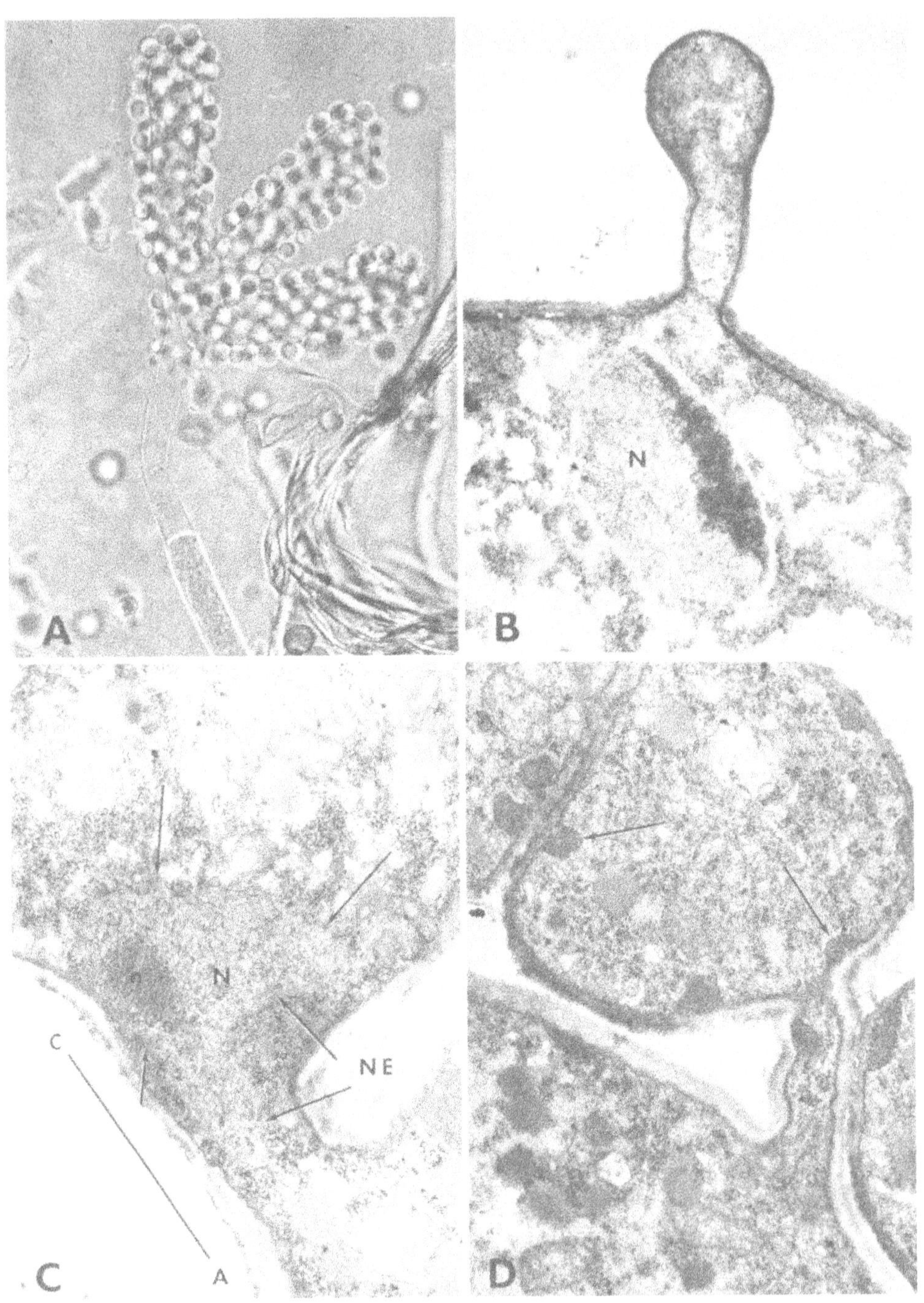


Figure 4.6 Conidium ontogeny in the *Chromelosporium* state of *Peziza ostracoderma;* see text. Reproduced by permission of the National Research Council of Canada from the *Canadian Journal of Botany*, 48 (1970): 361-6.

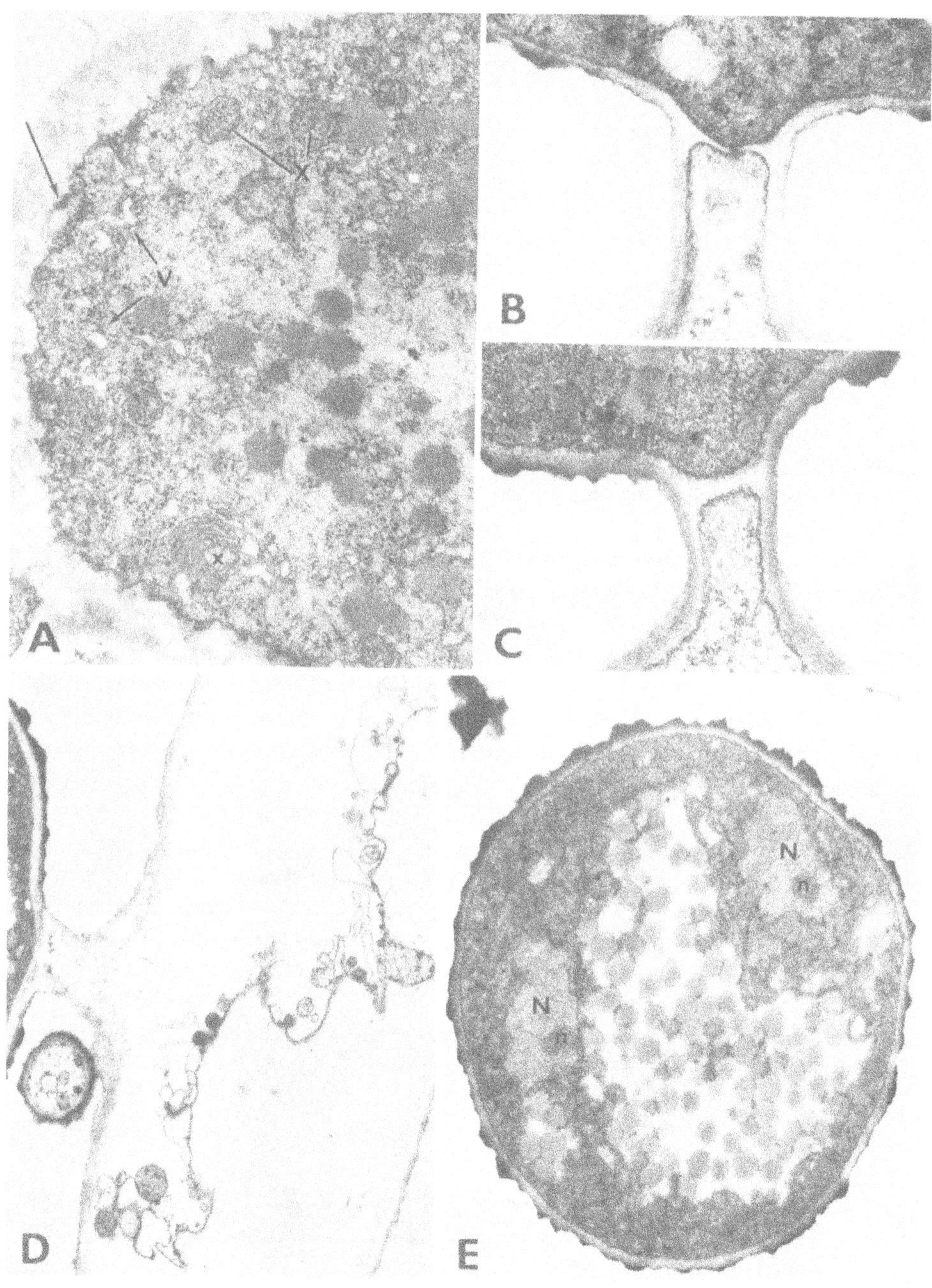


Figure 4.7 Conidium ontogeny in the *Chromelosporium* state of *Peziza ostracoderma*; see text. Reproduced by permission of the National Research Council of Canada from the *Canadian Journal of Botany*, 48 (1970): 361-6.

DR KENDRICK

Can you tell whether the septum which cuts off the conidium is a double septum or not?

DR G.C. HUGHES

I don't think it is double.

DR CARMICHAEL

This is the type of spore release I mentioned, in which the peg or connection is so narrow that the formation of another septum is redundant.

DR G.C. HUGHES

It is particularly redundant here because there's nothing living left in the ampulla.

DR CARMICHAEL

Even in those fungi which produce successive conidia, where contents must remain in the ampulla, the hole is so small that when they break off, a process which requires very little leverage, the wound can be plugged very easily. They don't require a double septum for release.

DR G.C. HUGHES

With all this talk about single and double septa, it has become apparent that septum formation is of considerable interest. In line with this I'd like to show you a few pictures of septa in conidiophores of an *Oedocephalum* and a *Chromelosporium*, both imperfect states of *Peziza* species.

Figure 4.8 A (X21,000) shows the septum of the *Oedocephalum* conidiophore. The septum grows inward from the wall of the conidiophore and is characterized by a spherical plug in the pore. In the *Chromelosporium* state of *Peziza ostracoderma* - Figure 4.8 B (X29,000) and C (X59,000) - the picture is very different. We find much more elaborate structures around the septal pores. B shows a very early stage in the development of this apparatus, and C shows the ultimate elaborations. We aren't really sure of the significance of these structures surrounding the septal pores in *Chromelosporium*, but they are interesting. We usually refer to the pores as "peristome teeth pores" because of the strong resemblance of the structures to the peristome teeth of mosses. It is of primary interest to us that in two imperfect states of *Peziza* whose conidium ontogeny is so similar (synchronous production of blastospores on denticles), there is such a striking difference in septal pore structures.

DR ELLIS

In the formation of transverse septa, it is often clear that it is only the inner wall of the conidiophore that participates in septum formation, and that often there is a double wall formed by inward growth from two circular zones of the inner wall, as shown in Figure 4.8 A, with a middle lamella between them, which may under some circumstances dissolve.

DR G.C. HUGHES

Yes, I've found that in everything I've looked at.

DR KENDRICK

This is an ordinary cross-wall, not one at which we expect dissociation of elements to occur. It now seems that the difference between an ordinary septum in

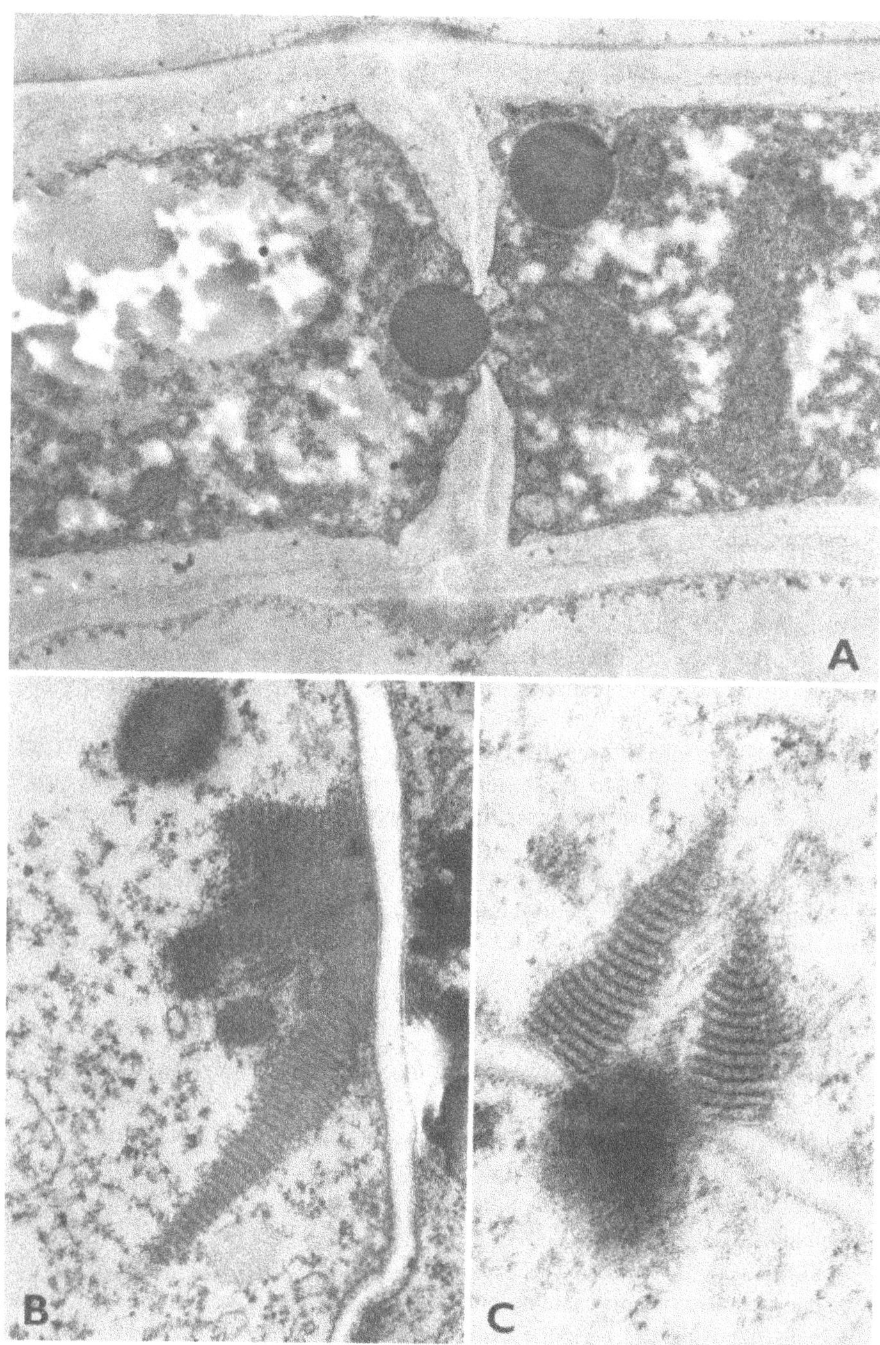


Figure 4.8 Septum formation in *Oedocephalum* and *Chromelosporium*; see text.

a hypha, and a "double septum" which enables conidia to be detached readily, may be one of degree rather than of kind.

DR S.J. HUGHES

In *Chromelosporium* the wall is hyaline and may be more plastic, while in *Gonatobotryum* it may be less extensible, so that when pressure is applied from within, it ruptures.

DR G.C. HUGHES

I started off with the assumption that *Chromelosporium* behaved like *Gonatobotryum*. It certainly looked that way, but it now appears that they behave differently.

DR KENDRICK

Perhaps the brown melanin-like pigment in the walls of *Gonatobotryum* has something to do with its lack of extensibility. This pigment may be an excretory product, but it is known to reduce water loss and protect from light. Perhaps it incidentally renders the wall less plastic also.

DR MÜLLER

I have difficulties even with the term spore. [*Laughter*] It is very hard to define, since any fungal cell can grow out and form a new mycelium. A spore is not like a seed. We have all intermediates between an ordinary fungal cell and the specialized cells we call spores. Most of the difficulties we are encountering in this discussion are caused by those intermediates. We must decide the limit of the term spore, but we must also realize that this necessary decision will be arbitrary.

DR KENDRICK

Do you like the idea of accepting Vuillemin's concept of conidiospores or "conidia vera," as opposed to those derived from pre-existing elements? (In other words, I need not give my paper on arthroconidia, because they are not really spores!)

MR NAG RAJ

In the imperfect state of *Ceratocystis paradoxa*, conidia produced as phialoconidia can later become thick-walled and pigmented resting spores. Even whole phialides and mycelial cells can become transformed. So here's a case of a fungus which can turn almost any part of itself into chlamydospores.

DR CARMICHAEL

Could we agree that in the production of conidia there is a gradation from one kind to another, and that to find a sharp distinction is next to impossible? There will always be doubtful or border-line cases. But there are also, of course, typical examples of different kinds. We must realize that these are simply high points in a series of intergrading forms. Agreement on this point might simplify future discussion.

DR KENDRICK

We must beware of two extreme positions: one in which we concentrate on the oddballs, ignoring the rest; the other in which we concentrate on the norms, ignoring the deviant. We have now seen so many strange intermediate forms that we should be prepared to admit the existence of a continuum. The majority of the fungi fall into the wide bulges in my diagram (Figure 4.9), but there are

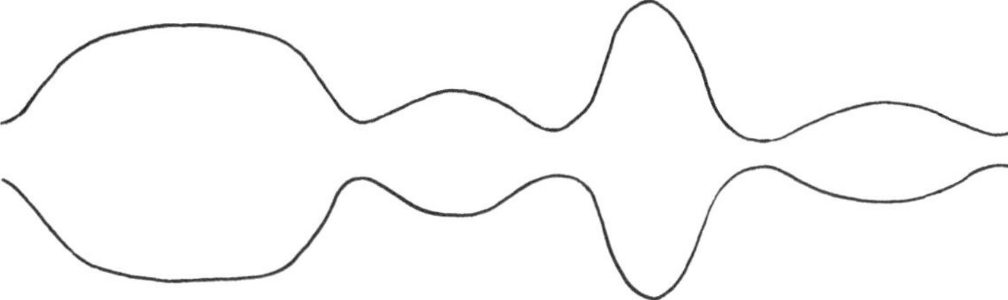

Figure 4.9 Diagram of the "fungal continuum"; see text.

forms that bridge the gaps between the bulges. Perhaps some of us thought until now that these gaps were complete gaps. Inevitably, we shall concentrate on the unusual or atypical at this meeting, since we are already aware of the major groupings, but we need to clarify our perspective of them, and perhaps trim their outlines more neatly. It is in the intermediate forms that the use of terms is especially critical.

DR ELLIS

I would agree that there are links between all spore types, with one exception, the phialoconidium. This is entirely distinct from every other spore form.

DR CARMICHAEL

My time-lapse pictures of *Aureobasidium pullulans* show repeated budding from the same point. Is this a phialide or not?

DR ELLIS

It is perhaps the simplest form of phialide. But *Aureobasidium* produces simple arthroconidia, blastic budding, and true phialoconidia. I'd like to see time-lapse studies of true *Aureobasidium pullulans.* There are many different concepts of *A. pullulans*, as Dr Cooke has pointed out, but in organisms conforming to *our* concept of the species, true phialoconidia *are* produced.

DR CARMICHAEL

I think phialides intergrade with sympodulae, and with successive blastoconidia. There is a typical phialide which is quite distinct; but you will also find organisms where the question arises: is this a phialide or an annellide?

DR ELLIS

I can't agree. I'm sure I can always distinguish a phialide.

DR CARMICHAEL

You work mainly with dematiaceous fungi that have nice thick walls. Perhaps they don't intergrade. But in the thin-walled hyaline hyphomycetes I have often been unable to decide just what was going on. There is considerable variation in a single fungus.

DR ELLIS

An accepted, up-to-date definition of what constitutes a phialide and a phialoconidium is very necessary and will, I hope, emerge later in this meeting. But I feel sure that the general concept has been quite distinct for a long time.

DR KENDRICK

We mentioned earlier the *Chalara* phialide with its long collarette within which there are several differentiated conidia. We wonder whether (i) differentiation began at the apex and proceeded downward until the upward movement of cytoplasm following the rupture of the phialide apex stabilized the situation, or (ii) the meristem was fixed from its inception, and the first few phialoconidia formed while the phialide was still elongating.

Let us assume that the first of those two hypotheses is true, the idea of an initial basipetal movement of the "conidiogenous locus," as I might call it. I can suggest that the first few spores are, in fact, endoarthroconidia, while those spores elaborated later by a fixed meristem, supplied with raw materials by an upwelling of cytoplasm from below, are true phialoconidia. We could be watching a re-enactment of the evolution of the phialide.

DR COLE

The problem is perhaps one of wall relationships. In the case of phialoconidium formation, a secondary wall is being formed, whereas in the production of meristem arthroconidia, as we have conceived them up to now, we are dealing with the entire wall of the conidiophore or fertile hypha. I prefer Dr Kendrick's second explanation of the phenomena occurring in the *Chalara* phialide of *Ceratocystis paradoxa*.

DR ELLIS

In all cases, the wall (or walls) of the phialoconidium is (are) formed de novo.

DR CARMICHAEL

It appears that general agreement on any of this terminology is not possible at present, and perhaps we should hear the rest of the talks and see the rest of the examples before we again try to reach an agreement.

EDITOR

*This topic was reopened during the terminology session. For further lively discussion, and the conclusions reached, see Chapters 15 and 16.*

# 5
# Porospores

M.B. ELLIS

The term porospore was coined by Dr Hughes in 1953 for a conidium which develops through a pore in the wall of the conidiophore. Among the examples of genera in which porospores are formed he gave *Helminthosporium, Torula*, and *Alternaria*.

In the past 16 years the term has gained wide acceptance, but the interpretation of what constitutes a porospore has varied. Subramanian in 1962 referred hyphomycetes producing porospores to the family Helminthosporiaceae, taking *Helminthosporium* (Figure 5.1) as the type genus for the family. He took *Torula*, however, as the type genus for the Torulaceae, a family which he considered to be characterized by the production of blastospores. He was familiar with and had described and figured *Torula herbarum*, the type species of *Torula.*

In 1963, Luttrell applied the adjective porogenous to conidia originating as protrusions through pores in the conidiophore wall.

Campbell, carrying out an electron microscope study of conidium structure and development in *Alternaria* at Bristol last year, showed that the conidia are formed by budding through a fairly well-defined channel in the outer wall of the conidiogenous cell, apparently induced enzymically. There is no sign of mechanical rupture. The inner or secondary wall of the conidiogenous cell becomes the outer or primary wall of the freshly developing conidium. The walls are not difficult to distinguish because the inner wall in *Alternaria* is electron-transparent whereas the outer wall has granular electron-dense material which may be concentrated in the outermost layer.

If it could be shown generally that, when porospores (or, as we may now prefer to say, poroconidia) develop, a channel in the outer wall of the conidio-

Commonwealth Mycological Institute, Kew, Surrey, England.

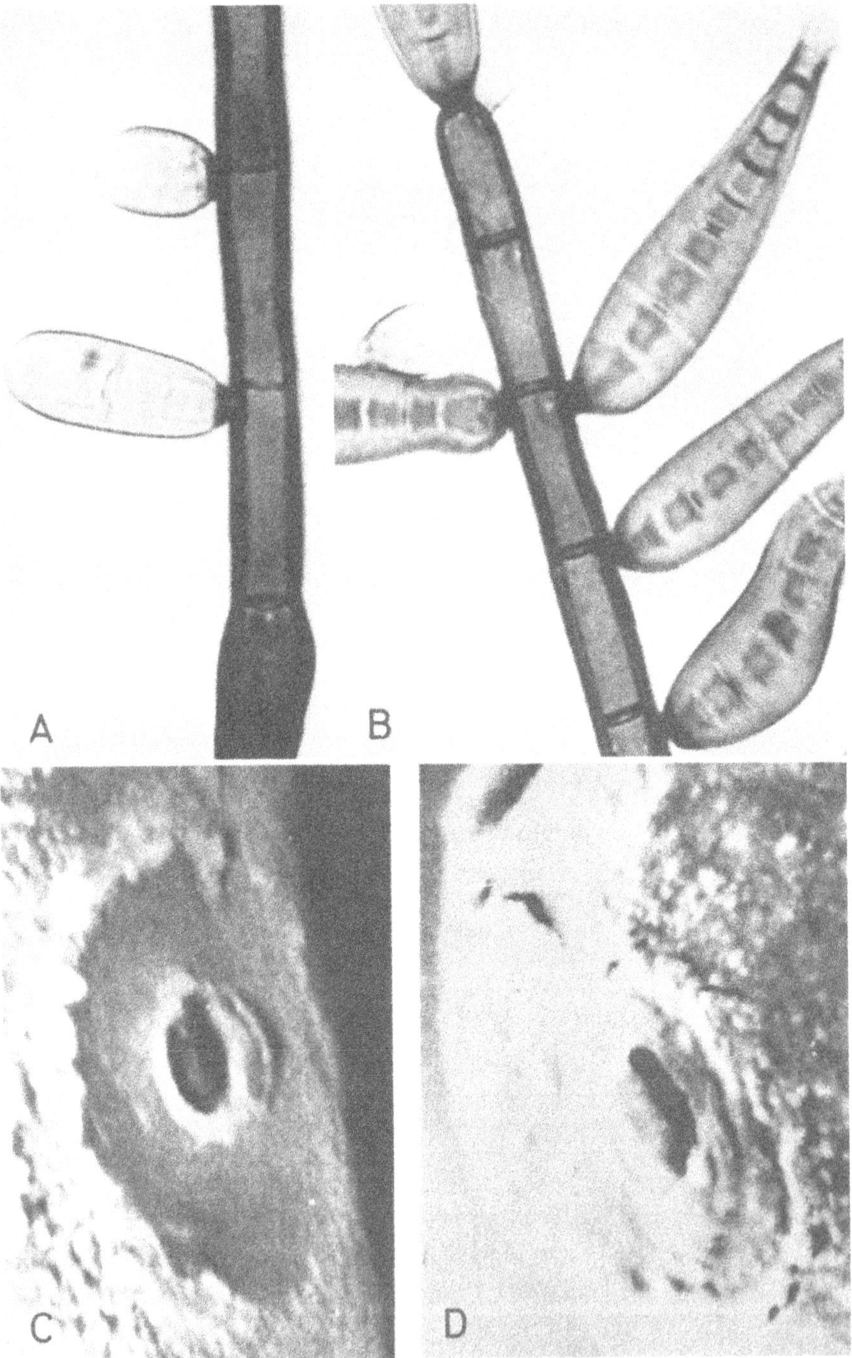


Figure 5.1 *Helminthosporium velutinum*. A and B (X1,000): A shows a channel passing through the thick outer wall of the conidiophore, and B shows conidia attached at the apex and in verticils below septa. C and D (X12,000; scan-

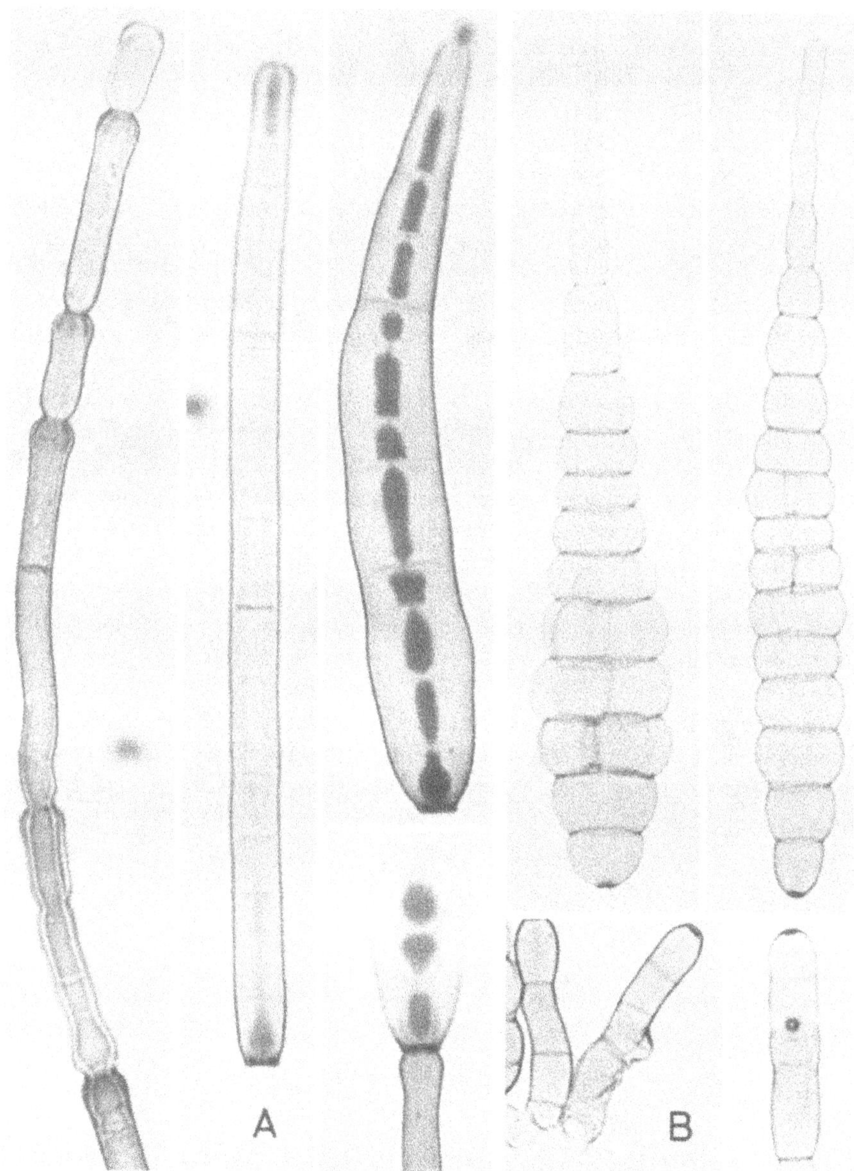


Figure 5.2 A: *Corynespora cassiicola,* X800; the conidiophore on the left shows repeated percurrent proliferation; the narrow cylindrical conidium, the young attached conidium, and the broad, obclavate conidium are all from the same colony. B: *Alternaria brassicae*, X500; the conidiophores illustrate sympodial development, and a scar sunken below the rounded contour is seen clearly at the apex; conidia are rounded off at the base and are very easily detached.

---

ning electron micrographs by C.V. Subramanian): C is a surface view of the end of a channel through which a conidium has developed, and D is part of the corresponding base of the conidium.

genous cell is induced enzymically, and that there follows protrusion of the inner wall through this channel, then poroconidia could be easily defined and distinguished, on the one hand from phialoconidia, which use neither wall of the conidiogenous cell, and on the other hand from blastoconidia, aleuriospores, and arthrospores, which make use of both.

Conidiogenous cells forming phialoconidia may be termed monophialidic and polyphialidic; those forming blastoconidia, monoblastic and polyblastic. When we come to poroconidia, however, although we could apply the term monoporous to the conidiogenous cells forming one conidium through a pore, the term polyporous could not be applied to conidiogenous cells which form several poroconidia because this adjective has been used in mycology in quite a different sense.

At the 10th Botanical Congress in Edinburgh, I suggested the term monotretic for a conidiogenous cell which forms one conidium (poroconidium) through a channel in its outer wall, and polytretic for a conidiogenous cell which forms conidia by protrusion through a number of separate channels in its wall. To date about a dozen genera have monotretic conidiogenous cells, and less than twenty have polytretic ones.

In *Corynespora* (Figure 5.2 A) and *Stemphylium* the monotretic conidiogenous cells are integrated, terminal, and percurrent; in *Spondylocladiella* they are discrete and determinate.

In *Helminthosporium* (Figure 5.1) and *Spadicoides* polytretic conidiogenous cells are integrated, terminal and intercalary, and usually determinate. More commonly, however, polytretic conidiogenous cells are sympodular as they are in *Alternaria* (Figure 5.2 B), *Curvularia, Drechslera, Exosporium*, and *Ulocladium.*

# 6
# Fine Structural Studies on "Poroconidium" Formation in *Stemphylium botryosum*

FANNY E. and G.C. CARROLL

Until recently the fine structure of conidium ontogeny in the Hyphomycetes has been a much neglected field. Several factors have probably contributed to the general discouragement of such studies. First, many hyphomycetous conidia and conidiophores are extremely resistant hydrophobic structures which have proved notoriously difficult to fix and embed satisfactorily for electron microscopy. Secondly, the diffuse and random orientation of conidiogenous loci in many hyphomycetes makes accurate location of the active sites in thin sections an arduous task. Finally, a tendency towards asynchronous sporulation gives rise to uncertainty in assigning a given image seen under the electron microscope to a known stage in conidium development.

As a result of such difficulties the developmental criteria proposed by Hughes (1953) for classifying the Fungi Imperfecti have not yet received adequate careful analysis at the fine structural level. With the exception of a single recent paper (Campbell 1969), to be discussed below, this situation holds for fungi producing poroconidia (Hughes's section VI) as it does for fungi in other sections of Hughes's system. In Hughes's initial characterization of section VI the presence of minute pores in conidiophores at the conidiogenous loci was cited as the sine qua non for the group. Luttrell (1963) later pointed out, however, that "there is no evidence as to whether porogenous conidia actually arise as protrusions of the protoplast through pores dissolved in the conidiophore wall and are therefore fundamentally different from blastogenous conidia or whether, as Ellis suggests, the 'pores' are merely 'thin areas' in the conidiophore wall that bulge outward to form the conidium initial in much the same fashion as in the formation of blastogenous conidia." The present study was undertaken to provide evidence on this question.

Department of Biology, University of Oregon, Eugene, Oregon, U.S.A.

 *Stemphylium botryosum* Wallroth was chosen as an experimental organism because it grows and fruits readily in culture, responds well to conventional fixation and embedment procedures for electron microscopy, will produce a dense palisade of short conidiophores in response to wounding, and can be induced to sporulate synchronously with the appropriate ultra-violet-light/temperature regime. It must be stressed that this work is still in progress, and the results presented here should be considered as tentative. We plan to supplement this work with light microscope studies using Nomarski phase-interference optics and fluorescence optics; at the fine structural level correlation of freeze-etch images with those obtained from conventional thin sections will be attempted. Where technical defects (knife scratches, chatter) are obvious in certain critical photographs here, replacements will be obtained.

## MATERIALS AND METHODS

About two weeks before fixation of conidiophores for electron microscope studies fresh cultures were started from single ascospore isolates (Leach no. 132) of *Pleospora herbarum* (Fr.) Rab., the perfect state associated with *Stemphylium botryosum*. These cultures were grown in a 12-hour light-dark regime at 21°C on 2 per cent malt agar plates. After 10 days, conidia from such cultures were inoculated onto Petri plates containing 2 per cent malt agar and incubated at 27°C in constant darkness. After 120 hours cores were taken from the centre of each colony with a no. 15 cork borer and transferred to small, sterile Petri plates. Synchronous sporulation was then induced according to methods formulated by Leach (1967, 1968). Cultures kept at 27°C were exposed to ultra-violet light ($\lambda$ = 320-420 m$\mu$) from a lamp (GE F15T8) suspended 30 cm above the plates. At the end of a 12-hour period all plates were returned to total darkness and held at 27°C while sporulation occurred. Samples were taken from the lateral surfaces of the cores and fixed at various intervals for electron microscopic study. Throughout the remaining discussion developmental stages are designated according to the time from the beginning of the ultra-violet treatment at which they were fixed.

Samples were fixed and embedded for electron microscopy according to the following procedure:

Fix in 2.5 per cent glutaraldehyde buffered at pH 7.8 with 0.1*M* s-collidine for 2 hours at room temperature (15 minutes in vacuo).

Wash several times in same buffer.

Post-fix in 1 per cent osmium tetroxide buffered at pH 7.8 with 0.1*M* s-collidine for 2 hours at room temperature.

Wash several times in distilled water.

Soak for 2 hours in 0.5 per cent aqueous uranyl acetate, pH 4.6.

Wash several times in distilled water.

Dehydrate in a graded ethanol series and wash several times in reagent grade acetone.

Infiltrate with graded acetone-plastic mixtures and finally embed in 100 per cent plastic. The following plastic formulation was used (Mollenhauer 1964): 62 ml Epon 812, 100 ml DDSA, 81 ml Araldite 6005, DMP-30 1 drop/ml plastic (accelerator).

Bubbles in the plastic were removed in a vacuum oven at 60°C, and polymerization of the plastic was carried out at 60°C for several days. Thin sections were cut on a Sorvall MT-1 ultramicrotome and mounted on 200-mesh copper grids. Sections were post-stained in methanolic solutions of uranyl acetate as described by Stempak and Ward (1964) followed by Reynolds lead citrate (Reynolds 1961), and examined with a Philips EMU-300 electron microscope.

## RESULTS

Fixations of conidiophores were started 14.5 hours after the beginning of ultraviolet induction when they had reached their maximum length of five or six cells; at this point they stopped growing and began maturation prior to sporulation. Observations on thin sections of apical portions of young conidiophores fixed at 14.5 hours reveal an undifferentiated cell wall 0.2-0.3$\mu$ thick containing a thin layer of some osmiophilic substance on its exterior surface (Figure 6.1). Within the cytoplasm, osmiophilic granules ranging from 20 to 100 m$\mu$ in diameter appear in association with internal convoluted membrane systems and convoluted areas of the plasma membrane. Although such granules are distributed thoughout the first cell of the conidiophore (Figure 6.1 A), they appear to be somewhat larger towards the base of this cell (Figure 6.1 C) than at the apex (Figure 6.1 B). Oblique sections of this portion of the conidiophore indicate that convoluted plasma membrane profiles visible in transverse sections represent complex outpocketings of the membrane (Figure 6.1 D). Figure 6.1 D shows numerous black granules lodged in the crypt-like areas between these outpocketings and the cell wall. In Figure 6.1 C the cytoplasm has pulled away from the wall during specimen preparation, leaving an electron-transparent area between wall and plasma membrane. We have noted repeatedly that the portion of the conidiophore in the immediate vicinity of the first septum is particularly prone to this sort of fixation damage.

At a somewhat later stage (15.5 hours) a second sort of inclusion is found in the older, basal cells of the conidiophore. Such inclusions appear as tenuous fibrillar elements within cytoplasmic vesicles (Figure 6.2) and as massive aggregations of fibrils between the plasma membrane and the cell wall (Figures 6.2, 6.3 A). Although the latter occur between outpocketings of the plasma membrane similar to those seen in younger portions of younger conidiophores, the plasma membrane in general is much less convoluted here than in the younger cells described above (compare Figure 6.4 A with Figure 6.1 C). In later stages differentiation of the lower conidiophore wall into three distinct layers has become apparent: an outer, extremely electron-dense layer about 200 m$\mu$ thick; a middle, electron-transparent area 30-60 m$\mu$ thick; and an inner, granular zone of intermediate electron density, 500-750 m$\mu$ thick (Figures 6.2, 6.3 A,B). Careful scrutiny of the inner wall layer reveals numerous fibrils oriented parallel to the longitudinal axis of the cell.

In the same developmental stages (15.5 hours) numerous lightly staining irregular bodies, bounded only by a darkly staining interface, arise in association with myelin-like swirls of membrane (Figure 6.3 B, 6.4 A, B). At this time such

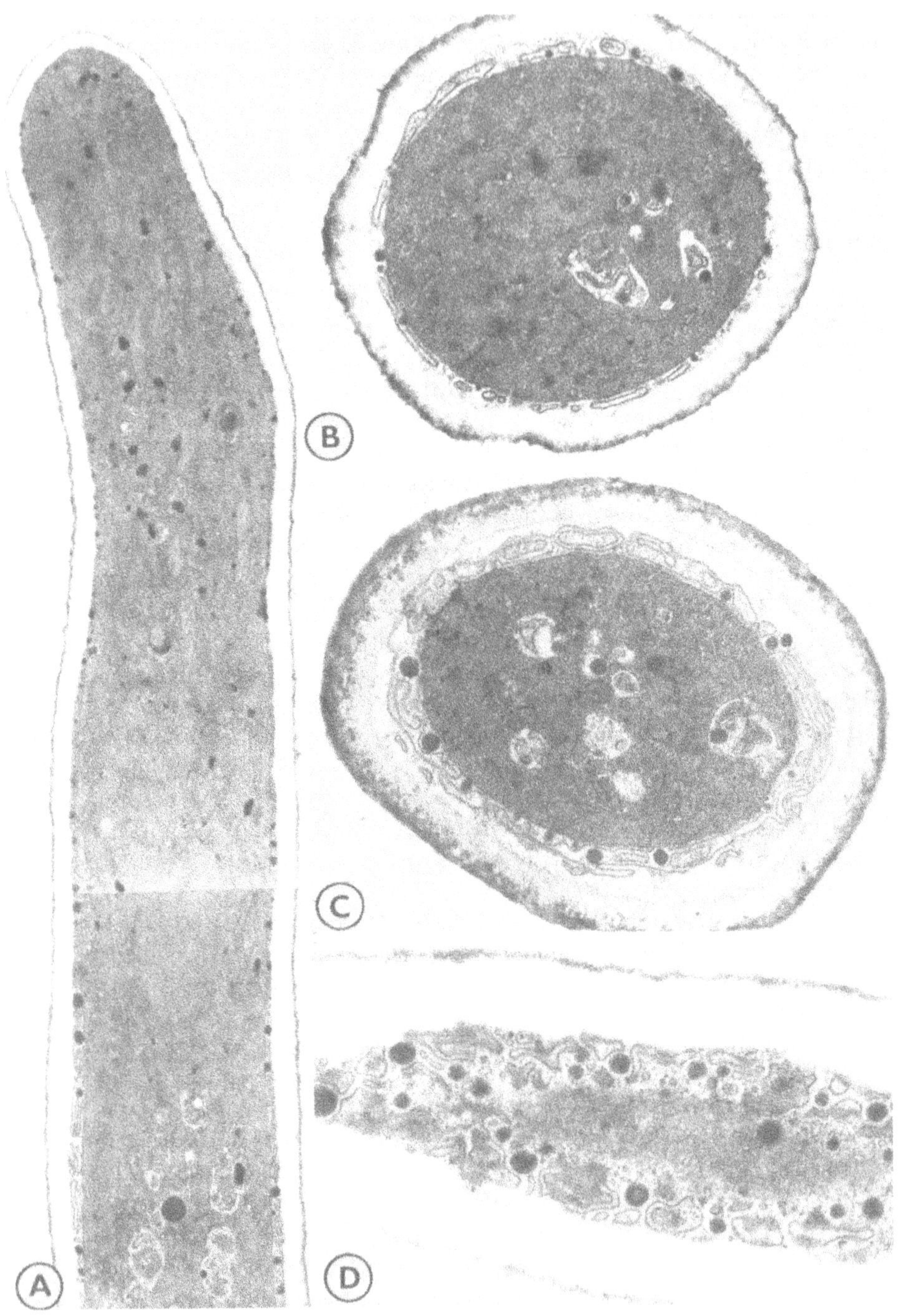


Figure 6.1 Sections of young conidiophores (14.5 hours): A, longitudinal sections of first cell at conidiophore apex, ×14,200; B, transverse section of conidiophores near apex, ×22,800; C, transverse section of conidiophore near first septum, ×21,500; D, oblique section of conidiophore near first septum, ×25,700.

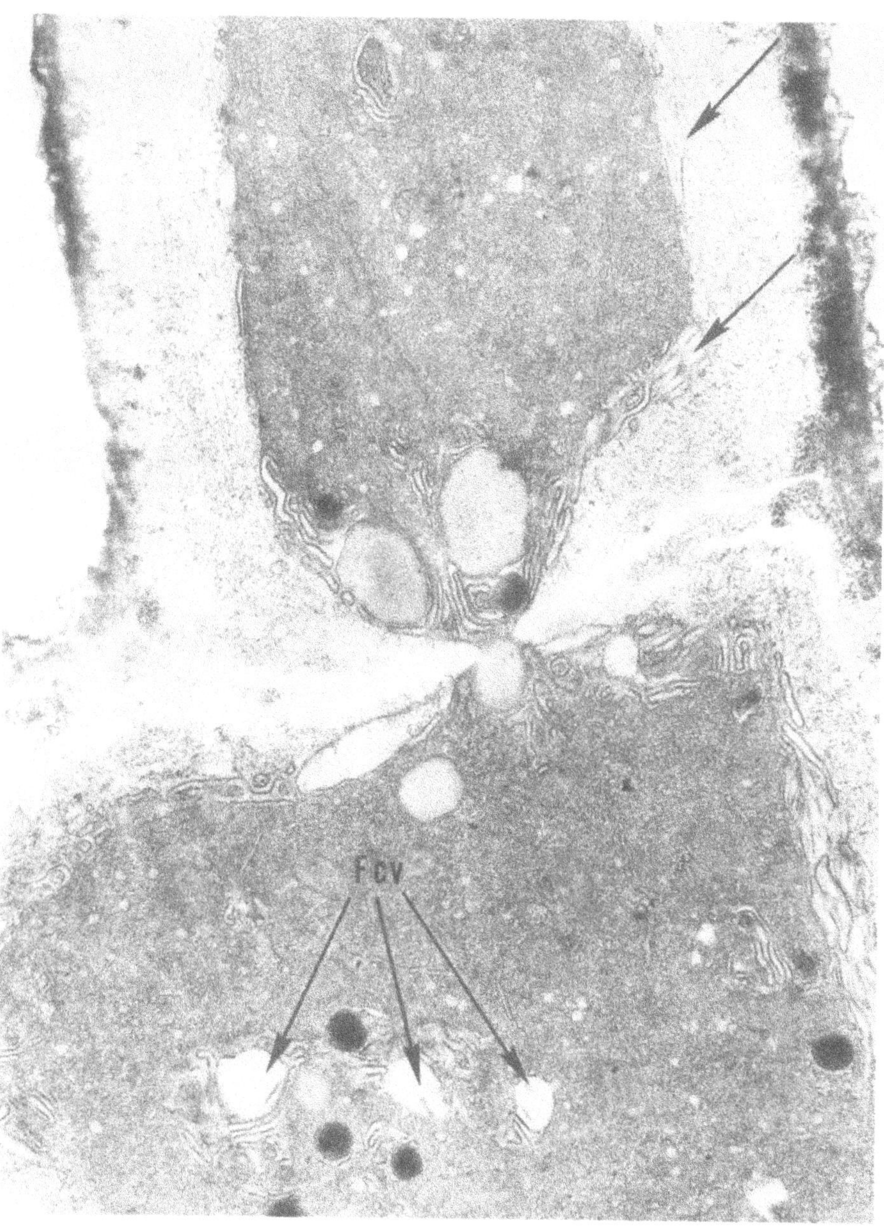


Figure 6.2 Longitudinal section of septum with pore, in conidiophore at point of branching near base (15.5 hours). Note crypt-like areas between plasma membrane and wall which contain fibrillar elements (arrows) and vesicles in cytoplasm containing similar elements. Fcv = fibril-containing vesicles. X35,000.

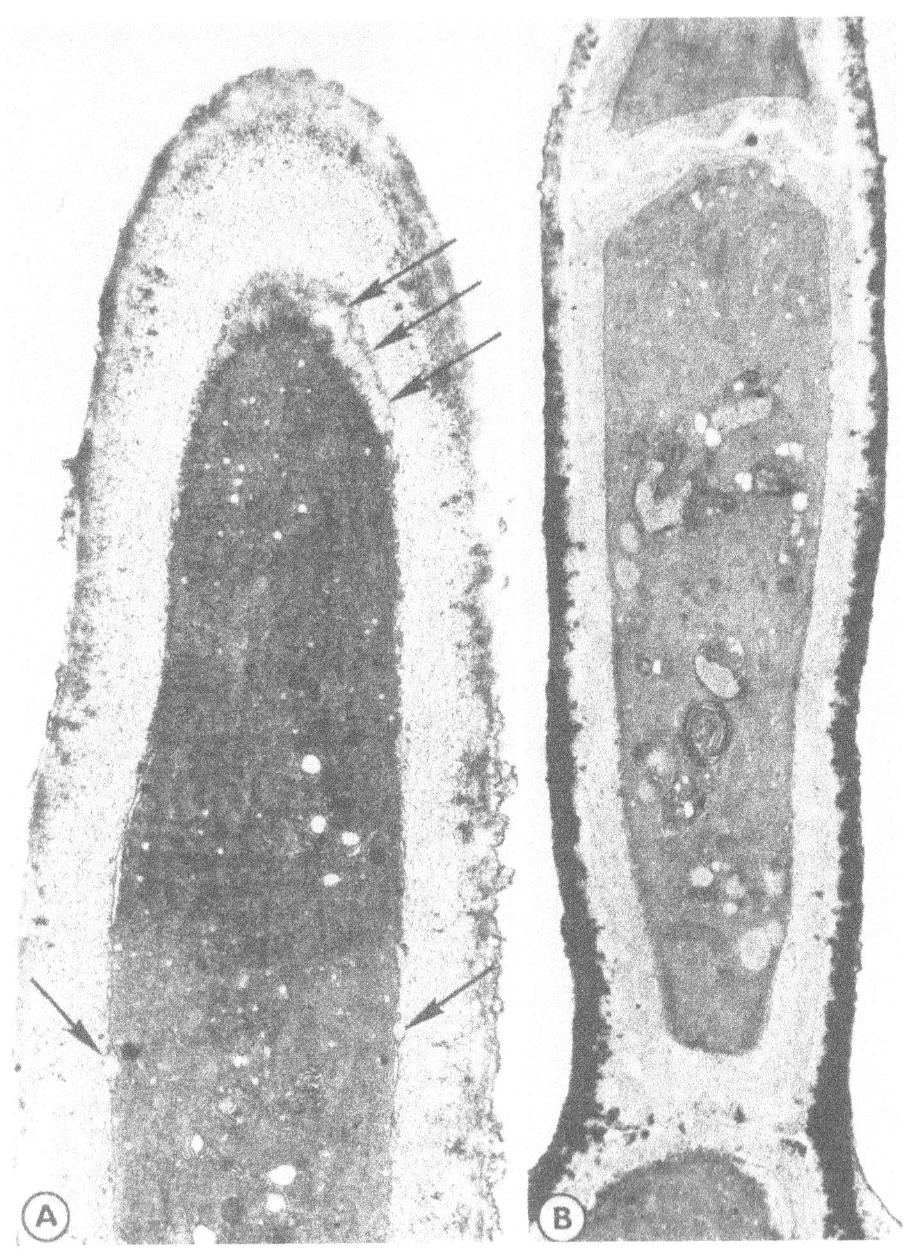

Figure 6.3 Longitudinal sections of conidiophores near base. A: second cell from base; arrows indicate fibrillar areas between plasma membrane and wall (15.5 hours); X21,500. B: basal cell of conidiophore (16 hours), X14,100.

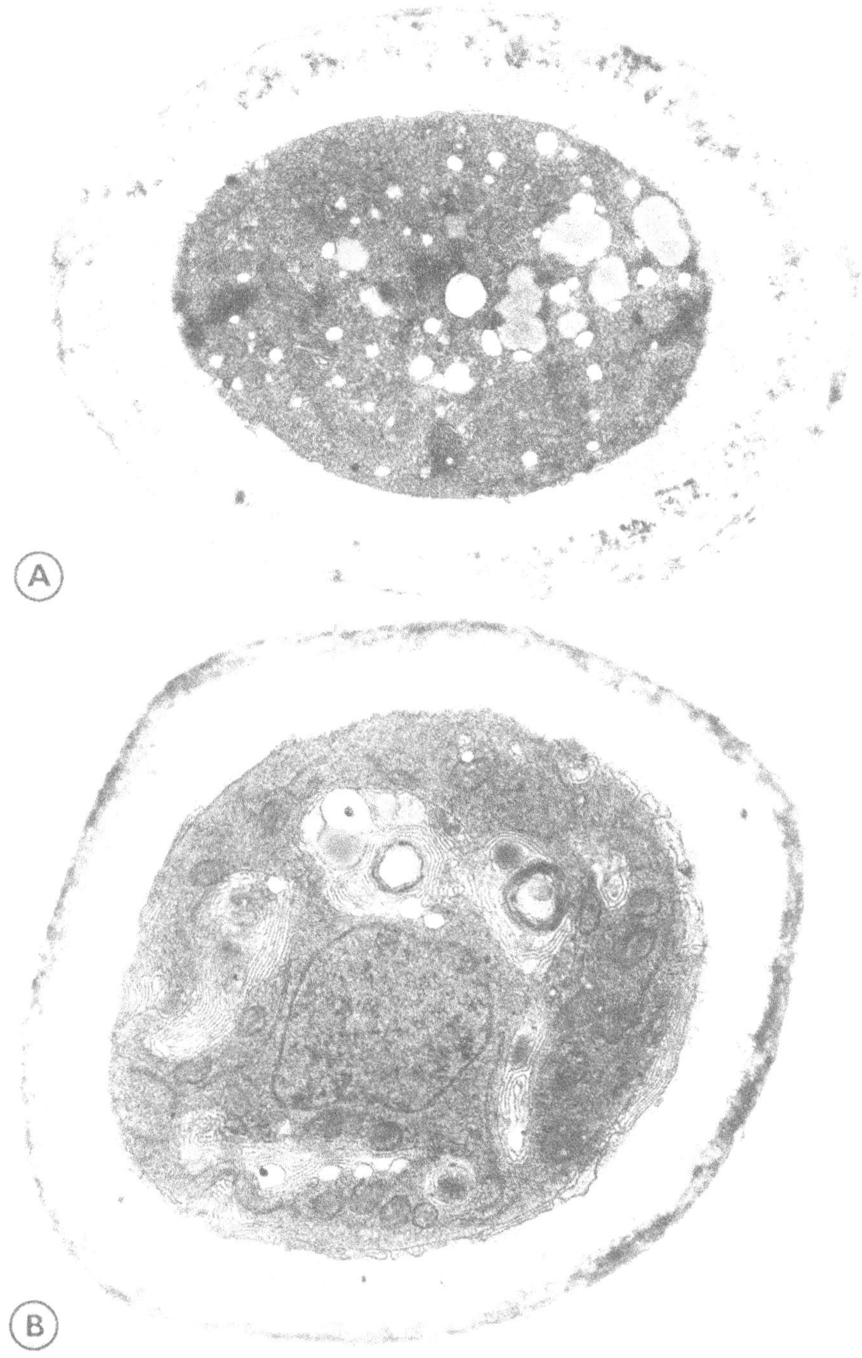


Figure 6.4 Transverse sections of conidiophores (15.5 hours): A, section near base of conidiophore, X22,200; B, section at a higher level, towards conidiophore apex, X22,200.

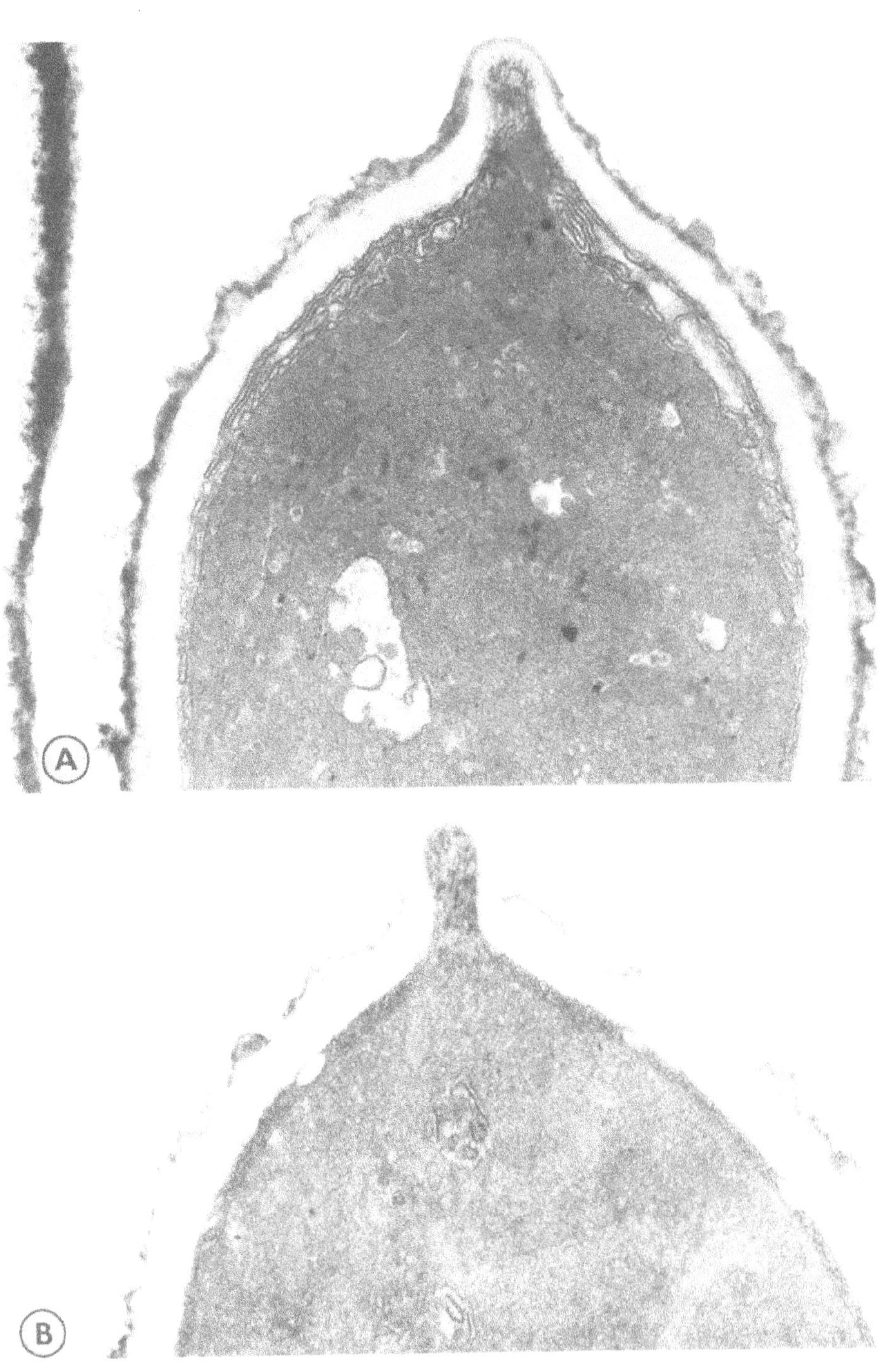


Figure 6.5 Longitudinal sections of successive stages in early conidium development at conidiophore apex (15.5 and 16 hours for A and B respectively), X32,000.

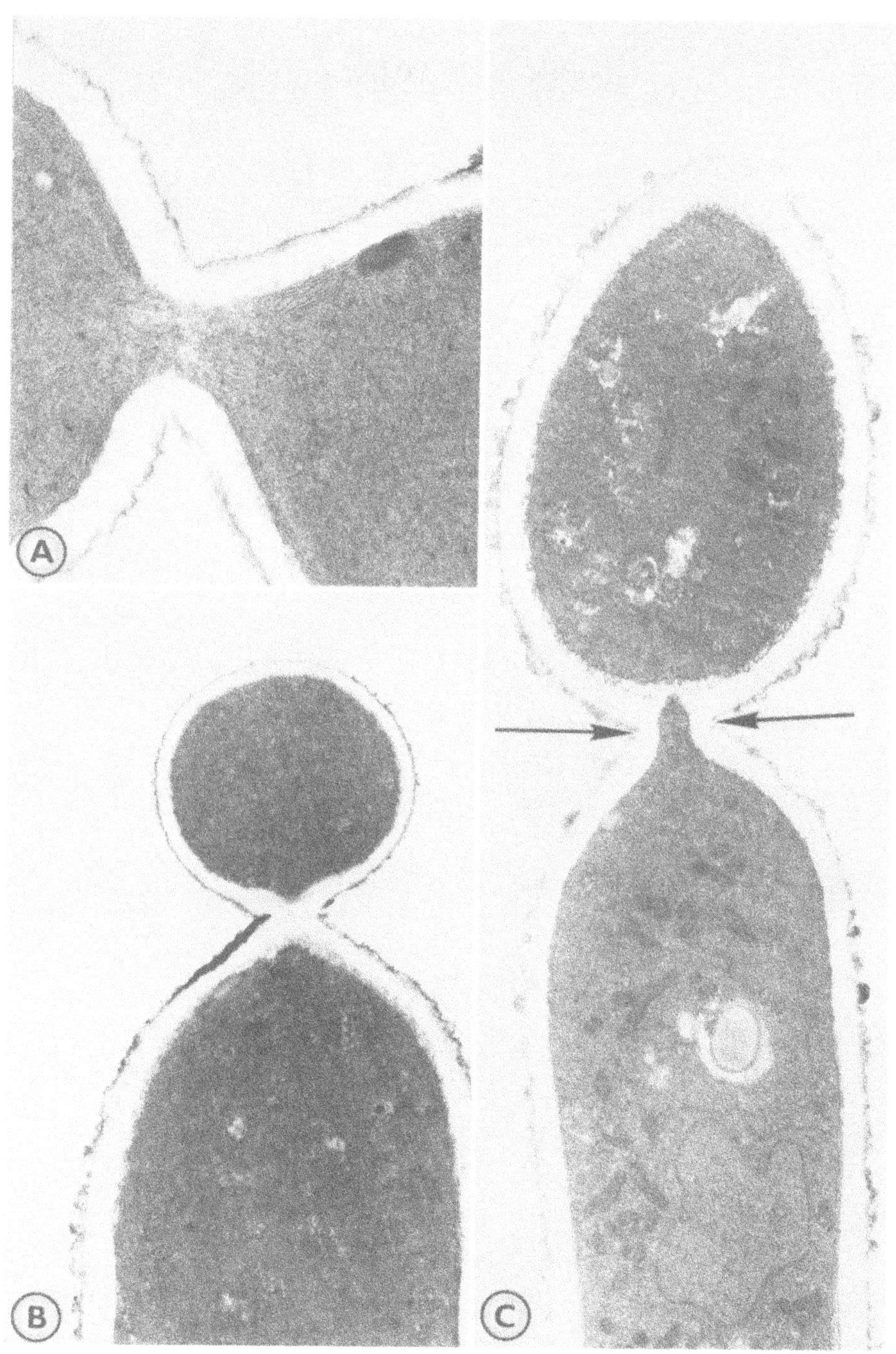


Figure 6.6 Longitudinal sections of junction between conidiophore and young spore: A (20 hours), X22,300; B (16.5 hours), X14,400; C (20 hours), arrows indicate eventual line of secession between conidiophore and conidium, X13,000.

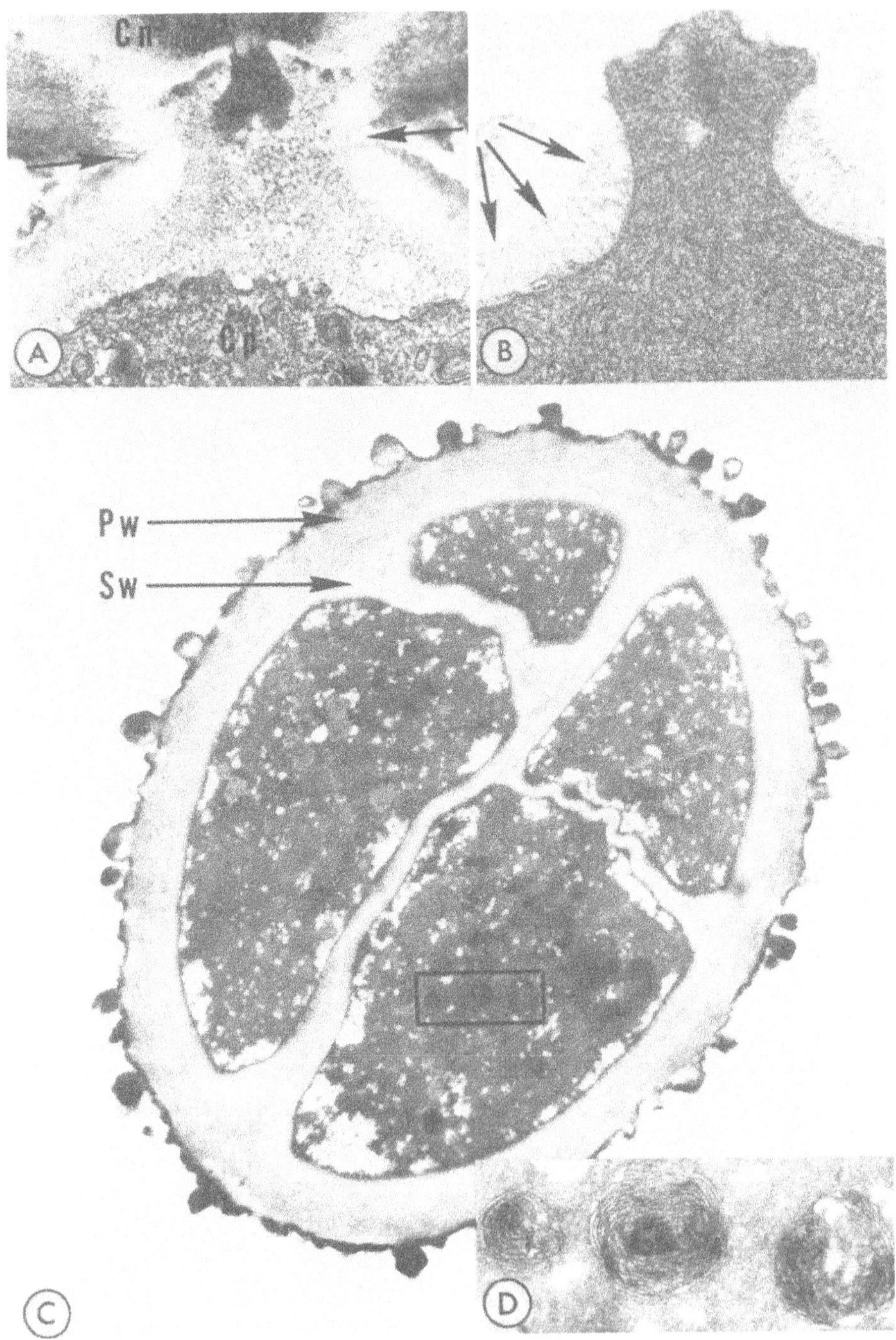


Figure 6.7 A: longitudinal-oblique section of junction between mature conidium (Cn) and conidiophore (Cp); arrows indicate line of secession (22 hours); X36,000. B: longitudinal section of conidiophore apex from which conidium has seceded; arrows show radial arrays of fibrils in inner wall of conidiophore (20 hours);

associations are particularly prominent in the middle cells of the conidiophore (Figure 6.4 B); in the basal cells the membrane swirls are much less in evidence and many of the light-staining bodies occur free in the cytoplasm (Figures 6.3 B, 6.4 A). Such "gray" inclusions are also found associated with complex involutions of the plasma membrane near septal pores (Figure 6.2). 

Sporulation commences at 15.5-16.0 hours. The conidium initial is first visible at the conidiophore apex as a simple bud-like protrusion whose wall is continuous with that of the conidiophore (Figure 6.5 A, B). The external osmiophilic layer of the conidiophore is at first thinner at the site of budding; in later stages, however, it becomes thicker over the spore initial and similar to that of the conidiophore itself. Indeed, at one point the spore initial appears indistinguishable from the conidiophore (Figure 6.6 A). The primary walls of the conidiophore and conidium remain continuous throughout conidium development (Figure 6.6 A, B, C), and only during the final stages of conidial maturation does a presumptive line of secession become visible at the apex of the conidiophore (Figure 6.7 A). Observations of the junction between conidium and conidiophore at these later stages suggest that this line of secession is not a clean, straight break across a narrow neck. Rather, the conidium-conidiophore junction appears to be a plug-and-socket arrangement in which a narrow channel of conidiophore cytoplasm protrudes into the conidium and abuts against a thin, perforate septum which is continuous with the inner part of the conidial wall (Figures 6.6 C, 6.7 A). The break occurs between the protrusion of the conidiophore cytoplasm and the conidial septum, leaving on the one hand a conidium with a crater-like secession scar, and on the other a conidiophore with an apical, crown-like cytoplasmic protrusion (Figure 6.7 B). A radial array of fibrils perpendicular to the plasma membrane is evident around the cytoplasmic neck in a mature conidiophore (Figure 6.7 B). At the same time wall differentiation proceeds in the maturing multicellular spore much as it does in the maturing conidiophore cells. An external, electron-dense, primary wall appears on the outside separated from a less electron-dense inner wall by a thin, electron-transparent zone (compare Figure 6.3 B with Figure 6.7 C). The individual cells are separated on their inside faces only by the electron-transparent zones between their secondary walls. Thus the primary wall acts as a sort of outer casing surrounding a number of separate cells each covered by its own secondary wall.

Aside from some slight vesiculation in the neck of the bud (Figure 6.5 A) and convolution of the plasma membrane at the apex of the conidiophore (Figures 6.5 A, B and 6.6 A, B) the micrographs presented here show little evidence of any special metabolic activity at the conidiogenous locus. As the conidial initial enlarges, convolution of the plasma membrane becomes less noticeable in the conidiophore and more evident in the initial (Figure 6.6 C); further, the conidial cytoplasm appears shrunken away from the wall in a manner similar to that described for the apical cell of the young conidiophore (compare Figure 6.1 A with

×28,700. C: longitudinal section of mature conidium (22 hours); Pw = primary wall, Sw = secondary wall; ×11,300. D: portion of C enlarged to show membranous nature of darkly staining bodies in lower right cell of conidium; ×45,000.

Figure 6.6 C). As maturation proceeds, myelin-like membrane swirls with associated "gray" bodies appear first in the conidiophore apex (not pictured) and then in the conidium (Figure 6.7 C, D). Finally, the conidial cytoplasm becomes extremely dense and, in addition to the inclusions noted above, an electron-transparent granular substance accumulates near the cell peripheries.

## DISCUSSION

Differentiation is evident at the fine structural level both in walls and in the cytoplasm during conidium formation and maturation in *S. botryosum.* Changes in the wall structure can be seen even with the light microscope during conidium formation in many hyphomycetes and are the basis for such terms as porospore (Hughes 1953) and distoseptate conidium (Luttrell 1963). In *S. botryosum* the most obvious such change appears as a gradual darkening of the wall during maturation. This correlates well with the thickening of the external osmiophilic layer with age that is observed at the fine structural level, and lends support to the idea that this layer is in fact composed of pigment. Further evidence for this comes from unpublished data from our laboratory on *Stemphylium floridanum* Hannon and Weber in which the lightly pigmented conidiophore apex and young conidia fluoresce brilliantly when stained with primulin and observed under a fluorescence microscope. Heavily pigmented cells in the lower part of the conidiophore and in the mature conidia do not fluoresce. However, when mature conidia are broken apart with pressure from the coverslip, the inner walls of the conidium are exposed and fluoresce. This would indicate that the lack of fluorescence in unbroken cells arises from the masking effect of an external dark pigment layer. Campbell (1969) has noted an electron-dense zone in conidiophores and conidia of *Alternaria brassicicola* (Schw.) Wiltshire and has also concluded that it is probably melanic.

The pore between conidiophore and conidium is exposed after conidium secession only as a small unpigmented area where the cytoplasm between conidiophore and conidium was previously continuous. Although the septum separating the mature conidium from the conidiophore is much thinner than the primary wall, it is doubtful whether a scar would be discerned here with a light microscope in the absence of any pigmentation in the wall of the conidiophore or conidium. In fact the vast majority of species described as having porogenous conidial development also have pigmented spores; in many hyaline-spored species a similar pattern may exist, as yet undiscovered because of the virtual invisibility of the pore areas.

In a number of species dark annular areas surrounding the pore in the conidiophore or conidium, or both, occur (Luttrell 1963). These have not been seen in *S. botryosum*, but do occur in *A. brassicicola*, where their ultrastructure has been noted (Campbell 1969). Apparently in that species such structures arise as extensive thickenings of the secondary wall surrounding the pore area within the conidiogenous cell. Eventual melanization renders them visible under the light microscope. Again, such structures may be present but invisible in hyaline-spored species.

In Hughes's initial diagnosis (1953) and in general discussions which have followed (Luttrell 1963; Simmons 1967) authors have stressed that although poroconidia might be formed by a budding process, a sharp discontinuity is maintained between the conidium and conidiophore walls. Our results contradict these statements. At least in *S. botryosum* the primary walls of the conidiophore and conidia are continuous until the time of conidium secession. Campbell (1969) claims that the secondary wall of the conidiogenous cell is continuous with the primary wall of the conidium. The situation in *A. brassicicola* seems to be similar to that in *S. botryosum* in many respects, however, and we would interpret Campbell's "primary wall" to be a superficial zone of pigment deposition within his "secondary wall" (compare our Figure 6.5 A, B with his Figure 17). Careful scrutiny of his Figure 17 reveals an external dark layer over a lighter layer in walls of both the conidiogenous cell and the budding conidial initial. In any case, conidial development in both his study and the present one appears to be fundamentally blastic. If this proves the case for other members of Hughes's section VI, then the difference between a "porospore" and a "blastospore" will become a tenuous distinction at best.

As aging of cells in the conidiophore proceeds, fibrillar elements appear in vesicles within the cytoplasm, in massive aggregates at the edge of the cytoplasm, and as isolated elements in the matrix of the wall itself. It appears likely that such elements are synthesized in cytoplasmic vesicles which migrate to the periphery of the cell and release their contents to the wall. Within the wall matrix such fibrils become reoriented and probably play an important role in secondary wall differentiation. In the lateral walls of the older cells of the conidiophore they are aligned parallel to the long axis of the cell; in the pore area they occur in radial arrays perpendicular to the plasma membrane.

Although cross-wall formation in the conidiophore and conidium has not been studied here in sufficient detail to record a developmental sequence, the fine structure of the mature cross-wall in *S. botryosum* appears similar to that of *A. brassicicola* and may arise by the process Campbell (1969) has suggested for that species. In *S. botryosum* the final result in conidiophore and conidium is an outer heavily pigmented wall encasing a number of cells each enclosed in its own separate secondary wall. This is precisely the situation described as distoseptate by Luttrell (1963). Its occurrence in a species in which individual cells do not readily separate when the outer wall of the conidium is crushed (as they do in, for example, *Helminthosporium sorokianum* Sacc. in Sorok.) suggests that the distinction between euseptate and distoseptate conidia as described by Luttrell (1963) may not be a clear one.

Evidence of changes in cytoplasmic structures and inclusions during conidial development seems conclusive. Even in the preliminary work presented here we find clear-cut cases of a restricted occurrence of certain cytoplasmic inclusions. During early stages of conidiophore maturation (15.5 hours) osmiophilic granules appear particularly numerous in cells near the tip of the conidiophore. Such inclusions may be similar to the smaller granules found in growing hyphal tips of *Ascodesmis sphaerospora* Obrist (Brenner and Carroll 1968). The composition and function of such granules is completely unknown. They may be some wall component which is synthesized in the cytoplasm and migrates to the wall; alternatively, they could contain enzymes important for the final stages of conidiophore growth or early stages of conidiophore maturation. Preliminary resolution of these questions must await histochemical work at the electron microscope level.

A second type of inclusion, here described as a "gray" body, arises in the basal cells of the conidiophore, later at the apex of the conidiophore, and finally in the cells of the conidium itself. Such bodies are always seen initially in connection with myelin-like swirls of membrane. Similar membrane swirls have been observed in yeast, in connection with both electron-transparent inclusions (Linnane, Vitols, and Nowland 1962) and inside vacuoles (Smith and Marchant 1968). Campbell (1969) has seen both "gray" bodies and membrane swirls, but not in close association. He suggests that the former may contain phospholipid and that the latter may be an artifact of glutaraldehyde fixation. Studies using permanganate-fixation and freeze-etch techniques to be carried out soon in our laboratory should at least confirm or deny the reality of these myelin-like figures in *S. botryosum.*

A third type of inclusion appears only in the mature spores; it shows up in cells stained with osmium tetroxide as peripheral electron-transparent granular aggregates. Campbell (1969) has noted similar inclusions in spores of *A. brassicicola,* and on the basis of preliminary paper chromatography of spore hydrolysates has suggested that they are some unknown polymer of glucose.

Other more general cytoplasmic changes occur during conidiogenesis in *S. botryosum*. A zone of plasma membrane convolution appears first in the middle cells of the conidiophore, then at the apex of the conidiophore, and finally in the conidium itself. Similar convolutions of the cell membrane have been noted by Campbell (1969) for *A. brassicicola.* It is suggested both in Campbell's paper and here that such membrane convolution is related to cell wall synthesis. Cytoplasmic shrinkage has been noted first in the conidiophore and later in the conidium here. Although this is essentially an artifact, it is an expression of a selective sensitivity of certain cells to the fixation methods used. As such, it can be taken as further evidence for a sequence of cytodifferentiation during development. Finally, changes in general cytoplasmic density occur during spore maturation. As the spore matures, the cytoplasm stains increasingly densely; in the fully mature spore very low contrast electron images are obtained, and organelles within the cytoplasm can be detected only with difficulty. Presumably this change in staining properties is related to dehydration of the cells.

Conidia in the Hyphomycetes develop as direct derivatives of a hyphal system; in many cases cells of the conidia differ but little from cells of the conidiophore. Thus it comes as no surprise to find that cytodifferentiation during conidium formation in *S. botryosum* involves similar changes in cells of the conidiophore and cells of the conidium itself. Waves of cytoplasmic change appear to move up the conidiophore to the conidium in acropetal sequence as the individual cells concerned differentiate. One expects that a detailed documentation of the fine structure and cytochemistry of such changes in a variety of hyphomycetes will prove a laborious task; clearly this area demands and deserves the interest of a number of careful workers.

ACKNOWLEDGMENTS

The authors wish to acknowledge and express thanks for support from the Graduate School of the University of Oregon and the Brown-Hazen Fund received while this work was in progress.

REFERENCES

Brenner, D.M., and Carroll, G.C. 1968. Fine-structural correlates of growth in hyphae of *Ascodesmis sphaerospora.* J. Bacteriol. 95: 658-71

Campbell, R. 1969. An electron-microscopic study of spore structure and development in *Alternaria brassicicola.* J. Gen. Microbiol. 54: 381-92

Hughes, S.J. 1953. Conidiophores, conidia, and classification. Can. J. Botany 31: 577-659

Leach, C.M. 1967. Interaction of near-ultraviolet light and temperature on sporulation of the fungi *Alternaria, Cercosporella, Fusarium, Helminthosporium,* and *Stemphylium.* Can. J. Botany 45: 1999-2016

- 1968. An action spectrum for light inhibition of the "terminal phase" of photosporogenesis in the fungus, *Stemphylium botryosum.* Mycologia 60: 532-46

Linnane, A.W., Vitols, E., and Nowland, P.G. 1962. Studies on the origin of yeast mitochondria. J. Cell Biol. 13: 345-50

Luttrell, E.S. 1963. Taxonomic criteria in *Helminthosporium.* Mycologia 55: 643-74

Mollenhauer, H.H. 1964. Plastic embedding mixtures for use in electron microscopy. Stain Technol. 39: 111-14

Reynolds, E.S. 1961. The use of lead citrate at a high pH as an electron opaque stain in electron microscopy. J. Cell Biol. 17: 208-13

Simmons, E.G. 1967. Typification of *Alternaria, Stemphylium,* and *Ulocladium.* Mycologia 59: 67-92

Smith, D.G., and Marchant, R. 1968. Lipid inclusions in the vacuoles of *Saccharomyces cerevisiae.* Arch. Mikrobiol. 60: 340-7

Stempak, J.G. and Ward, R.T. 1964. An improved staining method for electron microscopy. J. Cell Biol. 22: 697-700

## DISCUSSION

DR ELLIS

This seems to be a beautiful example of blastic development! [*Laughter*] Many of us have suspected for a long time that the conidia of *Pleospora herbarum* are not poroconidia in the sense that the conidia of *Helminthosporium velutinum* are poroconidia. There is often a swelling and attenuation of the wall at the apex of the conidiophore of *S. botryosum* before conidia are produced. This isn't like the tretic method of development found in *H. velutinum,* the type of the group, where a very narrow pore pierces a very thick wall. No electron micrographs of sections through pores of *H. velutinum* have yet been made, but they are badly needed. It is just conceivable that the pores of *H. velutinum* may be basically similar to those of *S. botryosum,* in which case we would no longer be able to use the term poroconidium in a developmental sense. Electron microscope work must be done on all fungi producing what have been called porospores, but *H. velutinum* should be investigated first. It has a thick, hard, inextensible wall which we believe must be dissolved before anything can come out; but perhaps electron microscope pictures will show it blowing out after all.

DR KENDRICK

The distinction which must be made is whether the outer wall of the conidiophore blows out and forms a continuous integument around the conidium, or whether it is dissolved away enzymically to allow the egress of an inner layer which will form the outer wall of the conidium. In the first case, demonstrated so well by Dr Carroll for *Stemphylium*, development is blastic. In the second, where the new conidium emerges from a narrow channel through a thick wall, development is tretic.

DR SUBRAMANIAN

I'm sorry Dr Madelin isn't here to speak about Campbell's work on *Alternaria.* The dark annulus, which often makes poroconidia easy to identify, is clearly visible in darkly pigmented fungi. Campbell also worked with an albino mutant. Wall relations are exactly the same as in the wild type, but no annulus is visible. Almost all recognized poroconidium-forming genera are dematiaceous. Perhaps some hyaline fungi produce conidia in the same way, but have not been recognized because they exhibit no easily visible annulus.

A second point. We have found that some *Drechslera* isolates produce mostly blastoconidia when incubated under certain conditions. Thus the two methods may be interchangeable. Perhaps the conditions under which Dr Carroll's *Stemphylium* were grown have had the same effect.

DR ELLIS

Did Campbell find the same method of conidium development at the apex of the conidiogenous cell and at the apex of the conidium?

DR SUBRAMANIAN

I'm not sure if he got the conidiophore apex, but he saw conidia germinating to produce germ tubes which then became conidiogenous. He found the same mechanism at these apices as at the apex of a conidium producing a secondary conidium directly.

DR KENDRICK

If *Alternaria* and *Stemphylium* really differ in this way, what does this imply for the close taxonomic relationships that have been assumed to exist between the two genera? After all, there are species of *Alternaria* with perfect states in the Pleosporaceae.

DR PIROZYNSKI

I wondered to what extent the degree of development of the conidiophore walls determined the type of conidium ontogeny. In *Spadicoides obovata* the apical conidium, which develops where the wall is still relatively plastic, is blastic, whereas the lateral conidia have to emerge through a thick, non-elastic, fully differentiated wall, and may have to seek other means of exit, such as dissolution of a pore.

DR ELLIS

The pore in the centre of a scar on a conidiophore is no indication that a poroconidium has been formed. Pores frequently appear in the centres of scars where a blastic type of conidium has been produced, and are there as a purely secondary phenomenon.

DR KENDRICK

The pore, then, may be a compensatory mechanism that is needed only when spore production is delayed, and must take place through a previously consolidated wall. The consensus of the group appears to be that conidium ontogeny in *S. botryosum* is clearly blastic, and that much more evidence, of the kind that Dr Carroll has provided, is needed before the terminological fate of the poroconidium or tretoconidium can be decided - whether it will be amalgamated with blastic spore formation, or will be maintained as a distinct process.

*So ended our first day. We had encountered some difficulties and disagreements, which we would try to resolve later in the conference, but we had also sensed some areas of general agreement. And there had been some surprises - notably Dr Carroll's revelations about the supposed porospores of* Stemphylium.

*Next morning, Dr Cooke read his succinct yet comprehensive summary of the first day's proceedings. Thus primed, we entered our second day of debate. I for one was already beginning to experience changes in my way of looking at the Hyphomycetes; the conference promised well.*

# 7
# The Phialide

C.V. SUBRAMANIAN

The main objective of this conference is to evolve an acceptable terminology supported by unambiguous definitions. Although the definition of the term phialide as originally used by Vuillemin (1910a, 1910b, 1911) was primarily based on the shape of the conidiogenous cell (phiala = cup or bowl), it is now clear that not all of the fungi which Vuillemin considered to produce phialides show the same pattern of development. The recent interest in conidium ontogeny, stimulated by the pioneering work of Vuillemin (1910a, 1910b, 1911), Mason (1928, 1933, 1937, 1941), and later Hughes (1953), has prompted several students of the Hyphomycetes to investigate problems of conidium ontogeny more thoroughly, using several distinct, but complementary, approaches: ordinary light and phase contrast microscopy, transmission and scanning electron microscopy, fluorescent antibody staining, time-lapse cinemicrography, etc. I shall review some of the more significant observations emerging from these studies.

Our knowledge of phialoconidium ontogeny is still inadequate, but it seems that variations in conidium ontogeny exist among the many fungi currently considered to produce phialides and phialoconidia. Some of these patterns in phialoconidium ontogeny can be summarized here, with special emphasis on lacunae in our knowledge, which should stimulate further work.

1 One such pattern is seen in *Phialophora lagerbergii* (Melin & Nannf.) Conant (Figure 7.1). From the time-lapse study of Cole and Kendrick (1969) on conidium ontogeny in this fungus, the sequence of events seems to be as follows. In the formation of the first conidium, a relatively narrow zone of the wall at the tip of a phialide blows out, and a conidium *with its own wall* is differentiated within the blown-out tip. This conidium is, therefore, endogenous, and the wall

Centre for Advanced Studies in Botany, University of Madras, Madras 5, India.

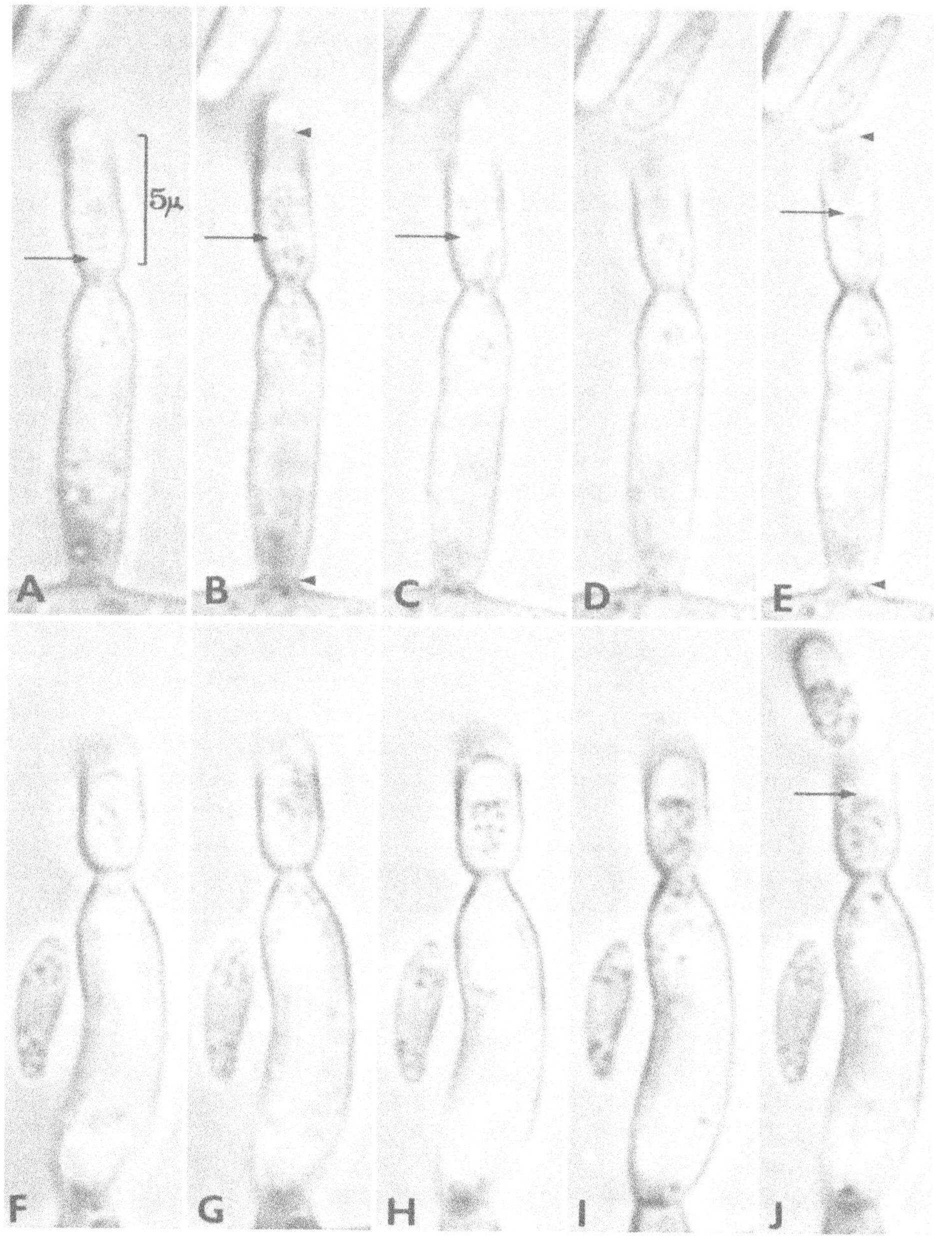


Figure 7.1 *Phialophora lagerbergii,* time-lapse sequence showing formation of phialoconidia. Arrows indicate apex of developing conidia. Arrowheads indicate common reference points (basal septum and upper limit of collarette). (From Cole and Kendrick 1969)

of the phialide does not seem to contribute to the wall of the conidium. When the first conidium has been differentiated, the apical extension of the phialide ceases growth, and a bulb-shaped protoplasmic protuberance appears immediately below the first conidium and touching its base. This is the initial of the second conidium, the further growth of which pushes the first conidium up so that it eventually breaks through the now inelastic wall of the phialide apex and emerges. Most of the extension of the phialide which originally enclosed this first conidium remains as a conspicuous cupulate collarette. The second conidium develops and is differentiated within the collarette. Conidia are produced one by one in succession from the open end of the phialide. Although each successive conidium develops from what appears to be a blowing out of a wall (or membrane?) distinct from and internal to the phialide wall, the genesis of this conidium wall is not clear. Also, the conidia do not show, under the ordinary light microscope, a basal attachment scar, and therefore it is possible that a new wall is laid down around each secondary conidium only after a protoplasmic mass has been completely extruded from the open tip of the phialide. Unfortunately, evidence from studies on fine structure is not available to settle this point.

The work of Lowry, Durkee, and Sussman (1967) and of Oulevey-Matikian and Turian (1968) on the fine structure of development of microconidia in *Neurospora crassa* Shear & Dodge suggest that walls of phialoconidia arise de novo. The microconidial apparatus in this fungus can be interpreted as a much reduced phialide similar to that described by Subramanian and Lodha (1964) for *Bahupaathra samala* Subram. & Lodha (Figure 7.14 C). In *Neurospora crassa,* formation of a protuberance from a cell of a microconidiophore presages the development of a microconidium (Figure 7.2 A). The cell wall of the conidiogenous cell apparently has several layers (Figure 7.2 A, B) and, as the protuberance enlarges, these cell wall layers are ruptured or dissolved following the blowing out of what is apparently a newly formed wall at the site where the protuberance appears (Figure 7.2 A) and this newly formed wall indeed seems to become the wall of microconidium (Figures 7.2 and 7.3). The newly formed wall is clearly seen in several of the electron micrographs of Lowry et al. (1967) and also of Oulevey-Matikian and Turian (1968). According to Dodge (1932) and Lowry et al. (1967), each conidiogenous cell may produce several conidia, but information on the fine structure of development of successive conidia is unfortunately still not available. How are conidia formed successively from the open end of the conidiogenous cell? Is it possible that, in the formation of the second and each later conidium, a mass of protoplasm is extruded from the open tip of the conidiogenous cell and is then surrounded by a wall formed de novo in each case?

The type of development described for *Phialophora lagerbergii* is quite common. The difference in the size and shape of the first-formed and later conidia from the same phialide seen in *P. lagerbergii* is known for several other genera and species also (Hughes 1953; Kendrick 1961b). However, in many forms the collarette is not as conspicuous as in *P. lagerbergii*. In several fungi the collarette is reduced to a small frill and it may be very inconspicuous or even not discern-

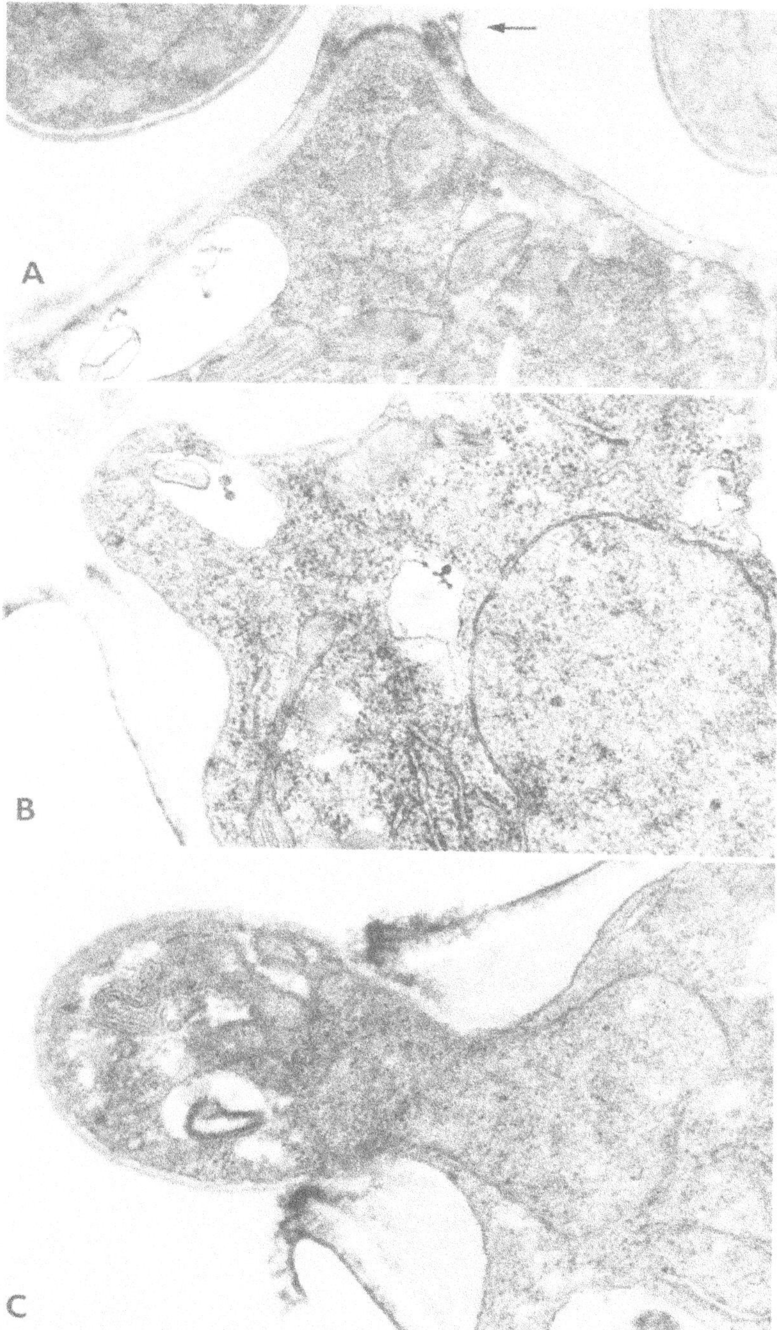


Figure 7.2 *Neurospora crassa*, electron micrographs of development of microconidia. In A, arrow points to protuberance which presages development of conidia. Note degeneration of outer hyphal wall in this region. (From Lowry, Durkee, and Sussman 1967)

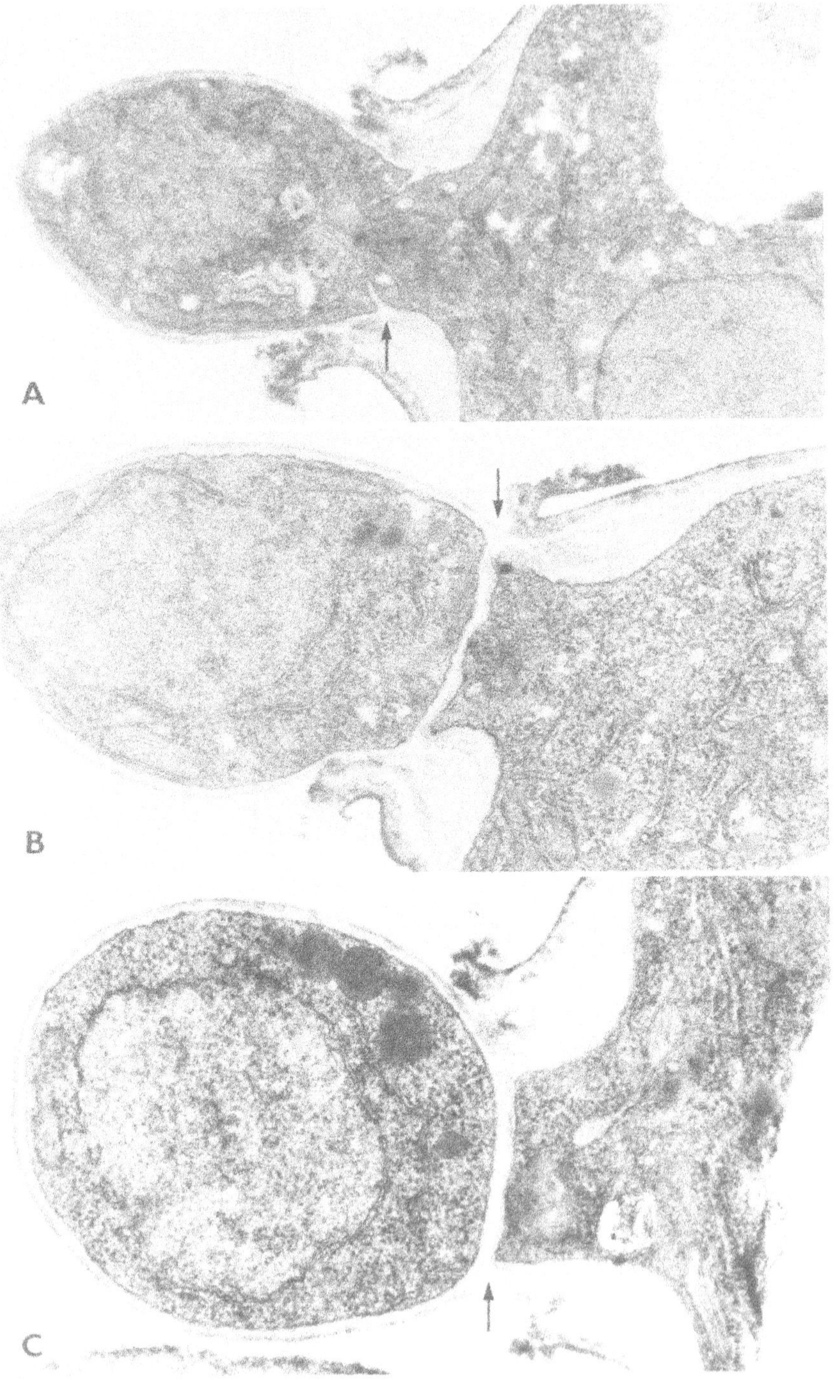


Figure 7.3 *Neurospora crassa*, electron micrographs of development of microconidia. Arrow in A points to formation of cross-wall cutting off the conidium at its base. Arrows in B and C point to breaking down of connection between conidium and conidiogenous cell. (From Lowry, Durkee, and Sussman 1967)

ible at all. Detailed developmental studies of these variations are few, but it would appear that in certain cases the wall at the tip of the phialide may be easily ruptured or dissolved enzymatically, allowing the initial of the first conidium to emerge from the open tip and continue development to maturity. When this happens the collarette may be inconspicuous or absent. In still other cases which may be interpreted as intermediate between these and *P. lagerbergii*, an apical protuberance of the phialide may develop to a greater or lesser degree before its tip is ruptured or dissolved, and the more delayed this rupture or dissolution, the more conspicuous the resulting collarette or collar. Another interpretation of the inconspicuous or conspicuous collar may be that in the former the wall of the protuberance breaks at its base and in the latter the break occurs at progressively higher levels. When the break takes place, part of the broken phialide wall may form a cap over the first conidium. 

In *Verticillium albo-atrum* Reinke & Berthold the first conidium develops within a protuberance at the tip of the phialide; the wall of this protuberance breaks before development of the first conidium is complete (Seshadri 1970c). The wall remaining below the break forms a distinct funnel-shaped collarette. The broken-off part of the wall of the protuberance remains at the apex of the developing conidium. Buckley, Wyllie, and DeVay (1969) recently published some electron micrographs of conidium formation in this fungus. Electron micrographs of the development of the first conidium were not given, and their Figures 7, 9 and 10 show the development of what may be the second conidium, since the phialide tip is already broken and possesses a distinct collarette (Figure 7.4 A, B, C, D). It is noteworthy that the wall of this conidium is formed de novo, and the conidium is cut off by a septum (? double septum). How the third and later conidia arise, and the genesis of the walls of these conidia, are not clear.

The macroconidia and the microconidia of *Fusarium* spp. develop in basically the same way as the conidia of *V. albo-atrum.* The first macroconidium is delimited within a relatively long and large blown-out apical extension of a phialide (Figure 7.5 A). Because of pressure from within, and prior to the formation of the second conidium, the wall of the blown-out part is broken off irregularly (Figure 7.5 A). Part of the blown-out wall of the phialide may often form a cap for the first conidium, a fact confirmed also by the technique of fluorescent antibody staining by Goos and Summers (1964) for *Fusarium oxysporum* Schlecht. ex Fries. A succession of conidia is produced from the open end of the phialide, the wall of each conidium being apparently formed de novo. From observations on *F. solani* (Mart.) App. & Wollenw. and *F. decemcellulare* Brick it would seem that a phialide may produce macroconidia or microconidia, but not both (Seshadri 1970c).

According to Kendrick (1963b), in *Phialocephala fusca* Kendrick a number of conidia may accumulate within the blown-out tip of the phialide, before the wall ruptures, and the resulting collarette is sometimes cupulate, but usually flaring and irregularly split (Figure 7.5 B). A critical investigation of the full sequence of events is needed.

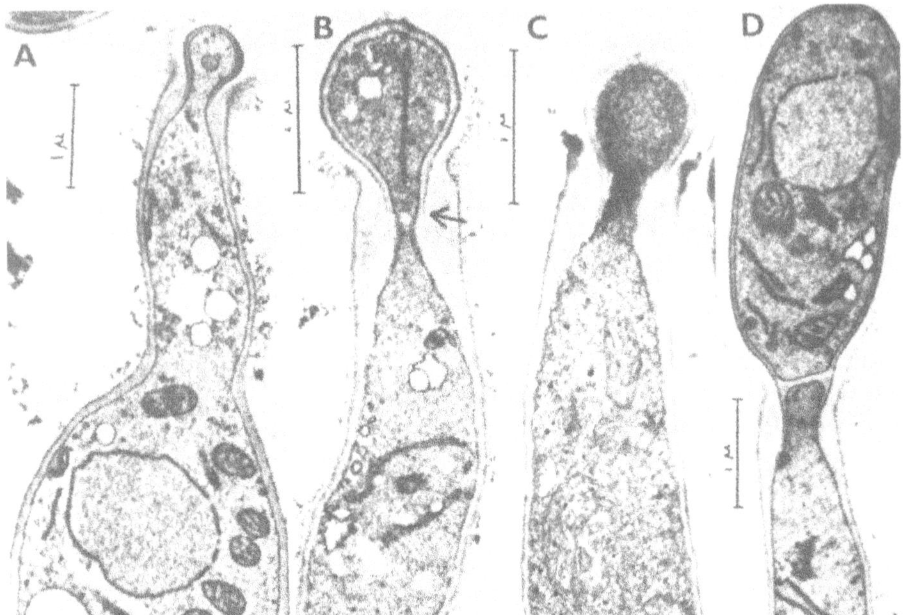


Figure 7.4 *Verticillium albo-atrum,* electron micrographs of stages in the development of phialoconidia. Note the collarette at the broken tip of the phialide, and the newly formed phialoconidium wall. Arrow in B indicates limit of new outer wall surrounding the emerging conidium. (From Buckley, Wyllie, and DeVay 1969)

In *Chloridium chlamydosporis* (van Beyma) Hughes a collarette remains at the tip of the phialide following the liberation of the first conidium, and appearances suggest that more than one conidium is produced concurrently from the open tip (Tubaki 1963, p. 41, Pl. I, H). This needs further elucidation, although the observations of Seshadri (1970a) indicate that conidium development in this species may not be concurrent but successive as in all phialides, the interesting feature apparently being the fact that the successive conidia seem to develop from different loci at the open tip of the phialide.

In *Stachybotrys chartarum* (Ehrenb.) Hughes, the first conidium may be delimited in a manner different from the types described so far. During initiation of the first conidium a narrow zone at the tip of the phailide blows out, and when the blown-out part attains its characteristic shape and size, it is cut off from the tip of the phialide and is thus the first conidium. When the first conidium has been shed, the diameter of the open tip of the phialide corresponds to the point of insertion of the first conidium. As far as we can tell from studies with the light microscope, the blown-out wall of the phialide itself is the wall of the first conidium in this case, or at least forms part of it (Seshadri 1970c). Subsequent conidia are then formed from the open tip of the phialide singly and

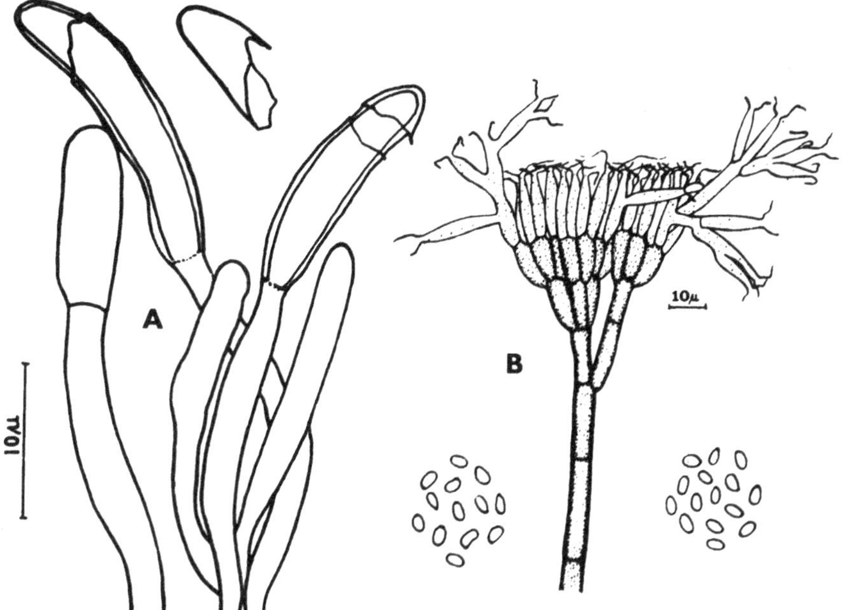


Figure 7.5 A: *Fusarium decemcellulare,* development of the first macroconidium. Note that the swollen apical part of the phialide, within which the conidium develops, is ruptured irregularly. B: *Phialocephala fusca.* Note that the collarettes are cupulate or flaring, and irregularly split. (After Kendrick 1963b)

successively, each having a wall apparently formed de novo. Details need to be investigated.

Successive conidia may become aggregated to form loose or fragile chains, e.g., *Phialocephala dimorphospora* Kendrick (Figure 7.6 A) and *P. bactrospora* Kendrick (Figure 7.6 B), or may form "false heads," e.g., *Cephalosporium acremonium* Corda and *Verticillium albo-atrum.* In certain species, the conidia may form fragile chains, or "false heads," or both, depending upon the conditions under which the fungus grows, e.g. *Gliomastix murorum* (Corda) Hughes.

2 Another pattern of phialoconidium development is seen in *Penicillium corylophilum* Dierckx and can also be illustrated from the time-lapse studies of Cole and Kendrick (1969) (Figure 7.7). In the development of the first conidium, the protoplast appears to burst through the outer wall of the phialide at its tip, leaving an inconspicuous collarette. The extruded protoplast is apparently surrounded by a membrane or wall and is the initial of the first conidium. When the first conidium has attained its final shape and size, a second initial appears below the first, then a third below the second, and so on. There is thus a basipetal succession of conidia developing from the open end of the phialide. The walls of the first and subsequent conidia in the chain appear to be a continuum, but not an extension of the (primary) wall of the phialide itself.

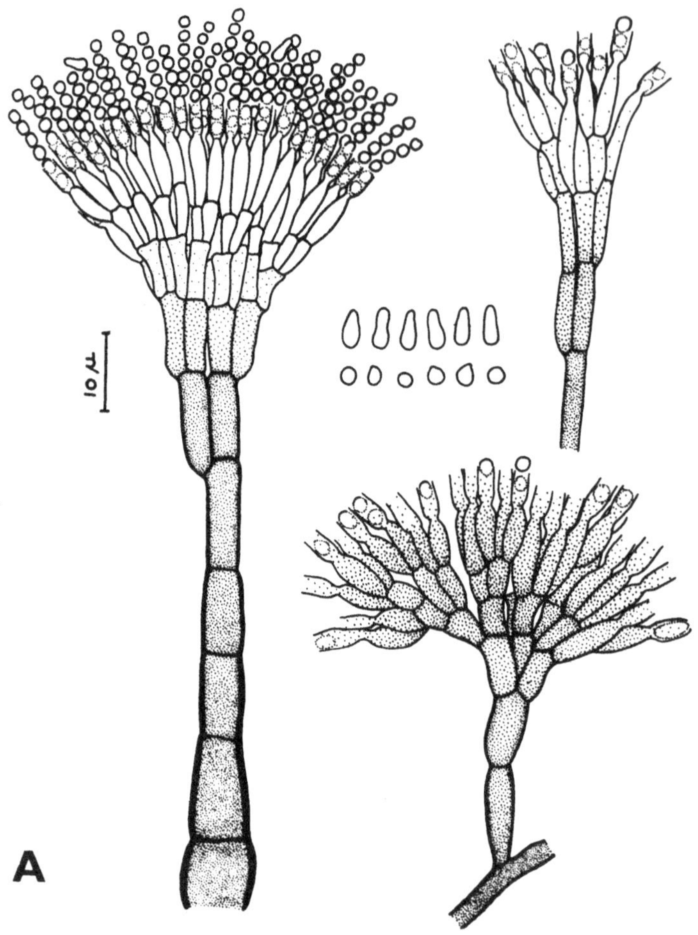


A correct understanding of the initiation of conidium development and the genesis of the conidial wall can be achieved only by studies of the ultrastructure of the entire process. Zachariah and Fitz-James (1967) described the fine structure of phialides and development of phialoconidia in *Penicillium claviforme* Bainer. Their electron micrographs show that the wall of a developing conidium is laid down internal to the phialide wall, at the apex of the phialide. As this wall material is extruded and expands to give the conidium its final form, a septum arising from the newly formed wall separates the conidium from the cytoplasm of the phialide. The process is repeated in the formation of each succeeding conidium. It is assumed that the phialide wall encases the first-formed conidium, and is ruptured by its expansion, but this has not been seen in electron micrographs yet. What is clear is that the phialoconidium wall is not merely an extension of the phialide wall (Figure 7.8).

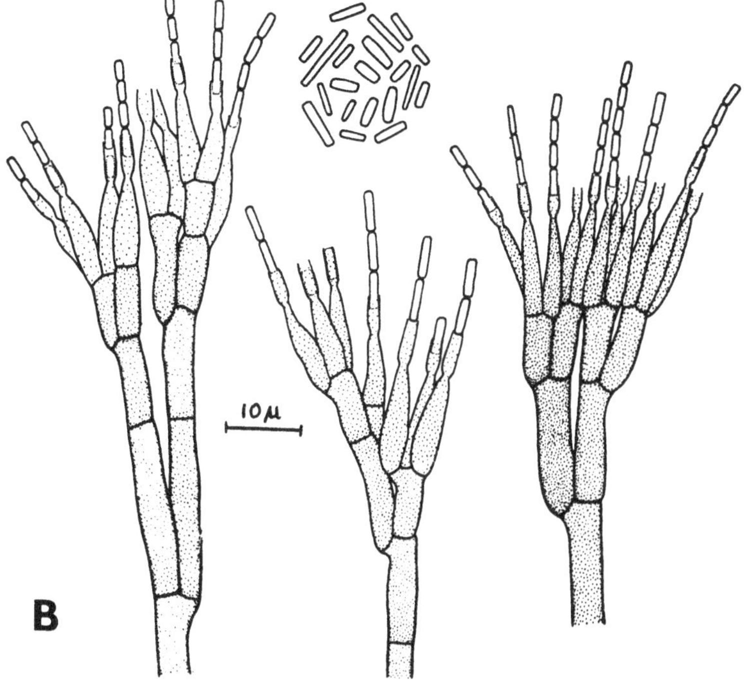


Figure 7.6 A (facing): *Phialocephala dimorphospora;* note prominent collarettes, fragile phialoconidial chains, and the difference in shape between the first conidium and the later-formed conidia. B: *Phialocephala bactrospora*; note phialides and fragile phialoconidial chains. (Both after Kendrick 1961b).

Somewhat the same type of conidium ontogeny seems to characterize *Aspergillus niger* v. Tieghem. From the electron micrographs of Tanaka and Yanagita (1963) it is clear that the phialide wall almost completely surrounds the first conidium except at the base, where it is broken off at a point that corresponds to the broken apex of the phialide. Also, the walls of the first and the second conidium initials are continuous both with each other and with the newly formed wall within the apex of the phialide. However, one cannot be sure if the individual conidia in a chain are delimited by a process of septum formation as in the case of *P. claviforme.* Electron micrographs of all phases of conidium ontogeny of *A. niger* are not available, but observations under the ordinary light microscope (Seshadri 1970c) show that the outer walls of the second and subsequent conidia in a chain are continuous, this wall also forming the "isthmus" that separates adjacent conidia (Figure 7.9). The newly formed inner wall of each conidium may be thicker than the common outer wall and sometimes becomes confluent with the outer wall so as to obscure the latter's presence. In fact, the thin outer wall may be clearly discernible only as isthmi between conidia.

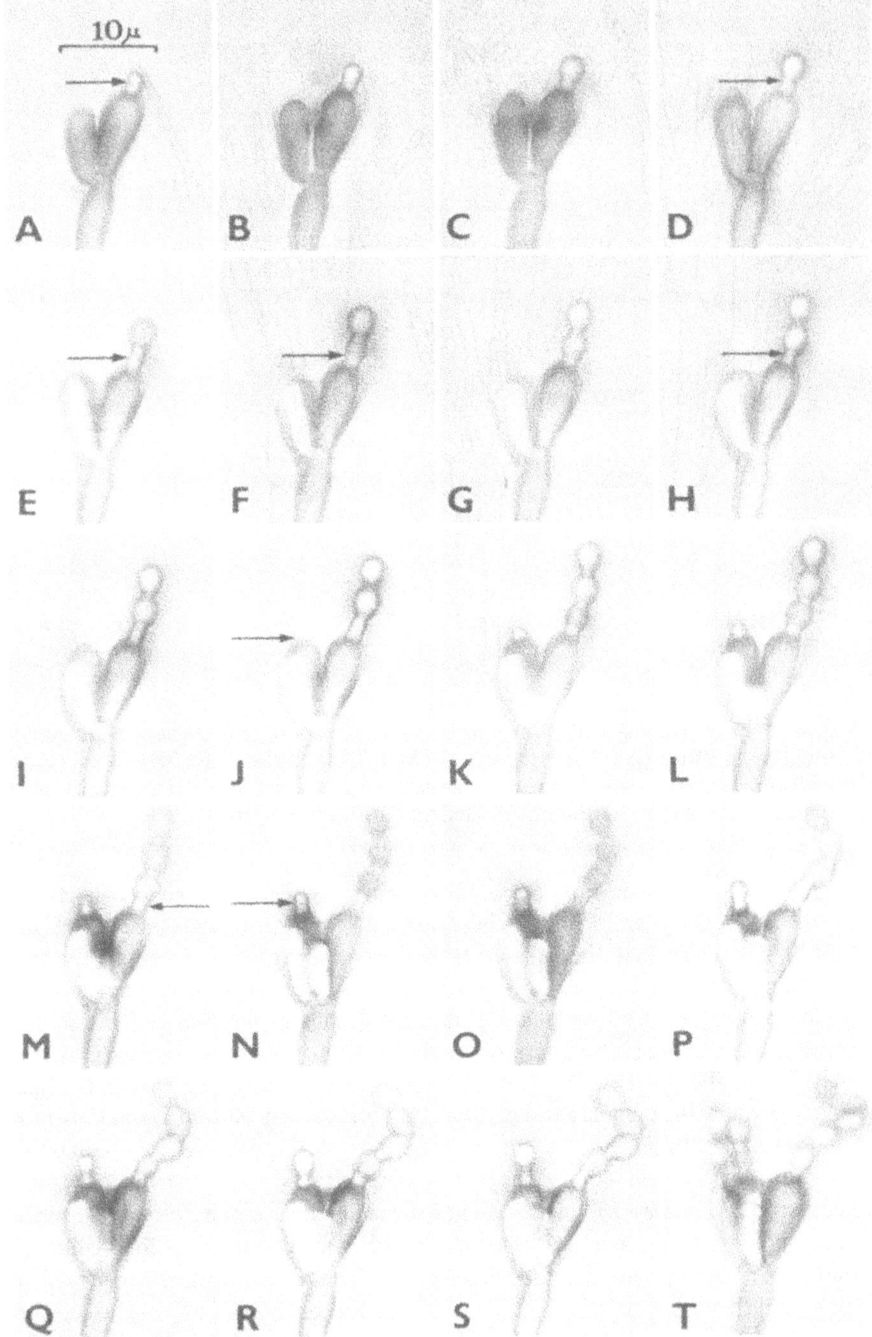


Figure 7.7 *Penicillium corylophilum*, time-lapse sequence of development of phialoconidial chains. Note that conidia are separated by "isthmi." (From Cole and Kendrick 1969)

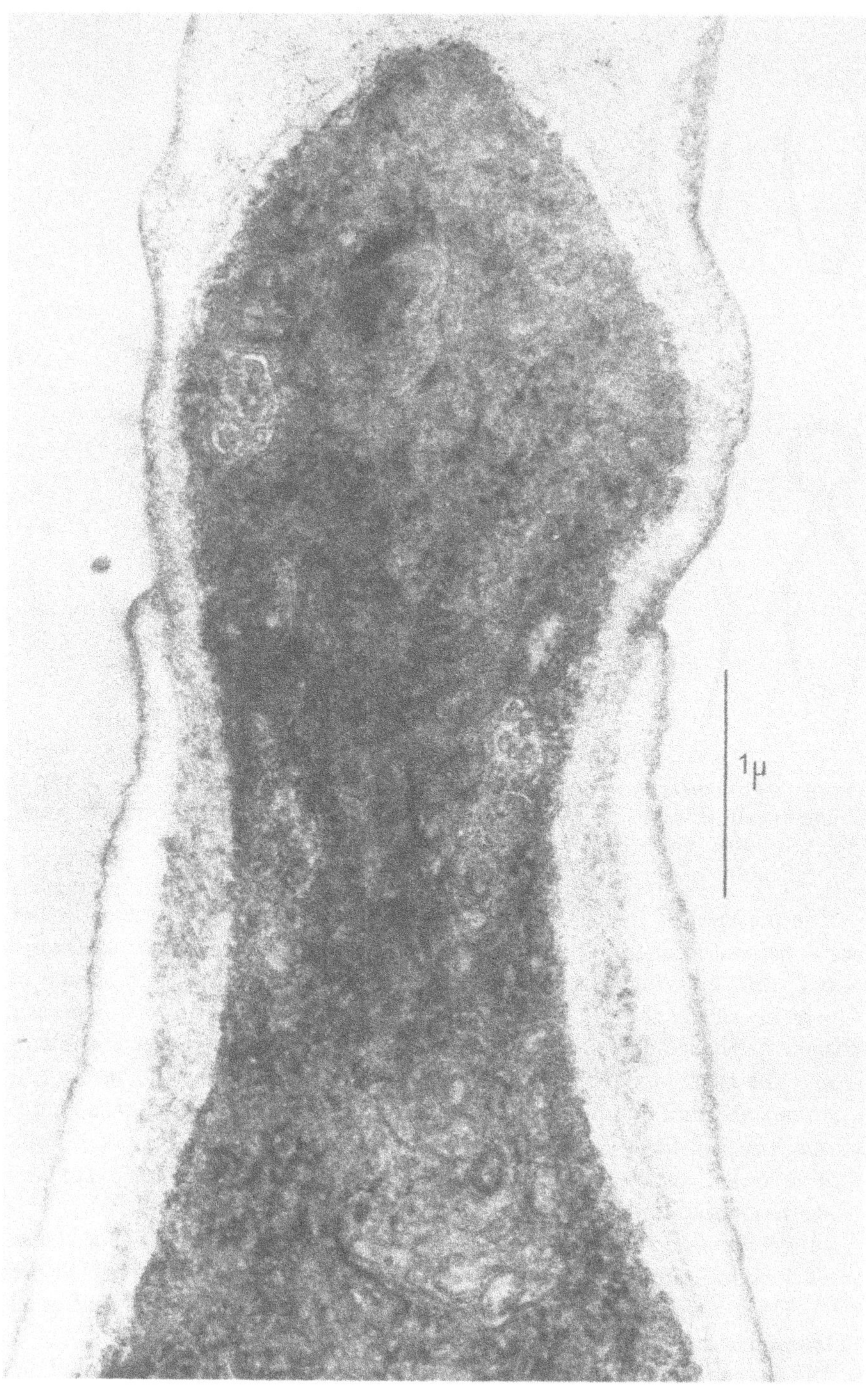


Figure 7.8 *Penicillium claviforme*, longitudinal section of phialide apex. Kellenberger-osmium fixation. (Reproduced by permission of Dr K. Zachariah)

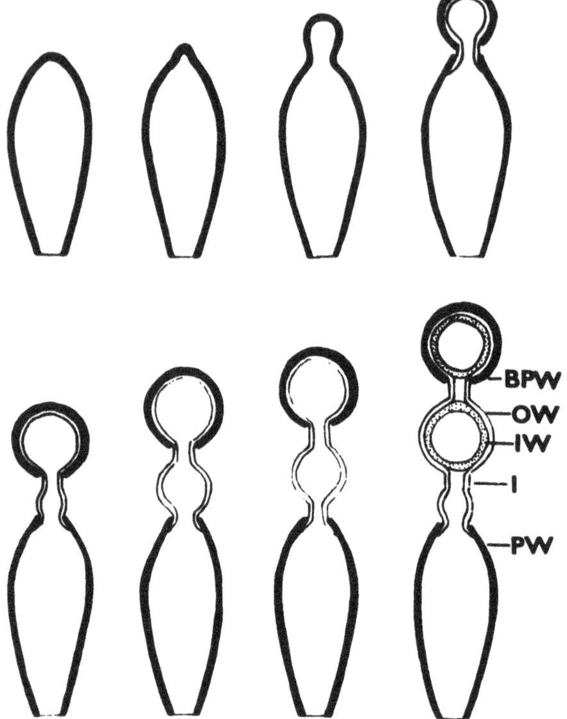


Figure 7.9 *Aspergillus niger,* diagrams illustrating conidium ontogeny. PW = phialide wall; BPW = broken-off part of phialide wall enveloping first conidium; OW = conidial outer wall; IW = conidial inner wall; I = isthmus.

The presence of intercalary isthmi in conidial chains of *Aspergillus* spp. has been observed frequently by workers in the past (Thom and Raper 1945, Figure 6, 63C). Some of Sugiyama's (1967) electron micrographs of conidial chains in *Aspergillus* clearly show these isthmi, which in certain species appear completely shrunken and thread-like, possibly an artefact of preparation for electron microscopy. The isthmi are very conspicuous in phialoconidial chains of *Phialomyces macrosporus* Misra and Talbot (Misra and Talbot 1964). In *Phialotubus microsporus* Roy and Leelavathy, the isthmi are long and tube-like (Roy and Leelavathy 1966). Typical isthmi are also seen in conidial chains of *Paecilomyces fusisporus* Saksena (Seshadri 1970c).

In *Memnoniella echinata* (Riv.) Galloway, conidia develop in basipetal chains as in *A. niger*, but no isthmi are seen between them (Seshadri 1970c) (Figure 7.10). Once again, the wall of the phialide seems to make no contribution at all to the wall of any of the conidia.

In *Aspergillus giganteus* Wehmer, the electron micrographs presented by Trinci, Peat, and Banbury (1968) indicate that the phialide wall has at least two electron-dense layers, often with an electron-transparent one sandwiched be-

Figure 7.10 *Memnoniella echinata,* development of phialoconidial chains.

tween them. In the formation of the first conidium both electron-dense layers apparently take part, and this first conidium is formed as a blown-out end from the phialide tip (Figure 7.11 A). Subsequent conidia are produced in basipetal succession and only the inner electron-dense layer seems to be involved in their formation (Figure 7.11 B). Presumably the outer layer breaks off when the first conidium is released. According to the authors, the conidia are delimited by a process which resembles septum formation in somatic hyphae. The cross-wall is initiated by invagination of the protoplasmic membrane of the phialide in the upper region of the tapered phialide neck. It could not be ascertained whether there was protoplasmic continuity between conidia.

3 A third type of phialoconidium ontogeny is found in conidial *Ceratocystis paradoxa* (Dade) Moreau (=*Thielaviopsis paradoxa* (de Seynes) v. Hoehnel) and has recently been described from time-lapse studies by Cole and Kendrick (1969). The phialide is long, cylindrical, and somewhat tapered. One to several conidia differentiate basipetally within the phialide. When the phialide stops growing, the pressure resulting from continued differentiation of a basipetal succession of conidia ruptures the outer wall of the phialide at its apex and facilitates the escape of the conidia. Within each phialide the zone where the conidia are differentiated becomes fixed sooner or later. According to Cole and Kendrick, the wall of the phialide from this point to its open tip is the collarette. The first-formed conidium is morphologically distinct from all other conidia, being rounded at the apex; later-formed conidia are flattened at both ends; all conidia are cylindrical. The observations of Seshadri (1970b) on an Indian iso-

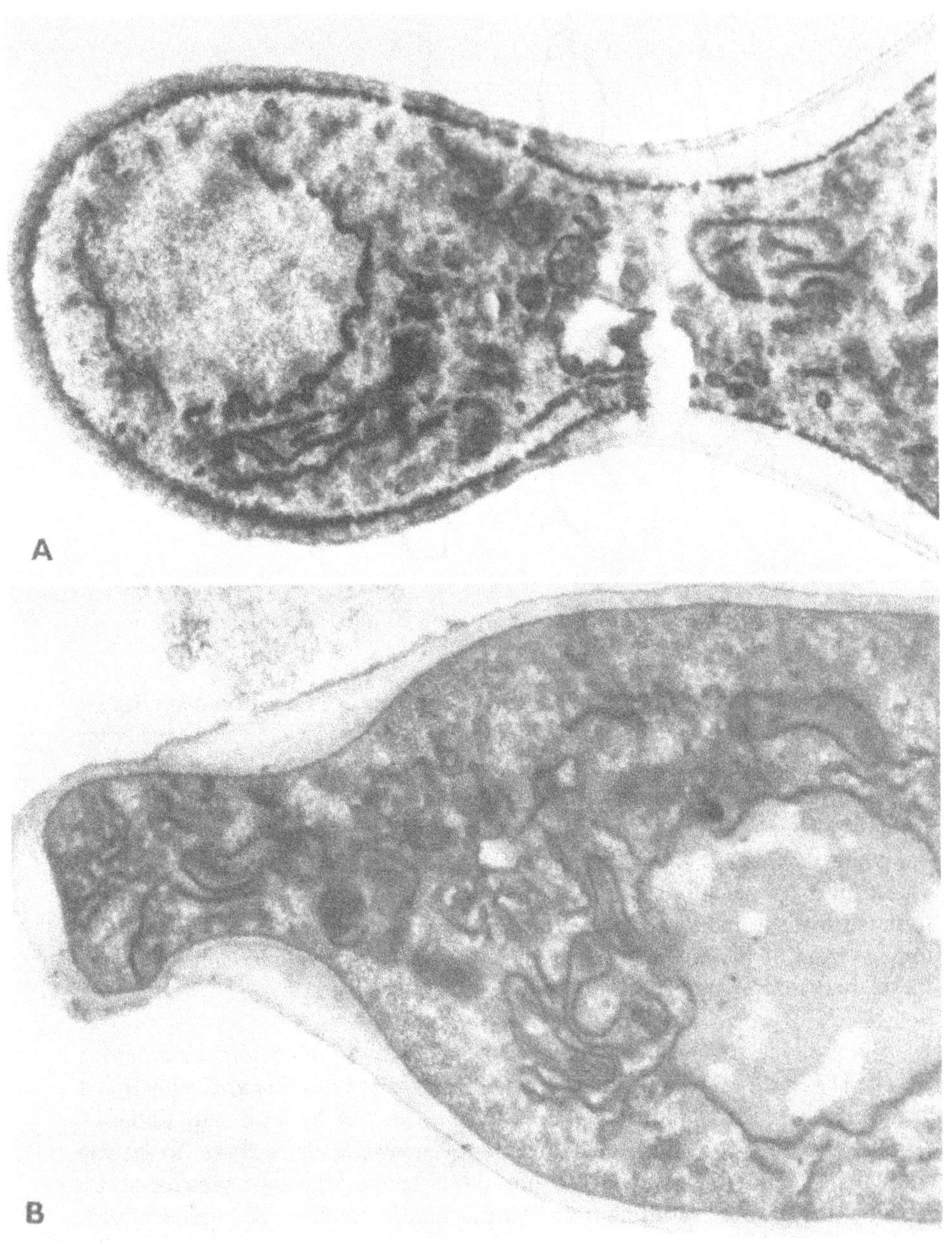


Figure 7.11 *Aspergillus giganteus*, electron micrographs showing development of the first (A) and later (B) conidia from a phialide. (From Trinci, Peat, and Banbury 1968).

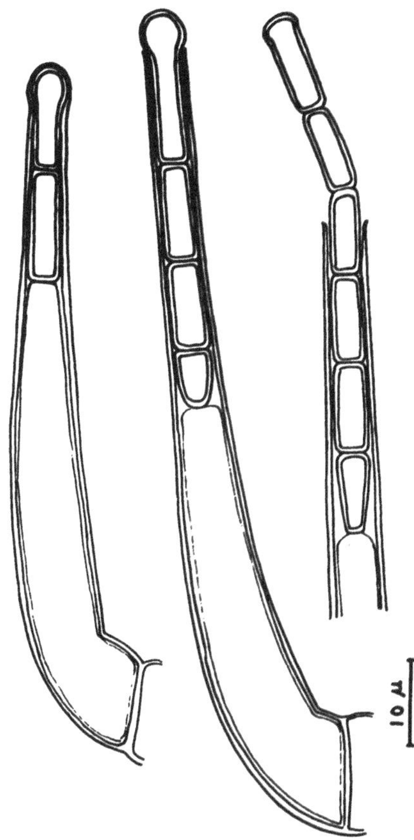

Figure 7.12 *Thielaviopsis paradoxa,* development of cylindrical phialoconidia from long phialides.

late of the same fungus under the ordinary light microscope confirm those of Cole and Kendrick (1969). In addition, it was seen that each conidium appears to have a thick hyaline wall all over, but it did not appear that these conidia were delimited from each other by a process of septation or double septation (Figure 7.12). Indeed, the recent report of DelVecchio, Corbaz, and Turian (1969) on *Thielaviopsis basicola* (Berk. and Br.) Ferraris adds support to this conclusion. From the electron micrographs presented by these workers it is evident that the wall of the phialide in *T. basicola* makes no contribution to the walls of the conidia. The individual conidia within the tubular phialide appear quite distinct from each other, each surrounded by its own wall, and have no organic connection with the wall of the earlier or later conidium or that of the phialide (Figure 7.13 A). The sequence of events in both *T. basicola* and *T. paradoxa* may be as follows. In the development of each conidium the first step appears to be protoplasmic cleavage followed by the development of a wall around the protoplasmic mass cleaved out. This cleavage seems to take place at the conidiogenous locus within the phialide, and successive cleavages followed by development of a totally new wall around each cleaved-out mass result in the production of several conidia in basipetal succession.

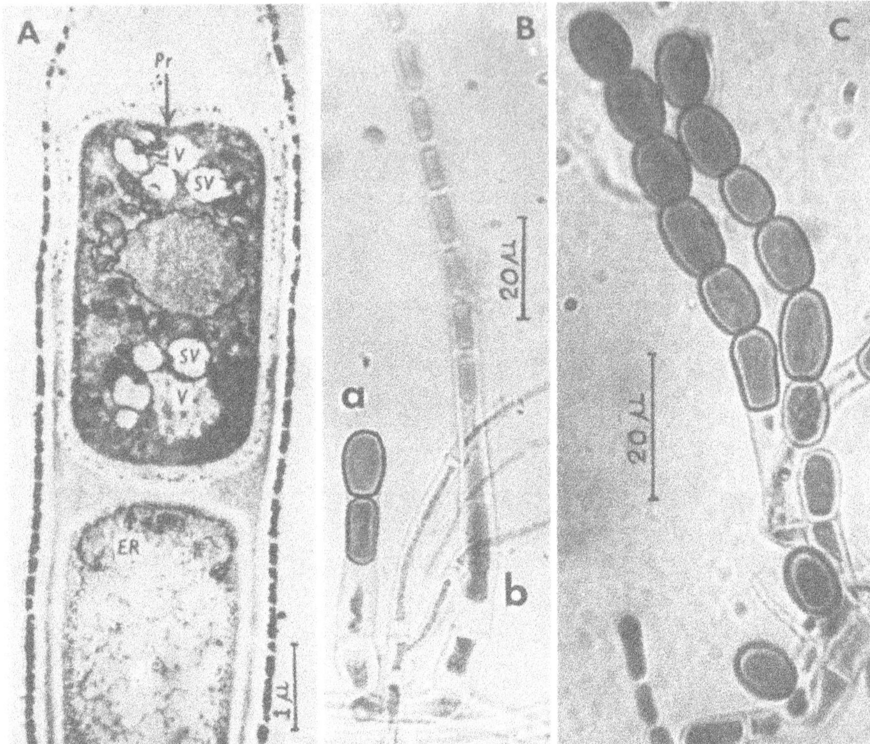


Figure 7.13 A: *Thielaviopsis basicola,* electron micrograph of a phialide with two phialoconidia. Note that the conidial wall is formed de novo. The wall of the earlier-formed conidium (the upper one) is better developed (with electron-dense particles) than that of the later-formed (lower) one. V = vacuoles; SV = storage vessels; Pr = protuberance; ER = endoplasmic reticulum. (From DelVecchio, Corbaz, and Turian 1969). B, C: *Thelaviopsis paradoxa.* Ba, short phialide abstricting dark, thick-walled, oval conidia from deep within; Bb, long phialide producing chain of cylindrical conidia; C, short phialides abstricting dark, thick-walled, oval conidia from near the tip.

Apart from the long, subulate phialides producing long, loose chains of cylindrical conidia (Figure 7.13 B(a)), the Indian isolate also produces considerably shorter phialides (Figure 7.13 B(b)) within which conidia are produced in the same way as in the long phialides. The conidia from the short phialides, though somewhat cylindrical in the beginning, soon assume an elongate-oval or oval to subglobose shape accompanied by thickening of the walls and deposition of dark pigment.

In yet another variation, phialides and conidia resemble the short phialides and their products described above, but the conidia are abstricted from near the apex of the phialide rather than from deep within it (Figure 7.13 C). Because these conidia are never restricted to the confines of a tubular collarette, they are not cylindrical even in the beginnning, but oval to subglobose. The two varieties of shorter phialide appear very closely alike and one may be only a variant of the other. Their ovoid, thick-walled, darkly pigmented conidia have been referred to as "chlamydospores" by some workers, but there is no doubt that they are produced from phialides. What appears most significant about these phialoconidia is the fact that they continue development and attain maturity after they are liberated from the phialide. 

It would seem that the proper elucidation of the relationship of the wall of the phialide to that of the conidia produced by it is the key to understanding phialoconidium ontogeny (or, for that matter, the ontogeny of any of the conidial types). The phialide wall (or its outermost layer according to another possible interpretation) does not contribute to the phialoconidium wall except in a few cases (e.g. *Stachybotrys chartarum*) where the primary conidium wall appears to be an extension of the phialide (primary) wall; however, these exceptions need further study. (In the case of the porospore (tretoconidium) also, the outer wall of the conidiophore does not contribute to the conidium wall; the conidium wall is an extension of an already existing inner wall of the conidiophore which blows out through a pore at the tip of the conidiophore, the pore apparently being the result of enzymatic dissolution.)

Successive production of conidia from a fixed conidiogenous locus is another feature of the phialide. This implies that the phialide shows no elongation once its tip is ruptured and it is actively producing conidia. Successive production of porospores from the same pore was reported by Luttrell (1963) for *Helminthosporium velutinum* Link ex Fries, but this needs further study. Comparison with true "annellospores" shows that the phialide can be distinguished by the fact that the successive conidia are delimited at the same level and the conidiogenous locus is fixed as long as a phialide is actively producing conidia. In the case of "annellophores," the conidiogenous cell, though often resembling the conventional phialide in appearance, produces the first conidium as a blown-out apex, then proliferates percurrently and transforms the new blown-out end into the second conidium at a higher level. A repetition of this process results in the production of successive conidia at higher levels, but each new proliferation produces only one conidium. Sutton and Sandhu's (1969) electron micrographs of several species producing annellophores are instructive in understanding conidium ontogeny in these annellophoric fungi. In *Cryptosporiopsis* sp., conidia are produced successively and at higher levels by a process of repeated proliferation; sometimes the conidia so produced may secede at a level lower than that at which the first conidium was shed. When this happens, annellations will not be seen under the ordinary light microscope and may pose a problem to the student of conidium ontogeny. However, typical abscission scars may be found on such conidia. No such scars are usually seen in phialoconidia; yet, in some species of *Phialocephala* (e.g. *P. repens* (Cooke and Ellis) Kendrick), each of the successive conidia which are liberated as discrete units has a distinct apiculate attachment scar (Kendrick 1963a). Since phialoconidia do not normally show a point of attachment or scar, what is the significance of this apiculus?

Each of the conidia successively produced from a phialide may be shed from the phialide as a separate unit, and these conidia may then slime down to form gloeoid masses at the tip of the phialide or may form loose or fragile chains. However, in some species, the successively produced conidia are held together by an outer wall common to all of them as in species of *Aspergillus, Penicillium, Paecilomyces, Memnoniella*; the conidia in these fungi are therefore seen to form *persistent* chains, in contrast to the *loose, fragile* chains seen in other fungi and so classified in this paper. In *Thysanophora penicillioides* (Roum.) Kendrick, as illustrated by Kendrick (1961a), the majority of conidia have a scar at each end, which suggests that they are held together in persistent chains; the conidia in this species are dry. Perhaps it is not a mere coincidence that all species known to produce persistent chains are dry-spored in the sense of Mason (1937), and those forming fragile chains are slimy-spored. Moreover, most species in which the conidia are known to slime down sometimes, and to form chains at other times (e.g. *Gliomastix murorum*), depending on conditions of growth, are those in which the phialoconidia are formed each as a separate unit, and this dual behaviour is not known in species producing persistent chains. It would be worth while investigating species producing phialoconidial chains to ascertain whether the chains are persistent or fragile as described here.

A few other features of the phialide remain to be discussed. In several species, the apex of the phialide may cease to produce conidia after a period of activity and then proliferate through the open tip and produce another fertile tip at a higher level. The new fertile tip produces a succession of conidia in the same way. This process may be repeated as, for example, in *Catenularia* spp. (Hughes 1965) (Figure 7.14 A). If the proliferation is followed by septation, the conidiogenous cell above can be interpreted as a new phialide. However, sometimes (e.g. *Catenularia, Fusarium*) no septa may develop (Figure 7.14 A), and in these cases the phialide may be interpreted as having given rise to a second conidiogenous locus at a higher level after proliferation.

In certain other fungi (e.g. *Codinaea* spp.; see Hughes and Kendrick 1968), phialides are seen with more than one open end (Figure 7.14 B). Hughes (1951), who observed them in *Lasiosphaeria hirsuta* (Fries) Ces. and de Not., referred to them as polyphialides. In the case of *Bahupaathra samala*, Subramanian and Lodha (1964) described the septate fertile branches as having funnel-shaped or cup-shaped collarettes from within which a succession of conidia are produced (Figure 7.14 C). The collarettes occur laterally and at right angles to the cells which bear them, except in the case of terminal cells of branches, where they are often terminal or subterminal. During development each collarette is at first borne terminally on the conidiogenous cell, but later it gradually assumes a lateral position owing to sympodial proliferation of the conidiogenous cell. A repetition of this process, coupled with progressive septation of the conidiogenous cell, results in numerous lateral collarettes borne on a septate "conidiophore." Where septa are not laid down, or septation is delayed, a single conidiogenous cell may have two or more open ends and be a polyphialide (Subramanian 1965). Whether only the most recently formed conidiogenous locus in a polyphialide is functional needs further study.

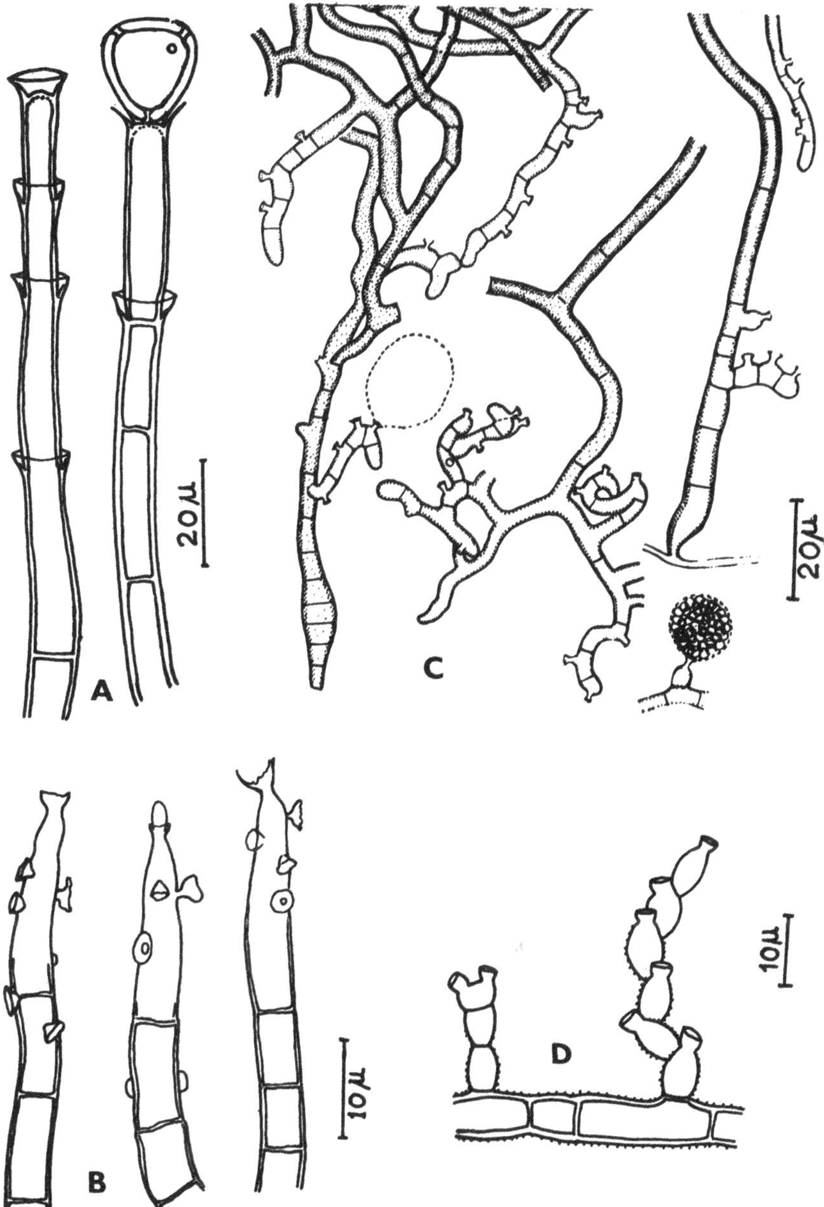


Figure 7.14 A: *Catenularia* state of *Chaetosphaeria novae-zelandiae* Hughes and Shoemaker, phialides showing percurrent proliferation (after Hughes 1965). B: *Codinaea fertilis* Hughes and Kendrick, polyphialides (after Hughes and Kendrick 1968). C: *Bahupaathra samala*, conidiophores and development of phialides (after Subramanian and Lodha 1964). D: *Capnophialophora* state of *Strigopodia batistae*, sympodially developing succession of phialides (after Hughes 1968)

Another interesting feature of some phialides is lateral proliferation to produce not merely a collarette but a new phialide, and this process may be repeated (Figure 7.14 D), resulting in a succession of sympodially developed phialides, e.g. the *Capnophialophora* state of *Strigopodia batistae* (Sacc. and Bres.) Hughes, as illustrated by Hughes (1968).

It is now clear that the definition of a phialide should be based on the following features relating to the development of phialoconidia:

(a) The phialide wall does not contribute to the phialoconidium wall, with a few possible exceptions which need further study; the phialoconidial wall is apparently formed de novo, possibly in one of three ways as already described, and typical phialoconidia often do not show clear attachment scars.

(b) The phialide produces a succession of conidia from a fixed conidiogenous locus without increasing in length in the process; conidia may be produced individually as discrete units, and then may form fragile or loose chains or "false heads"; in other cases, the successive conidia may be surrounded by a common wall and then they form persistent chains.

(c) The phialide may have a distinct collar or collarette, or a collarette may be inconspicuous or not discernible; the collar or collarette may be defined as that part of the phialide wall from the conidiogenous locus to the open tip.

(d) A phialide may proliferate percurrently and produce conidiogenous loci at higher levels; or a succession of phialides may sometimes be produced in a sympodial manner.

Conidiogenous cells which show these features may be considered to be phialides and they show an extraordinary range in shape.

ACKNOWLEDGMENT

I should like to conclude with an expression of appreciation and gratitude to Dr Kendrick for having given me the opportunity to participate in this meeting and for financial support to enable me to come.

REFERENCES

Buckley, P.M., Wyllie, T.D., and DeVay, J.E. 1969. Fine structure of conidia and conidium formation in *Verticillium albo-atrum* and *V. nigrescens*. Mycologia 61: 240-50

Cole, G.T., and Kendrick, W.B. 1969. Conidium ontogeny in hyphomycetes. The phialides of *Phialophora*, *Penicillium*, and *Ceratocystis*. Can. J. Botany 47: 779-89

DelVecchio, V.G., Corbaz, R., and Turian, G. 1969. An ultra-structural study of the hyphae, endoconidia and chlamydospores of *Thielaviopsis basicola*. J. Gen. Microbiol. 58: 23-7

Dodge, B.O. 1932. The non-sexual functions of microconidia of *Neurospora*. Bull. Torrey Bot. Club 59: 347-60

Goos, R.D., and Summers, D.F. 1964. Use of fluorescent antibody techniques in observations on the morphogenesis of fungi. Mycologia 56: 701-7

Hughes, S.J. 1951. Studies on micro-fungi. XI. Some hyphomycetes which produce phialides. C.M.I. Mycol. Pap. 45

- 1953. Conidiophores, conidia, and classification. Can. J. Botany 31: 577-659

- 1965. New Zealand fungi. 3. *Catenularia* Grove. N.Z. J. Botany 3: 136-50
- 1968. *Strigopodia*. Can. J. Botany 46: 1099-1107
Hughes, S.J., and Kendrick, W.B. 1968. New Zealand fungi. 12. *Menispora, Codinaea, Menisporopsis*. N.Z. J. Botany 6: 323-75
Kendrick, W.B. 1961a. Hyphomycetes of conifer leaf litter. *Thysanophora* gen. nov. Can. J. Botany 39: 817-32
- 1961b. The *Leptographium* complex. *Phialocephala* gen. nov. Can. J. Botany 39: 1079-85
- 1963a. The *Leptographium* complex. *Penicillium repens* C. and E. Can. J. Botany 41: 573-7
- 1963b. The *Leptographium* complex. Two new species of *Phialocephala*. Can. J. Botany 41: 1015-23
Lowry, R.J., Durkee, T.L., and Sussman, A.S. 1967. Ultrastructural studies of microconidium formation in *Neurospora crassa*. J. Bacteriol. 94: 1757-63
Luttrell, E.S. 1963. Taxonomic criteria in *Helminthosporium*. Mycologia 55: 643-74
Mason, E.W. 1928. Annotated account of fungi received at the Imperial Bureau of Mycology. List II (Fascicle 1). C.M.I. Mycol. Pap. 2
- 1933. Annotated account of fungi received at the Imperial Bureau of Mycology. List II (Fascicle 2). C.M.I. Mycol. Pap. 3
- 1937. Annotated account of fungi received at the Imperial Mycological Institute. List II (Fascicle 3 - General Part). C.M.I. Mycol. Pap. 4: 69-99
- 1941. Annotated account of fungi received at the Imperial Mycological Institute. List II (Fascicle 3 - Special Part). C.M.I. Mycol. Pap. 5: 101-44
Misra, P.C., and Talbot, P.H.B. 1964. *Phialomyces*, a new genus of hyphomycetes. Can. J. Botany 42: 1287-90
Oulevey-Matikian, N., and Turian, G. 1968. Controle métabolique et aspects ultrastructuraux de la conidiation (macro-micro-conidies) de *Neurospora crassa*. Arch. Mikrobiol. 60: 35-58
Roy, R.Y., and Leelavathy, K.M. 1966. *Phialotubus microsporus* gen. et sp. nov., from soil. Trans. Brit. Mycol. Soc. 49: 495-8
Seshadri, K. 1970a. Development of conidia in *Chloridium chlamydosporis* (unpublished)
- 1970b. Ontogeny of conidia of *Thielaviopsis paradoxa* (unpublished)
- 1970c. Studies on the phialide (unpublished)
Subramanian, C.V. 1965. Spore types in the classification of the Hyphomycetes. Mycopath. Mycol. Appl. 26: 373-84
Subramanian, C.V., and Lodha, B.C. 1964. Four new coprophilous hyphomycetes. Ant. van Leeuwenhoek 30: 317-30
Sugiyama, J. 1967. Mycoflora in core samples from stratigraphic drillings in middle Japan. II. The genus *Aspergillus*. J. Fac. Sci. Tokyo Univ., Sect. 3, 9: 377-405
Sutton, B.C., and Sandhu, D.K. 1969. Electron microscopy of conidium development and secession in *Cryptosporiopsis* sp., *Phoma fumosa, Melanconium bicolor,* and *M. apiocarpum*. Can. J. Botany 47: 745-9
Tanaka, K., and Yanagita, T. 1963. Electron microscopy on ultrathin sections of *Aspergillus niger*. II. Fine structure of conidia-bearing apparatus. J. Gen. Appl. Microbiol., Tokyo 9: 189-203
Thom, C., and Raper, K.B. 1945. A manual of the Aspergilli. Williams and Wilkins, Baltimore, Md.
Trinci, A.P.J., Peat, A., and Banbury, G.H. 1968. Fine structure of phialide and conidiophore development in *Aspergillus giganteus* "Wehmer." Ann. Botany 32: 241-9
Tubaki, K. 1963. Taxonomic study of hyphomycetes. Ann. Rep. Inst. Fermentation, Osaka 1: 25-54

114 Vuillemin, P. 1910a. 'Matèriaux pour une classification rationelle des Fungi Imperfecti. C. R. Acad. Sci. Paris 150: 882-4
- 1910b. Les Conidiosporés. Bull. Soc. Sci. Nancy, Ser.3, 11 (2): 129-72
- 1911. Les Aleuriosporés. Bull. Soc. Sci. Nancy, Ser. 3, 12 (3): 151-75
Zachariah, K., and Fitz-James, P.C. 1967. The structure of phialides of *Penicillium claviforme*. Can. J. Microbiol. 13: 249-56

## DISCUSSION

DR KENDRICK

In typical phialides, the first phialoconidium develops within the intact phialide wall. In Figure 7.15 A, *Phialophora,* that wall breaks at or near the apex (arrows), so that only a "cap" is lost and we are left with a deep collarette that remains for the rest of the life of the phialide. In *Aspergillus niger* (Figure 7.15 B), the outer wall breaks much lower down towards the "neck" of the phialide (arrows); thus the first conidium is encased in the phialide's wall as well as in its own, and we are left with almost no collarette. No phialide wall covers subsequent conidia in either case. We are dealing with essentially the same phenomenon in both fungi; there is merely a different point of weakness in the outer wall.

The secondary septation of polyphialides, which Dr Subramanian illustrated in *Bahupaathra*, reminds me of the problem presented by the tulasnellaceous basidium. The basidiomycetologists ask themselves whether the "septum" which develops at the base of each of the four "epibasidia" of *Tulasnella* is a secondary phenomenon. I believe that most of the contents of the "probasidium" migrate into these structures, and the formation of the cross-walls at the base may happen subsequent to much of the migration. Is this a basidium with four sterigmata, or a supporting cell which gives rise to four unicellular basidia? Upon the answer to this question rests the decision whether the basidium of *Tulasnella* is a holobasidium or a septate heterobasidium. There is a somewhat analogous problem in the phialide that is subdivided into many cells. Is it essentially one conidiogenous unit, or do the septa count? Are they laid down as "primary" or "secondary" septa?

Dr Ellis considers the phialide to be unique. Could we perhaps have his reasons at this point?

DR ELLIS

The only thing that delimits the phialide from all the other groups is the production of both outer and inner walls of the phialoconidia as distinct from the walls of the conidiogenous cell. In other words, the wall of the phialoconidium is formed de novo, except for the first phialoconidium, which carries away a part of the conidiogenous cell wall at dehiscence.

DR KENDRICK

We've had great trouble with first conidia in several groups because they aren't typical.

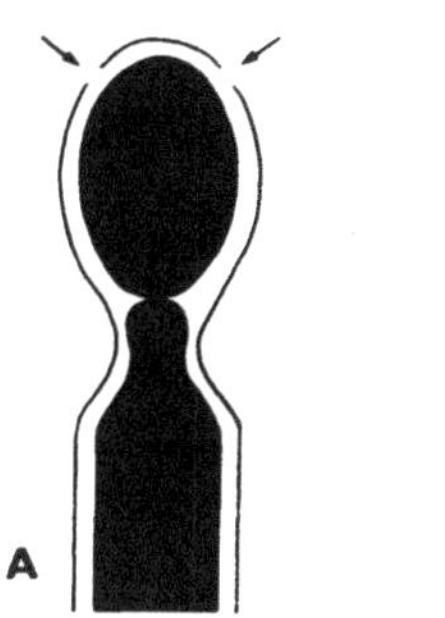

Figure 7.15 Diagrams comparing release of first phialoconidium in *Phialophora lagerbergii* (A) and *Aspergillus niger* (B) and its effect on the subsequent conformation of the collarette.

DR HENNEBERT

Dr von Arx does not make any distinction between phialides and annellophores. His keys to genera of Hyphomycetes are based on the assumption that for all practical purposes there are no distinguishing characters.

SOME MEMBERS

Oh! Oh!

DR HENNEBERT

He will assert that, since the conidiogenous cell of *Cryptosporiopsis* looks like a phialide in ordinary light microscopy, and has long been called a phialide, he will maintain that usage. Remember also that we may in future decide that poroconidia are really only blastoconidia, and that many phialides are really annellides when we look at their internal arrangements. So von Arx may have some reason to ignore the fine differences between phialides and annellides. We must be careful in making keys. They are only tools, not classifications. They must be practical. Not everyone can look at fungi with an electron microscope to identify them.

DR KENDRICK

Dr von Arx obviously wants a morphological, not a developmental classification. Judging from our recent discussion of annelloconidia, I hope that we may be able, by hindsight, to provide morphological clues to the different ontogenetic processes. Dr Hughes threw us a curve in 1953 when he established an experimental classification based on developmental characters, because, as we all know, it takes much more of an expert to work his scheme than to work Saccardo's scheme. And so there are still mycologists who use Saccardo's scheme when making keys for identification. The fact that Saccardo's scheme does not work for a number of fungi is unfortunate. It means that we have the responsibility of producing keys that incorporate as much ontogenetic information as possible without becoming so esoteric and arcane as to be unworkable.

DR SUBRAMANIAN

The annellophore produces only one conidium from each growing point. The phialide, by definition, has a fixed meristem and produces more than one conidium from it. But whether all the structures we have taken for granted as being phialides *are* phialides, I don't know.

EDITOR

*At this point the discussion was illustrated by projection of sequences from the Cole-Kendrick time-lapse film, "Conidium Ontogeny in Hyphomycetes," showing conidiogenesis in* Phialophora lagerbergii, *and the* Chalara *state of* Ceratocystis paradoxa.

DR KENDRICK

If the first few conidia in *Thelaviopsis paradoxa* phialides are delimited by a downwardly mobile zone of differentiation, they aren't typical phialoconidia - they could be called arthroconidia. In the light of what has already been said at this conference (which is just over a day old), that no longer surprises me. [*Laughter*] When a hypha is growing, the meristem is at the apex; when spore delimitation begins, I think the zone of active differentiation merely moves gradually down until it stops at a pre-determined point, where it remains thereafter. As the film sequence shows so beautifully, there is then a balance between the influx of cytoplasm and its differentiation into conidia.

DR CARMICHAEL

This particular problem bears out my feeling that phialoconidia of all kinds, including those from annellated phialides, are simply a kind of successive fission reproduction, or successive blastoconidia. In both fungi, the wall of the collarette seems to play an active part in expelling the conidia. It looks as if the wall has muscles in it!

DR SUBRAMANIAN

In phialides like that of *Phialophora,* there seems to be no attachment scar - a wall is laid down de novo all around the spore. In *Chalara*-type phialides, there may be protoplasmic cleavage. What about phialoconidia which have an apiculus? Is there fission here?

DR KENDRICK

These are differences of degree rather than of kind. The point at which the wall is being laid down could be up at the apex, or much lower down the phialide, or anywhere between, depending on the conidium. It's a matter of the location of the fixed meristem. If the wall is being actively laid down only as far down as the bottom of the conidium, then when the conidium is being cut off, the apex of the protoplast is apiculate, but it then retracts by surface tension. It probably *has* no wall at that point (Figure 7.16). It certainly doesn't look as if a double septum is involved in such cases.

DR ELLIS

In *Wallemia ichthyophaga,* all conidia have a common outer wall, but individual inner walls. This could be an extension of what is found in phialides of *Ceratocystis paradoxa.* Each conidium may be considered to be two-walled, so in *C. paradoxa* there are four wall layers - two for the conidiogenous cell, two for the conidium. In *Wallemia* they come out as a long tube, far beyond the end of the

Figure 7.16 Diagrams of suggested mechanism for production of phialoconidia with pointed abstriction scar. Wall-setting does not extend down to the protoplast of the phialide, which resumes the hemispherical shape dictated by surface tension as soon as the conidium has been abstricted.

phialide. At least I think it is a phialide. If it isn't, the conidia would be some kind of endoarthroconidia.

MME NICOT

Vuillemin considered the conidiogenous cell of *Sporendonema sebi* as a phialide.

DR S.J. HUGHES

In the Cole-Kendrick film, the first thing seen at the meristematic zone is the transverse septum. What we can't see is whether this occurs more or less simultaneously like a cleavage around the spore as in, say, a merosporangium. We must await electron microscope studies for that.

DR G.C. HUGHES

About the meristem. I'd be very surprised if ultrastructural studies don't show this to be a zone rather than a point. From the standpoint of organization, organelles concerned with this area will occur lower down in the phialide - mitochondria becoming geared up for the job - though, of course, the actual laying down of the wall may occur over a very short distance.

DR HENNEBERT

I think I should present to the meeting the new descriptive phialide terminology proposed by Dr W. Gams. He distinguishes phialides by their orientation with regard to the supporting hypha: (i) the orthophialide, growing at right angles to the supporting hypha, common in *Acremonium* (which he considers the correct name for *Cephalosporium* species); (ii) the plagiophialide, borne just below a septum of a supporting hypha, not at right angles to it, but inclined towards the direction of growth of the hypha; such phialides may be solitary or in verticils, and phialides may bear phialides; common in *Fusarium, Verticillium, Tilachlidium;* (iii) the schizophialide, a branched phialide without a septum at the base of each branch, in fact, a phialide with several functional openings; (iv) the pleurophialide, several openings on an intercalary hyphal cell (?), as in conidial *Pleurage verruculosa* and in *Cladorrhinum foecundissimum*; (v) the polyphialide, any cell that bears more than one collarette.

DR S.J. HUGHES

Do you mean, regardless of whether the collarettes arise in succession? I proposed the term 20 years ago for the phialides of *Lasiosphaeria hirsuta,* and I don't remember whether I was looking for a succession of collarettes or not.

DR HENNEBERT

Gams doesn't consider the question of a succession of collarettes. He is interested in the branching, and the relation to the supporting hypha.

DR KENDRICK

In our paper on *Codinaea,* Dr Hughes and I applied the term polyphialide specifically to a sympodially proliferating phialide with a succession of conidiogenous apices, only one of which, the most recently produced, is functional at any given time.

DR ELLIS

For descriptive purposes it is logical to describe a phialidic conidiogenous cell with only one opening as monophialidic, and one with two to many openings as polyphialidic.

DR KENDRICK

I am very concerned about Dr Gams's suggestions. I believe this kind of terminology could proliferate endlessly. If you want to describe every kind of branching in every kind of conidiophore in every kind of hyphomycete, you'll finish up with a million terms; and I don't think they will have any real significance in taxonomy, because many of them will describe mere environmentally imposed variations. The terms which already exist are perfectly adequate.

I would also like to suggest an extension of the term polyphialide to cover all sympodially proliferating phialides, whether or not they become septate subsequently. I have become convinced - perhaps only this morning - that there is a difference between septa laid down very near the tip of a hypha while it is growing, and those which are laid down later, as an afterthought. This concept might also solve the problem we had with the *Monilia* state of *Sclerotinia fructigena.* It was budding at the tip, but the septa were not laid down till much later. If the septa had been laid down at the beginning, the conidia would have been blastoconidia. As it is, they are laid down much later, and the conidia are arthroconidia. Perhaps we can now differentiate between "primary" and "secondary" septa in hyphomycetes.

DR CARMICHAEL

I also consider Dr Gams's new terms completely unnecessary.

DR KENDRICK

Dr Cole and I (1969) have discussed the idea of the variation in morphology of phialides, and have come to the conclusion that (and I quote) "in all these morphological variants, the existence of a localized, endogenous meristem producing a basipetal succession of phialospores through a rupture of the primary wall, has persuaded us that the over-riding criterion used in defining the phialide must be the method of conidium formation, not the morphology of the sporogenous cell." It would be impossible to have an unambiguous definition of the phialide based on traditional morphological characters. Early definitions *were* morphological, and for that reason they fell down.

DR COOKE

Some proliferation of terms may be necessary - something may appear that calls for it.

DR KENDRICK

That may be true. But to me the whole of science is a process of generalization, and the more our terms express an understanding of general underlying similarities rather than superficial differences, the more predictive they will be, and the better they will stand the test of time.

DR GOOS

Mason (1937) used the term prophialide for the cell subtending what he called the phialide of *Zygosporium.* Perhaps this would be a useful term for the metula in *Aspergillus* and *Penicillium.*

DR ELLIS

The conidiogenous cell of *Zygosporium* is not, in fact, a phialide. The single conidium is blastic.

DR GOOS

That is so, but perhaps the term prophialide may be a useful one to replace "metula."

DR ELLIS

I don't like the term. I think these are all branches, of several different orders - primary, secondary, tertiary branches of the conidiophore. There is usually in these cases a stipe with a branched head. The penicillus is a branched head.

DR KENDRICK

Dr Cole and I simultaneously thought of another objection. In basidiomycetes, the probasidium *becomes* a basidium, it doesn't *bear* a basidium.

MRS POLLACK

The first conidium that some of these conidiophores bear is so distinct from subsequent conidia that we need to take account of these two processes in our classification. We should also consider the position of the meristem when the first conidium is formed, and where it is subsequently: in the same place, above, or below.

DR FUNK

Dr Tubaki has applied some good descriptive terms to the origin of various phialoconidia: endogenous and subendogenous.

DR KENDRICK

You refer, perhaps, to the enormous difference between, say, *Ascoconidium,* with which you have worked, and, say, *Penicillium corylophilum.* If you put the two side by side, you could not connect them in any way. But Dr Cole and I would hesitate to return to the use of the term endoconidia. These are all phialoconidia, just as a bat and a buffalo are both mammals. The similarities are more basic than the differences.

*For a resumption of discussion on the phialide, and the conclusions we reached, see Chapters 15 and 16.*

# 8
# The Organization of the Penicillus of *Penicillium claviforme* Bainier

K. ZACHARIAH and P.O. METITIRI

The penicillus, or conidial apparatus of the penicillia, comprises two or more different kinds of cellular elements which develop sequentially from the apex of the conidiophore. Whorled branching of the elements at one or more steps in the sequence produces the broom shape of the penicillus. The most distal cells are the conidiogenous ones, the phialides.

In this report we describe three mutants in which the functioning of components of the penicillus has been variously altered; this work is the first stage of a project the purpose of which is to analyse conidium formation by study of appropriate mutants.

*Penicillium claviforme* Bainier belongs to the section Asymmetrica (asymmetrical penicillus), subsection Fasciculata (aggregated conidiophores) and possesses a growth habit which is striking and unmistakable (Raper and Thom 1949). The penicilli are restricted to the apices of coremia, where they form a smooth, continuous spore-bearing layer on the surface of the head. Thus, cellular changes which alter either the structure or proliferative activity of a sufficient number of penicilli can be detected by macroscopically examining the surface of the coremial head. We have exploited this feature in selecting mutants in which penicillus organization has been affected genetically; so far more than seventy mutants of this kind have been isolated and maintained in culture. On the basis of macroscopic characters they can be classified in several groups for convenience. Here we describe the morphology of penicilli and conidia of the wild type strain (WT hereafter) and members of the three mutant groups listed below:

BW group: coremium heads are compact, smooth, and white.
MH group: coremium heads are loose, fluffy, of irregular shape, and white.
BG group: coremium heads are compact, smooth, and green.

Department of Biology, University of Waterloo, Waterloo, Ontario, Canada.

The WT used, Q.M. no. 7504, was obtained through the kindness of Dr Emory G. Simmons of the Quartermaster Research and Engineering Centre, Natick, Mass., U.S.A. Mutagenesis was carried out by exposing germ tubes to a dilute solution of N-methyl-N-nitro-N-nitrosoguanidine for 12 hours. Selected mutants were grown on Difco Lima bean agar, and coremia of suitable ages were excised and processed for sectioning, for both light and electron microscope. Full details of all methods used will be published shortly (Metitiri and Zachariah, in preparation).

## THE WILD TYPE

In the hymenium of WT, each conidiophore gives rise to a relatively unordered zone of intertwined, anastomosed branches; these, in turn, produce metulae, which are slightly swollen cells, more regularly disposed than the branches. Next, the metulae each produce three to five parallel phialides; finally, the phialides produce conidia in basipetal succession (Figure 8.1 a). Metulae, phialides, and conidia are uninucleate.

Conidium formation in WT (Figure 8.1 a-j) provides a remarkable example of unequal cell division. First the wall of the developing conidium swells out at the apex of the phialide. Next the phialide nucleus divides mitotically with the spindle oriented parallel to the long axis of the phialide. The distal daughter nucleus migrates into the conidium and finally the complete conidium is cut off by a septum. The process is repeated rapidly many times. The depth of the conidial masses on mature coremium heads may be from 750 to 1,000 $\mu$, indicating that approximately 250-300 conidia are produced by a single phialide. Electron microscope studies of phialides have clearly shown that the wall of each conidium is elaborated on the inner surface of the apical region of the phialide, and that the cell membrane is highly convoluted and folded in this region, and in places is included in the formation of conspicuous mesosomes. These are believed to take part in the formation of the new membrane and wall material which is required for conidium formation (Zachariah and Fitz-James 1967).

## THE MUTANTS

The three kinds of mutant strains described in this report all belong to the large class of penicillus mutants, i.e., in all features of vegetative development they are virtually indistinguishable from WT. The differences become apparent only when their coremia reach the last stage of development. Macroscopically, at this stage, their heads fail to turn dark green and to show the other signs of active conidium formation, such as the appearance of compact masses of aggregated conidial chains. Light and electron microscopy show that, at the cellular and subcellular levels, the reasons for these gross alterations are not the same in the different types.

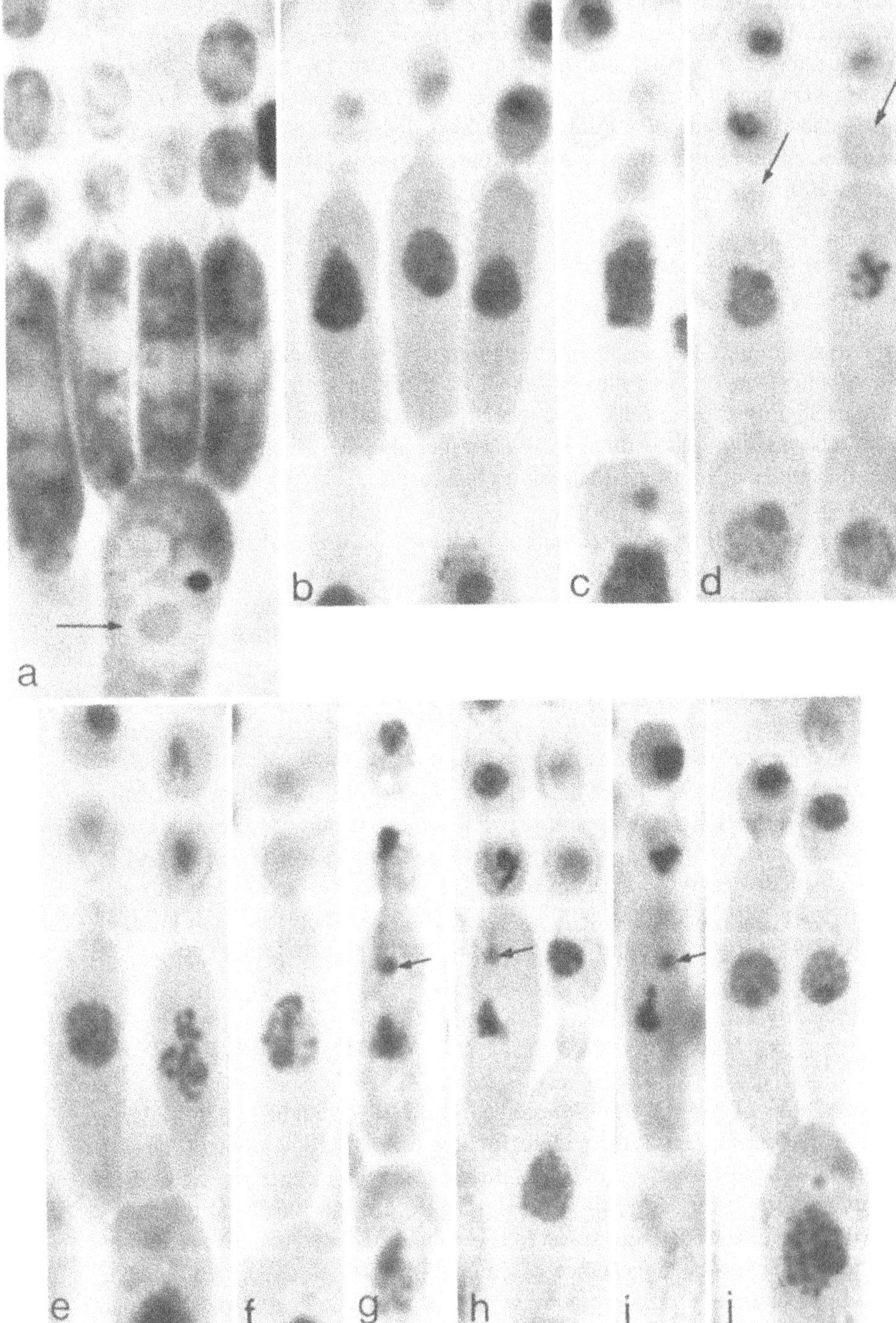

a
b
c
d
e
f
g
h
i
j

*BG group*

These mutants are characterized by coremium heads which are smooth and pale green. Microscopic examination shows that the phialide layer is usually indistinct because its cells are of various shapes and sizes and sometimes irregularly disposed. These mutants consistently produce very few conidia, usually only one per phialide. This feature is most clearly seen in mutant BG 95; the phialide layer is more regular than in other BG mutants, and each phialide produces only one indehiscent conidium. Appropriate staining techniques reveal two interesting facts; the conidia, though of approximately normal size, have very thick walls, and therefore only a small lumen (Figure 8.2 a-e); also, although each phialide possesses a nucleus, only a very few of the conidia do (Figure 8.2 f,g). These properties are shared by some of the other BG mutant strains, and it appears that the block to normal penicillus development occurs when the phialide starts to produce the wall material of the first conidium; in some way the massive abnormal structure that results prevents both the repetition of this step as well as the initiation of the dependent step of nuclear division.

*MH and BW groups*

The coremia produced by these mutants are similar in that the heads are always white, though the surface is rather more smooth and compact in the BW group. Microscopically too, the heads are similar in that they are composed of only crowded hypha-like elements which bear no conidia, but detailed examination shows significant differences. In MH strains, the strands are almost indistinguishable from the apices of vegetative hyphae; one difference is that they contain conspicuous vacuoles along their lengths (Figure 8.3 d). In contrast, close examination of the strands of any BW mutants shows that they are actually made up of files of phialide-like cells with distinct constrictions between them (Figure 8.5 a, b); and whereas the strands in MH mutants are merely continuations of the hyphae of the coremium stalk, those in BW mutants always arise (as do the phialides of WT) from a clearly defined zone of metulae (Figure 8.3 a). Furthermore, there is a striking difference in the distribution and behaviour of nuclei: in MH mutants, the cells of these aerial hyphae contain variable numbers

Figure 8.1 *Penicillium claviforme*, phialides of wild type (WT), all X3,000. (a) Toluidine blue; the single nucleus of the metula (arrow) and of each phialide contains a slightly darker nucleolus. (b-j) HCl-Giemsa (b) Nuclei of phialides in interphase; the first two phialides have conidia at a very early stage of development attached to their apices. (c) Phialide nucleus in early prophase; chromatin strands are just visible in it. (d) The nucleus of the second phialide is in metaphase, the interphase nucleus of the first phialide has a clear nucleolus; note the difference in size of the two most proximal conidia (arrows). (e,f) Phialide nuclei in metaphase. (g, h, i) Phialide nuclei in anaphase and early telophase; the arrows show the nucleolus, about to disappear. (j) Late telophase to early anaphase; the septum has been laid down in the second phialide but not in the first.

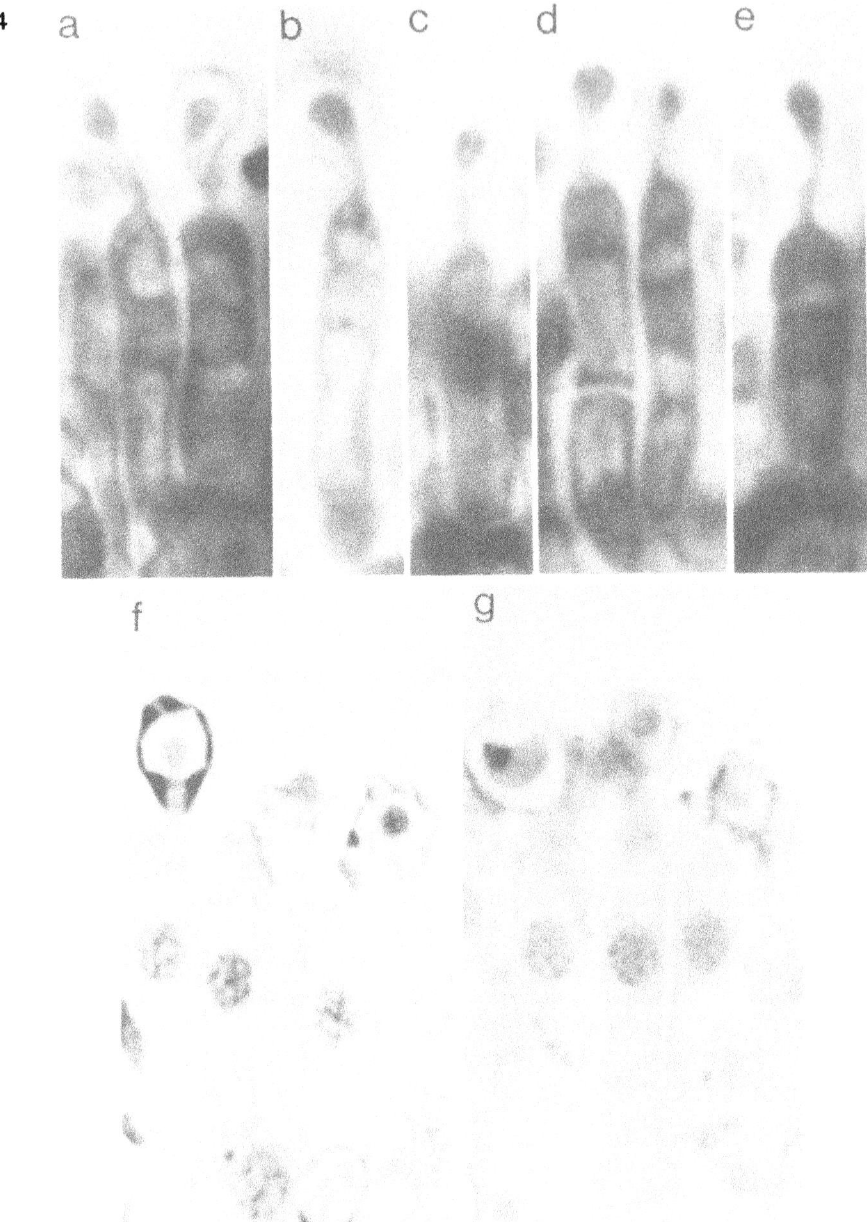


Figure 8.2 *Penicillium claviforme,* phialides and conidia of BG95, all X3,000. (a-e) Toluidine blue. (f,g) HCl-Giemsa.

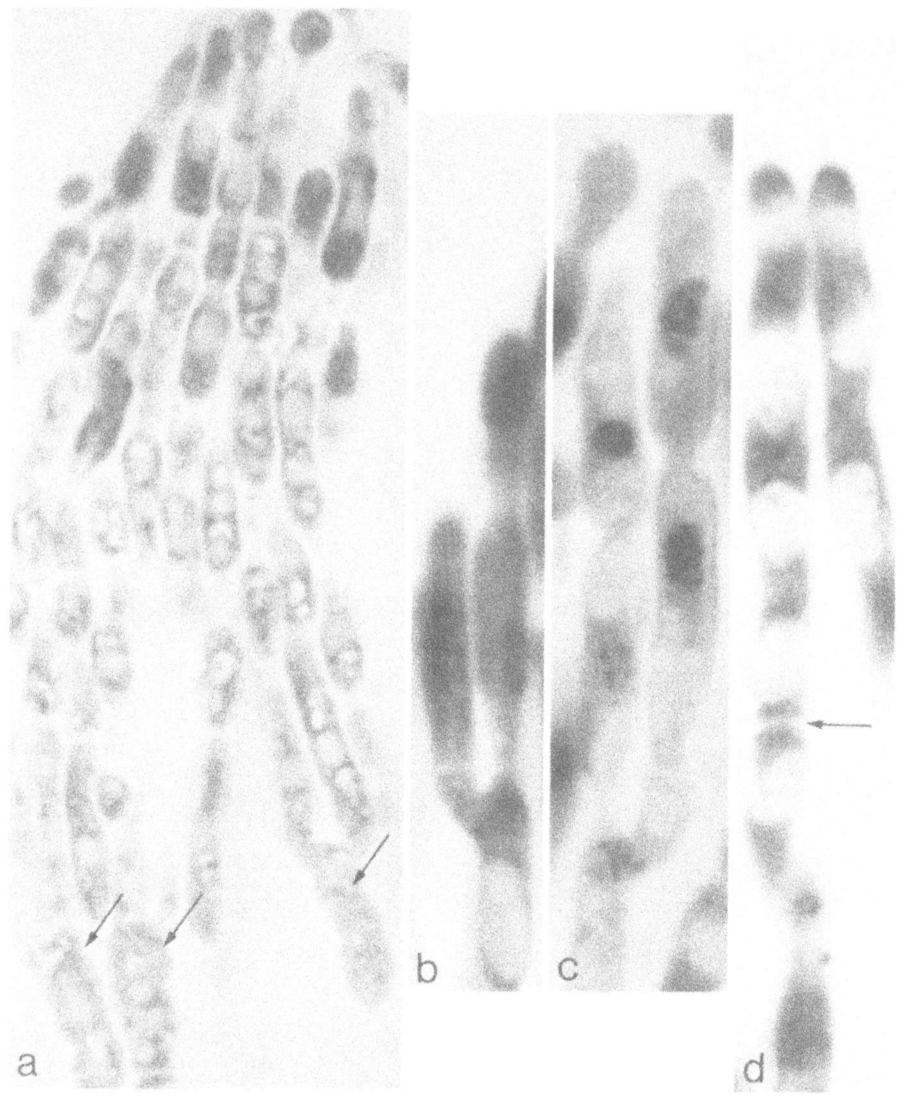


Figure 8.3 *Penicillium claviforme.* (a) Modified phialides of BW1 to show three metulae (arrows); toluidine blue, X1,700. (b) Branching of a modified phialide of BW38; toluidine blue, X3,000. (c) Branching of a modified phialide of BW38; HCl-Giemsa, X3,000. (d) Hyphae of the coremium head of MH62; arrow indicates septum; toluidine blue, X3,000.

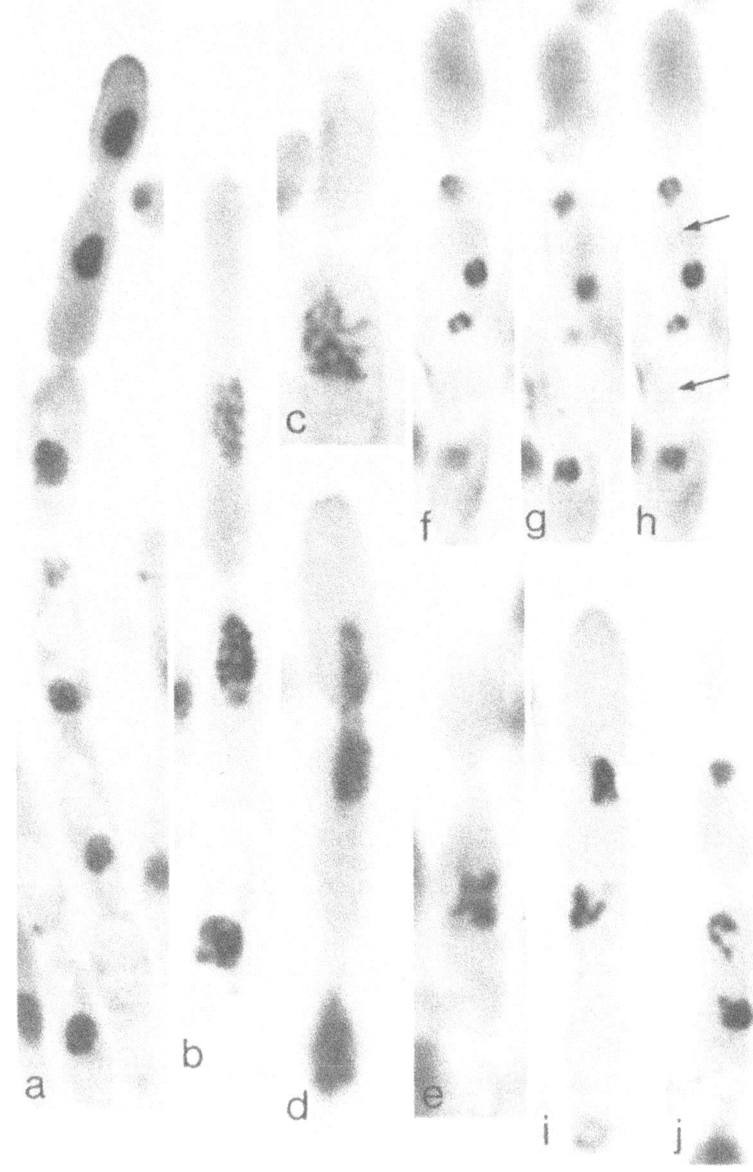


Figure 8.4 *Penicillium claviforme*, modified phialides of BW38; HCl-Giemsa; approximately ×3,000 except where stated otherwise. (a) The two most distal cells have the densest nuclei and do not show any cytoplasmic vacuoles, ×2,300. (b) The two most distal nuclei are entering prophase. (c) Late metaphase, ×3,600. (d) A migrating nucleus, ×3,600. (e) Early anaphase. (f, g, h) Three levels of focus to show two synchronous mitoses, each at telophase: the arrows indicate the equator of each spindle. (i, j) Later stages of telophase; the nuclei are less condensed than in f-h.

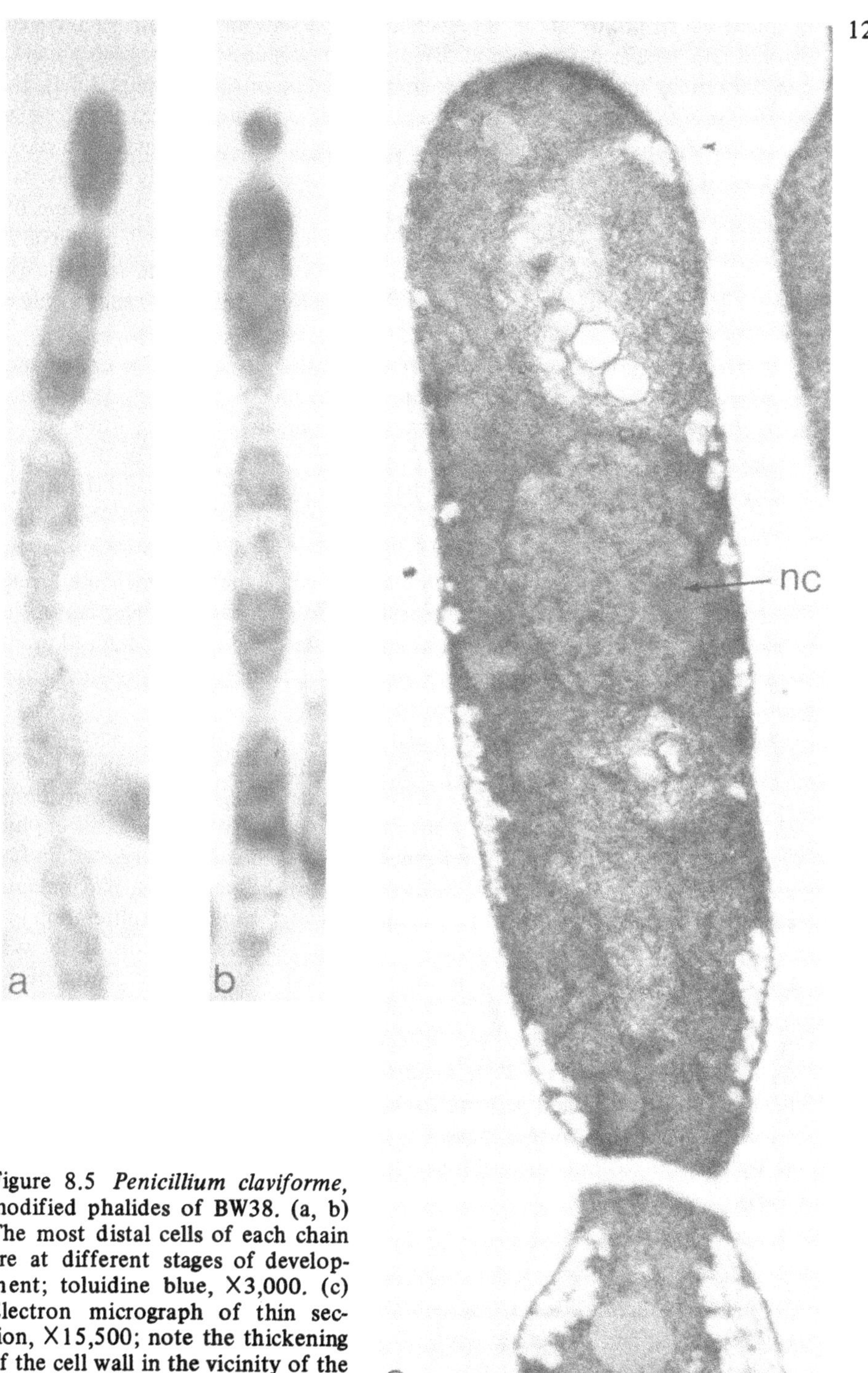


Figure 8.5 *Penicillium claviforme,* modified phalides of BW38. (a, b) The most distal cells of each chain are at different stages of development; toluidine blue, X3,000. (c) Electron micrograph of thin section, X15,500; note the thickening of the cell wall in the vicinity of the septum. nc = nucleus.

of nuclei which proliferate in no recognizable pattern; in BW mutants there is a distinct and regular pattern. In all BW mutants examined so far, each phialide-like cell usually has one nucleus, less commonly two or three (Figure 8.4 a), and only those nuclei in the most distal cells of each chain are capable of mitosis, as shown by Figure 8.4, in which prophase and early metaphase nuclei are recognized by their more or less distinct chromosomes (b, c) and early anaphase and telophase nuclei by their very condensed chromatin figures (e-j). In some BW mutants, two nuclei in distal cells of a chain may undergo mitosis simultaneously (f-h; in j one of the four daughter nuclei was not included in the section). Figures which we interpret as migrating nuclei (Figure 8.4 d) are also found only in the distal cells of the chains.

This karyological polarity is correlated with the distribution of cytoplasmic basophilia. When chains are stained with toluidine blue at pH 4 (to reveal ribonucleo-protein), the most intense coloration is found in the two or three cells at the apex of the chain (Figure 8.5 a, b). We believe this pattern results from the inversion of a cytoplasmic gradient, the effects of which are manifested in the unequal cell division a normal phialide undergoes in producing a conidium. That is to say, in WT phialides, the distal daughter nucleus of each mitosis loses the ability to divide (until conidium germination) while the proximal one divides repeatedly; in BW mutants, this situation is reversed. Such a polar control of nuclear activity and the correlated switch from basipetal to acropetal cell proliferation in certain *Aspergillus* mutants were first postulated by Thom and Raper (1945), and have now been confirmed by our study.

Two final points about these catenulate cells in BW mutants are worth noting. First, in some strains the cells form branches, which are usually initiated at the distal end of the cell (Figure 8.3 b, c). This feature, though never seen in normal WT phialides, is a characteristic means of proliferation of certain kinds of phialides, viz., "polyphialides" (Hughes and Kendrick 1968). Secondly, certain features of the cell wall are similar to those of normal phialides. In BW mutants, there is, as we have already noted, a complete switch from cell proliferation in a basipetal manner to cell proliferation in an acropetal manner. Thus the locus at which cell wall material is made is no longer fixed, but advances continuously. One might imagine that in this conversion the active cells lose all the characteristics of normal phialides, and in fact electron microscopy shows that the locus of proliferation in BW mutants does not involve two distinct cell wall layers (a phialide wall and an inner conidial wall) but only one. However, in this region the single wall layer is markedly thickened, as is the wall at the apex of a normal phialide, so that some trace of the normal organization remains (Figure 8.5 c).

The desirability of studying the epigenesis of phenotypic characters in problems of fungal taxonomy was first adumbrated by S. J. Hughes in a classic paper (Hughes 1953), and one tactic directed towards this end has been skilfully and successfully applied to sporulation in hyphomycetes by Kendrick and Cole. Our objective is an understanding of the mechanisms of cell division and differentiation. Our results are only preliminary, but they indicate the potential usefulness of morphological mutants in illuminating taxonomic relationships. It would be very interesting, too, to examine the regions of spore wall growth, perhaps with the use of fluorescent-labelled antibodies, in species with acropetal proliferation (*Cladosporium, Gonatobotryum*) and then to compare these patterns with those in mutants of basipetal species, such as our BW types. This might advance our understanding of the evolutionary relationship between phialoconidia and blastoconidia. Or again a cytological comparison of the spores of BG mutants with the solitary, thick-walled, indehiscent spores that are frequently found in normal cultures of *Scopulariopsis brevicaulis* might shed light on variations in annelloconidium structure and formation (Cole and Kendrick 1969).



Genes control phenotypic characters through the complex organization of the protein-synthesizing mechanism in the cytoplasm (the "epigenetic action system of the cell" of Waddington 1962). In only one of the mutants described here (namely BW types) have we found any indication of the involvement of components of this system in producing phenotypic alterations. But this connection is clearer in B mutants, a class of lytic mutants (Zachariah and Metitiri 1970), and recently we have discovered that the lower cells of the penicillus of mutant BG95 contain an abundance of dense virus-like particles which may be related to similar particles isolated from strains of *Penicillium stoloniferum* (Hollings and Stone 1969). It is not known at present how, if at all, they affect cell differentiation in BG95.

## ACKNOWLEDGMENTS

This research was supported by grants from the National Research Council of Canada and the Ontario Department of University Affairs.

## REFERENCES

Cole, G.T., and Kendrick, W.B. 1969. Conidium ontogeny in hyphomycetes. The annellophores of *Scopulariopsis brevicaulis*. Can. J. Botany 47: 925-9

Hollings, M., and Stone, O.M. 1969. Viruses in fungi. Sci. Progr. 57: 371-91

Hughes, S.J. 1953. Conidiophores, conidia, and classification. Can. J. Botany 31: 577-659

Hughes, S.J., and Kendrick, W.B. 1968. New Zealand fungi. 12. *Menispora, Codinaea, Menisporopsis*. N.Z. J. Botany 6: 323-75

Raper, K.B., and Thom, C. 1949. A manual of the Penicillia. Williams and Wilkins, Baltimore, Md.

Thom, C., and Raper, K.B. 1945. A manual of the aspergilli. Williams and Wilkins, Baltimore, Md.

Waddington, C.H. 1962. New patterns in genetics and development. Columbia University Press, New York

Zachariah, K., and Fitz-James, P.C. 1967. The structure of phialides in *Penicillium claviforme*. Can. J. Microbiol. 13: 249-56

Zachariah, K., and Metitiri, P.O. 1970. The effect of mutation on cell proliferation and nuclear behaviour in *Penicillium claviforme* Bainier. Protoplasma 69: 331-9

DR MÜLLER
Do we have the right to call nuclear division in such cases mitosis? The two daughter nuclei are not equivalent. One is responsible for the formation of the spore, and all the features that implies, and perhaps the other is responsible for the morphology of the phialide.
DR KENDRICK
But aren't we talking here about the turning "on" and "off" of various genes, rather than a lack of equivalence in their distribution? After all, cell differentiation goes on even after normal mitosis.
DR TUBAKI
*Cladosarum olivaceum*, a mutant of *Aspergillus niger,* behaved just like Mr Metitiri's mutant.
DR KENDRICK
An event of the kind induced in *Penicillium claviforme* by mutagenic agents might be just a freak. But such an event suggests that the evolution of the basipetal method might have been by a reversal of the polarity of the nuclei involved in the acropetal method. In other words, the event reported in *Cladosarum* and *P. claviforme* might really be a "reversion to type." The acropetal method of growth is the more primitive of the two - its logistics are less straightforward - all food material has to pass right through the chain of conidia to reach the new one at the tip.
DR ELLIS
The acropetal method of development is more widespread than the basipetal: more lines of evolution, more genera.
DR PIROZYNSKI
This fits Willis's "age and area" hypothesis which postulates that, in general, the commonest genera are the oldest rather than the best adapted. This seems to be true in the Uredinales and the Erysiphaceae - *Erysiphe,* the oldest and most primitive genus, is the commonest.
MR BHATT
In most lists of soil fungi, you will find *Penicillium* and *Trichoderma* and other basipetal spore formers occurring with very high frequency. Acropetal spore formers like *Cladosporium* are much less common.
DR ELLIS
This isn't a question of numbers of individuals, but rather an extension of different forms, a vast number of different kinds of fungi producing spores acropetally. There are only a limited number of genera producing phialides.
DR KENDRICK
But does that necessarily imply that they are the newer of the two groups? In many other groups of organisms, the older members are now the ones present in smaller numbers of genera and species. Without a good fossil record in the Fungi, we find it very hard to draw definite conclusions. Nevertheless, I agree with Dr Ellis. I am convinced that the acropetal method is the older and more primitive

of the two, but I base my arguments on the logistics of the spore-forming processes, not on occurrences or distribution. I don't know when the basipetal method developed, or whether it did so in the way I suggested. But it is obviously a phenomenon that is determined by nuclear behaviour. There's a lot to be said about the logistics of conidium formation in the Fungi Imperfecti in relation to their evolution; perhaps we're not yet ready to say much, but the time will come.

DR ELLIS

There is a fossil record of some blastic forms - *Clasterosporium* is an old genus, at least Tertiary.

# 9
# Annellophores

S. J. HUGHES

The successive proliferation of a cell, with the total involvement of each broad, transverse septum exposed by the liberation of a conidium or by the emptying of some other fruiting structure to produce a single, similar reproductive unit at the apex of each proliferation, is a common phenomenon in fungi, algae, and some mosses. It occurs in many groups of phycomycetes, in a few examples of basidiomycetes, and is a well-established method of producing a plurality of conidia in ascomycetes. An illustrated account of such proliferations in the Fungi as a whole, in algae, and in some mosses is presented elsewhere (Hughes 1971). The annellophores of ascomycetous Fungi Imperfecti are examples of cells producing percurrent proliferations, and this brief review is restricted to a consideration of these sporogenous cells.

In 1953 I proposed the term annellophore (Hughes 1953b) for such proliferating sporogenous cells and applied the term annellation* to each increment of the sporogenous cell as a result of successive proliferation and conidium production: it was not known to me at that time that some annellophores produce annellospores at the same level or even below the level of the original scar. Luttrell (1963) proposed the term percurrent for "conidiophore proliferations growing straight through the tip of the conidiophore, usually through the terminal conidial scar"; the term percurrent proliferations is appropriate for the sequence of events occurring in annellophores.

The term annellophore was originally applied to the conidiophores of *Annellophora* and *Sporidesmium,* to the short, flask-shaped sporogenous cells of *Spi-*

Plant Research Institute, Canada Department of Agriculture, Ottawa, Canada.

*The somewhat similar term annulation has been used in the description of "Hydroida," e.g. by Nutting (*Proc. U.S. Nat. Mus.* 21 [1899], 741-53) for corrugations on the pedicels supporting the "hydrothecae."

*locaea* and *Cephalotrichum* (*Doratomyces*), and to the sporogenous cells of *Melanconium* and of other form genera. In *Annellophora* the terminal cell of a conidium may also function as a sporogenous cell and sporulate by percurrent proliferations. An annellophore usually bears only a single sporogenous site. In *Spilocaea oleaginea*, however, the short, flask-shaped annellophores may bear two or even three necks, at the ends of which annellospores are produced (Hughes 1953a). Saccas (1944) illustrated annellophores in two other species of *Spilocaea* (as "*Fusicladium pyracanthae*" and "*F. dendriticum*") with annellations and a terminal proliferation bearing two broad projections each of which had sporulated. In one of the species ("*dendriticum*") an annellophore had bifurcated prior to producing any conidia.

Smith (1961) described the form genus *Polypaecilum* for two species described originally in *Scopulariopsis.* In the type species, *P. insolitum,* the conidiophores are irregular structures which bear terminal and sometimes lateral extensions; the distal ends of the main axis and laterals are extended into "more or less cylindrical outgrowths which are not individual cells but are continuous with the supporting branch. Each of these outgrowths is an annellophore, producing chains of spores and gradually elongating in the process." Smith used the term compound annellophore for these structures. His illustration of *P. insolitum* indicates that a single compound annellophore can bear up to nine sporulating projections, and some of the latter give the appearance of having arisen as a dichotomy. Such a single cell sporulating at multiple sites by percurrent proliferations is apparently uncommon. It was, however, illustrated by Saccas (1944) in his studies of *Spilocaea pomi* (as "*Fusicladium dendriticum*") in pure culture. His remarks and figures show that non-septate lengths of hyphae, particularly of the distal ends, could expand and produce many short lateral branches which sporulate at their cylindrical extremities.

Streiblová and Beran (1963) investigated the types and characteristics of scars formed as a result of budding, and the liberation of the buds, in a number of yeasts (see Figure 2.6). They listed some which multiplied by multilateral budding; in these, buds are formed progressively at different sites on the mother cell, and in *Saccharomyces cerevisiae* the maximum number of bud scars found was 25. I believe that such yeast cells which bud successively at new sites would rank as sympodulae. In the second group these authors included those yeasts which reproduce exclusively by fission and these belong to the genus *Schizosaccharomyces.* It is the third group of Streiblová and Beran which concerns us here; they stated that "reproduction by the intermediate process is characteristic of the tribe of the Nadsonieae and of two genera of asporogenous yeasts." By this method the first daughter cells are formed at each pole; "further daughter cells are formed at the poles in the same way from the same site." These scars were called "multiple scars" and because the growing number of scars results in the elongation of the polar areas of the cells, apiculate forms develop. The polar "excrescences" on the apiculate forms could be as long as the original cell. The subsequent application of electron microscope studies to the apiculate yeasts by Streiblová, Beran, and Pokorný (1964) showed that "multiple scars" have the character of concentric funnels with parallel or non-parallel rims. Still further studies by Streiblová and Beran (1965), using, among other techniques, ultra-thin sections, led the authors to remark that "the individual scar margins protrude clearly above the surface of the wall. The scars do not merge with one another continuously, the individual funnels remaining adjacent." "The highest individual reproduction ability of a *Saccharomycodes ludwigii* cell so far ascertained was 16 daughter cells out of a single mother cell, the maximum of 11 arising from one pole." These illustrated accounts indicate that "multiple scars" result from percurrent proliferations with total involvement of the exposed septum so that the cell in *S. ludwigii* and other apiculate yeasts is itself an annellophore, and furthermore an annellophore with two fertile areas.

Streiblová et al. (1964) remarked that earlier studies by other workers on apiculate yeasts "did not lead to the discovery of multiple scars, mainly because the investigators did not consider the possibility of their existence." The discovery of annellations in yeasts has at least explained the distinctive form of apiculate yeast cells, which gave rise to this appellation.

Sutton and Sandhu (1969) carried out electron microscope studies of conidium development on *Melanconium bicolor, M. apiocarpum, Cryptosporiopsis* sp., and *Phoma fumosa.* In the species of *Melanconium* they found that the annellophores increased in length as each successive terminal conidium was cut off from each successive percurrent proliferation, at a higher level than the previous conidium. This was illustrated and described by Hughes (1953b) in the *Melanconium* states of *Melanconis stilbostoma* and *M. juglandis*; in the last-named, the pale brown annellations are apparently attached to one another very tenuously because they readily become detached. In *Cryptosporiopsis* sp., Sutton and Sandhu (1969) found that the sporogenous cells were also annellophores but that the points of secession of the conidia are retrogressive or at approximately the same level as the first conidium; their illustrations strongly suggest that the successive annellations are expanded by pressure of the new proliferations formed within. In *Phoma fumosa* they found that the evidence was not so clear but nevertheless it "suggested an affinity with annellophoric development." Sutton and Sandhu implied that because new conidia may be formed almost at the same level as the first conidium or even lower down within the conidiophore, then such annellophores, under the light microscope, might be mistaken for phialides.

Cole and Kendrick (1969) recorded, by means of time-lapse photomicrography, successive conidium development and concomitant elongation of the annellophore in *Scopulariopsis brevicaulis* (Figure 9.1).

Preliminary unpublished results by Corlett and Hughes of electron microscope studies with *Cephalotrichum* (*Doratomyces*) *stemonitis* have confirmed the presence of distinct annellations in the sporogenous cells. Line drawings, prepared from electron micrographs by Dr M. P. Corlett, are shown in Figure 9.2, and the sequence of events in the production of conidia may be described as follows (see also Figures 19.3 and 19.4).

The first conidium initial arises as the distended apex of the sporogenous cell; the wall layers of the two structures are continuous. A transverse septum, de-

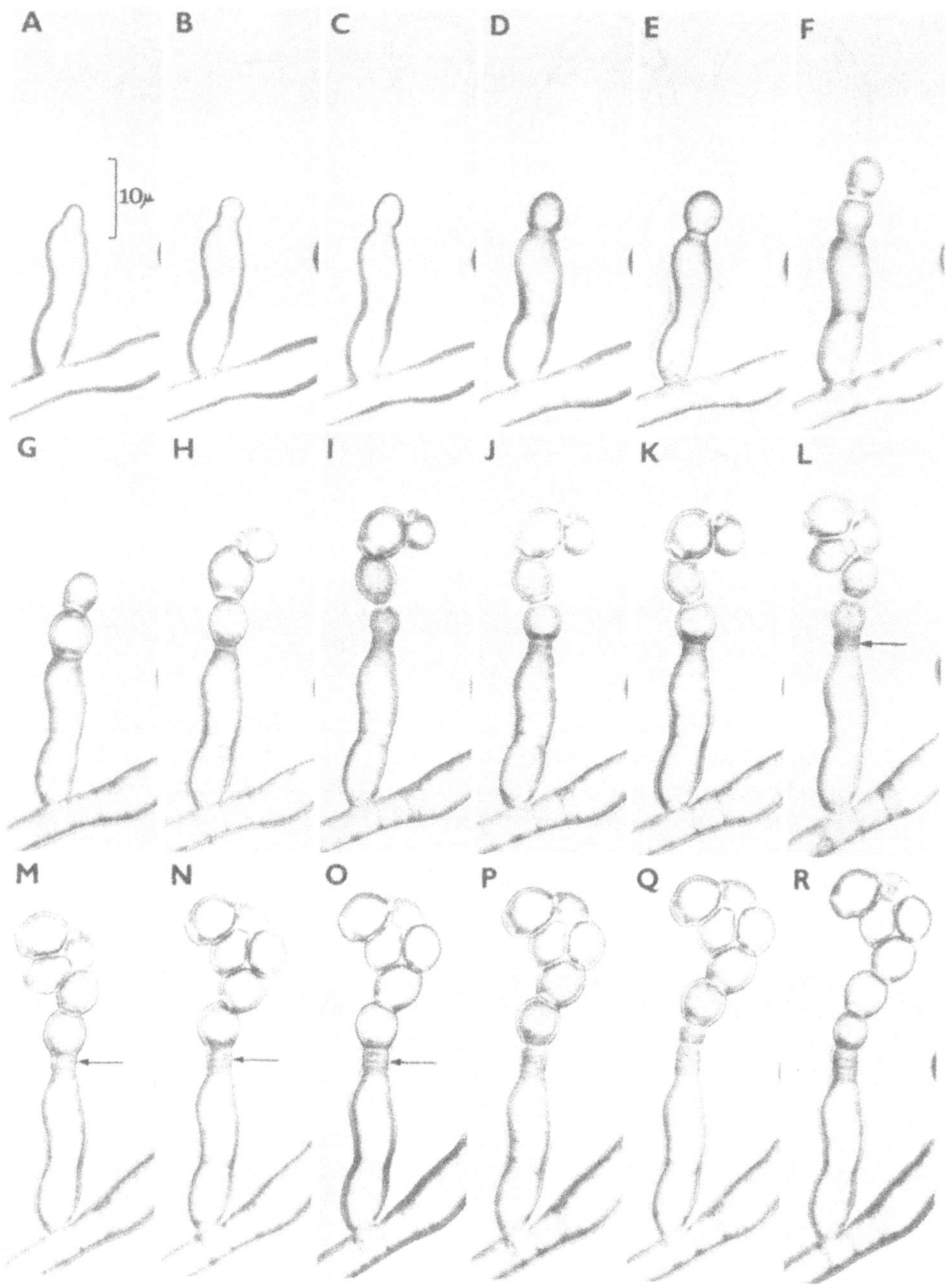


Figure 9.1 Time-lapse sequence of annelloconidium formation by *Scopulariopsis brevicaulis* (from Cole and Kendrick 1969).

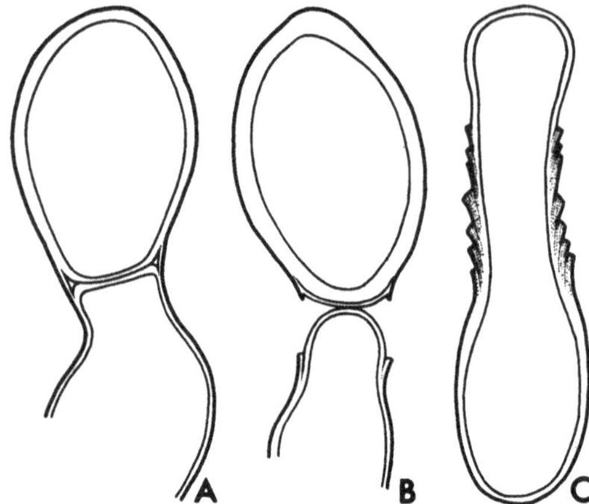


Figure 9.2 Percurrent proliferations in *Cephalotrichum* (*Doratomyces*) *stemonitis*, line drawings, from electron micrographs by Dr M.P. Corlett: A, first conidium delimited, and splitting at septum under way; B, first conidium seceded, conidium scar evident, and proliferation of annellophore started; C, annellophore with seven scars and developing apical conidium. ×ca. 5,000 (DAOM 57668).

rived only from the inner wall, delimits the conidium. The septum is of a double nature, and soon shows evidence of separating, especially adjacent to the outer wall where, in sections, a characteristic triangular area becomes evident. The transverse splitting of the double transverse septum and the circumscissile rupture of the outer wall, usually at the level of the septum, release the conidium. Probably the thrust of the base of the turgid conidium against the newly established apex of the the turgid conidium against the newly established apex of the equally turgid sporogenous cell assists in the rupture of the outer wall. The broken edge of the outer wall manifests itself as a conspicuous frill around the base of the spore, and its counterpart constitutes the scar on the sporogenous cell. The distal, exposed wall of the sporogenous cell, which is a part of the septum which delimited the first conidium, now becomes totally involved in the proliferation; the new apex extends and then distends into another terminal conidium initial. The outer layer of the wall of the proliferation and conidium initial becomes modified into a new outer wall and at the same time new wall material is added within. A transverse septum is formed as before, usually just above the level of the scar left by the previous conidium. This second conidium is released as before, by a transverse splitting of the septum and a circumscissile rupture of the outer wall. The process is repeated, resulting in a basipetal succession of conidia, each conidium upon liberation leaving a scar at the apex of each annellation.

The annellophoric sporogenous apparatus may be characterized as follows. Annellophores may be "free living," as in apiculate yeasts, but usually they are borne on a hyphal thallus, in which case they are cylindrical to subulate to flask-shaped, sessile or on stalks which are sometimes almost setose. They generally sporulate at one site but annellophores with two or more annellated projections are known. Annellophores are hyaline or variously coloured and may be aggregated in penicillate heads, produced on synnemata or on tuberculariaceous stromata, in acervuli or in pycnidia. 

Annellations vary considerably in length and shape in different species: they may be cylindrical and occasionally medially narrowed, to oval obpyriform, or they may resemble nested funnels. In some species they are of uniform length but in others the extent of proliferation before the onset of successive conidium development and delimitation can vary; in still other species, series of more or less equidistant conidium scars may be separated by inordinately long proliferations. Annellations are generally parallel but occasional ones may be oblique. They are usually of the same width in the same annellophore but may be successively narrower. The "annellations" of some enclosed annellophores are not evident because conidia can originate and secede at the same level as, or even at a lower level than the previous conidium. Annellations are usually concolorous with the rest of the annellophore although they may be proximally or distally darker; in hyaline annellophores the annellations may be difficult to observe with a light microscope but fluorescence microscopy holds out great promise.

Annellospores may be continuous, with one to many transverse septa, with longitudinal as well as transverse septa, or they may be variously branched or appendaged. They can develop in slimy heads, in dry chains, or they may fall away as soon as they are released. They often bear a conspicuous basal frill corresponding to the ruptured outer wall of the annellophore or its successive proliferations; the scars are usually more or less of the same width, but when annellations are successively narrower, so are the scars on the corresponding conidia. Like the annellophores, the conidia may be hyaline or variously coloured.

Records of the occurrence of the annellophoric method of producing conidia have now expanded considerably and include examples wherein annellations are obvious, even under a light microscope, but also those in which electron microscope studies are needed to show convincingly that this method is followed. Present evidence suggests that there is an intergradation from one extreme form to the other.

It has been pointed out elsewhere (Hughes 1971) that not all percurrent proliferations in Fungi Imperfecti involve the half-septum exposed by the secession of the terminal conidium. This stresses the need for electron microscope studies to determine the behaviour of existing or newly developed wall layers during conidiogenesis and conidium release. Perhaps the term annellophore should be used only for those structures in which each proliferation arises from the exposed half-septum, as for instance in *Cephalotrichum* and *Spilocaea* among the ascomycetous Fungi Imperfecti. The sporangiophore of *Albugo* (Phycomycetes) and the subtending cells of propagules in some species of *Sphacelaria* (Algae) are also apparently the morphological equivalents of annellophores and require close inspection.

In Chapter 2 some speculations were made on the derivation of the phialidic method of conidium production. It is tempting now to make conjectures concerning the derivation(s) of the annellophore. Four possible derivations may be put forward.

1. Annellophores may have arisen from the subtending cell of an ancestral sporangial apparatus in which the sporangium seceded as a unit and sporangiospore development was delayed and reduced to one or omitted: secession of the unit by a schizolytic separation (cf. Hughes 1971) followed by percurrent proliferation of the subtending cell and the production of another "sporangium" at the apex would establish an annellophore. This is essentially the derivation of the percurrently proliferating sporangiophore of *Albugo* given by Hughes (1971).

2. Annellophores may have arisen from phialide-like structures. This could have occurred by the secession of the first-formed endogenous phialospore complete with that part of the wall of the sporogenous cell that surrounded it; secession would be accomplished by a schizolytic separation of the spore from the cell below, and by a circumscissile rupture of the outer wall at the level of the septum. The subtending cell would then proliferate percurrently by the extension of the exposed wall and another terminal spore would form in the same manner at a higher level.

Although annellophores and phialides are common kinds of sporogenous cells in ascomycetes, records of their occurrence together in the same life cycle seem to be rare. However, Schol-Schwarz (1968) recorded "annellophore-like spore-mother cells" and phialides occurring together in cultures of *Rhinocladiella mansonii* which was revised to include *Trichosporium heteromorphum, Torula jeanselmei,* and others. She likened the annellate state of *R. mansonii* to the multiple yeast scars described by Streiblová et al. (1963, 1964). What significance can be attached to the rare association of annellophores and phialides remains to be seen.

3. Annellophores may have arisen from the subtending cell of a terminal chlamydospore which was capable of seceding schizolytically: percurrent proliferation, involving the half-septum of the subtending cell, and the production of a further terminal chlamydospore would establish an annellophore.

4. Annellophores may have arisen de novo from the subterminal cell of a hypha if the terminal cell separated schizolytically and proliferation followed as in 3 above.

Answers to many of the problems concerned with the structure and activity of annellophores, and perhaps some clues to their ancestry, may rest with electron microscope studies.

The possession of annellophores is currently being used as a key character in the circumscription of numerous form genera of Moniliales, Melanconiales, and Sphaeropsidales. There are, however, a few examples which have been described with conidia produced by different methods on the same conidiophore. For instance, in *Parodiopsis lophirae* Deighton (Moreau and Moreau 1959) sporogenous cells which produce staurosporous macroconidia may be annellated (presumably annellophores) or may produce macroconidia successively and sympodially:

Moreau and Moreau also illustrated a sporogenous cell which had at first proliferated percurrently and then sympodially, each proliferation having terminated in a macroconidium. Similar examples in ascomycetous Fungi Imperfecti are uncommon but they need close examination: in this irregularity they have their counterparts in the percurrent and sympodial proliferations associated with sporangium production in some phycomycetes, and with the production of gemmae in some mosses (Hughes 1971). But in spite of a few apparent intermediate examples between annellophores and sympodulae (sensu stricto) the production of the first kind of sporogenous cell provides a useful generic character throughout the ascomycetous Fungi Imperfecti. 

REFERENCES

Cole, G.T., and Kendrick, W.B. 1969. Conidium ontogeny in hyphomycetes. The annellophores of *Scopulariopsis brevicaulis*. Can. J. Botany 47: 925-9

Hughes, S.J. 1953a. Some foliicolous hyphomycetes. Can. J. Botany 31: 560-76

- 1953b. Conidiophores, conidia, and classification. Can. J. Botany 31: 577-659

- 1971. Percurrent proliferations in fungi, algae, and mosses. Can. J. Botany 49: 215-31

Luttrell, E.S. 1963. Taxonomic criteria in *Helminthosporium*. Mycologia 55: 643-74

Moreau, C., and Moreau, M. 1959. Champignons foliicoles de Guinée. 1. Ascomycètes du *Lophira alata* Banks et leurs parasites. Rev. de Mycol. 24: 324-48

Saccas, A. 1944. Etude morphologique et biologique des *Fusicladium* des Rosacées. Librairie le François, Paris

Schol-Schwarz, M.B. 1968. *Rhinocladiella*, its synonym *Fonsecaea* and its relation to *Phialophora*. Ant. van Leeuwenhoek 34: 119-52

Smith, G. 1961. *Polypaecilum* gen. nov. Trans. Brit. Mycol. Soc. 44: 437-40

Streiblová, E., and Beran, K. 1963. Types of multiplication scars in yeasts, demonstrated by fluorescence microscopy. Folia Microbiol. 8: 221-7

- 1965. On the question of vegetative reproduction in apiculate yeasts. Folia Microbiol. 10: 352-6

Streiblová, E., Beran, K., and Pokorný, V. 1964. Multiple scars, a new type of yeast scar in apiculate yeasts. J. Bacteriol. 88: 1104-11

Sutton, B.C., and Sandhu, D.K. 1969. Electron microscopy of conidium development and secession in *Cryptosporiopsis* sp., *Phoma fumosa*, *Melanconium bicolor*, and *M. apiocarpum*. Can. J. Botany 47: 745-9

## DISCUSSION

DR S.J. HUGHES

Could we have the views of this group on the terms annellophore and annellation? Do you consider "annellation" a suitable term for the proliferation of both *Sporidesmium* and *Spilocaea*?

DR SUBRAMANIAN

I would restrict the term annellation to the closely spaced, very regular proliferations found in the single conidiogenous cell of *Scopulariopsis*, *Doratomyces*, etc., which could also be called annellophores. We could simply say that those fungi whose conidiogenous cells proliferate irregularly have "percurrently proliferating" conidiophores.

DR KENDRICK

I am not convinced that these things are different enough to separate them like that. First, there are all the intermediate conditions. Second, a phialide is called a phialide whether it proliferates percurrently, sympodially, or not at all. Surely what we are after is the method. These things bear "annellations," and should be kept together.

DR PIROZYNSKI

Many conidiophores have periods of rest. In *Annellophora africana* closely spaced sets of annellations are succeeded by periods of vegetative growth which may reflect conditions unfavourable to sporulation. We should not attribute to these a significance they do not have.

DR ELLIS

The terms annelloconidium and annellospore have both been used at this conference. I am very much in favour of using the ending "-conidium," as suggested by Dr Kendrick, for whatever kind of spore we are considering, if it is produced by one of the Fungi Imperfecti.

DR KENDRICK

I prefer not only to use the ending "-conidium," but also to speak of "conidiogenous cell" and "conidiogenous hypha." We might as well be consistent.

DR HENNEBERT

I approve of the use of "conidium," but I foresee some difficulties if we eventually include in the Fungi Imperfecti other kinds of fungi, like Ustilaginales, Uredinales, and Phycomycetes, which have well-developed terminologies of their own.

DR KENDRICK

It will be quite a while before anyone gets around to working on these other groups from our developmental point of view. This should not deter us from setting our own house in order. There seems to be a consensus in favour of using "conidium."

*The discussion of percurrently proliferating conidiogenous cells and the conidia they produce is continued and concluded in Chapters 15 and 16. The karyology of the annellides of* Scopulariopsis brevicaulis *is discussed by Kendrick and Chang in Chapter 18, and their wall relations by Cole and Aldrich in Chapter 19.*

# 10
# The Sympodula and the Sympodioconidium

G.T. COLE

In 1953 Hughes included in section II of his experimental classification those hyphomycetes in which "the conidia are produced as blown-out ends singly at the apex of the conidiophore and of the successive new growing points which develop just to one side of the previous terminal conidium. At maturity, therefore, a conidiophore or sporogenous cell producing conidia in this way possesses a number of scars; each one of these was in turn terminal ... before being pushed aside by the development of a new growing point. Such conidiophores usually show a perceptible increase in length with the development of each conidium; but in some species ... the apex becomes swollen with the development of successive conidia."

At the time Hughes wrote this description no precise term had been proposed for this particular kind of conidiogenous cell. In 1933 Mason had referred to conidia which terminate the growth of a cell as "terminus phialospores"; unfortunately, he used the term phialospore in the wide sense originally proposed by Vuillemin in 1910. Hughes, on the other hand, restricted the term phialospore to denote the conidia produced successively from a specialized conidiogenous cell which does not increase or decrease in length during conidium formation.

In 1962 Kendrick proposed the name sympodula for the conidiogenous cell of Hughes's section II because it is a summation of the growth of many successive apices. He also suggested that the conidia be termed sympodioconidia but stated that "any term intended to describe the characteristics of Section II should refer to the sporogenous cell, as this and not the spore displays the diagnostic features of the group."

*Tritirachium album* Limber, which Hughes accommodated in his section II, can serve as an example of a sympodula-forming genus. The time-lapse sequence

Department of Biology, University of Waterloo, Waterloo, Ontario, Canada.

 of photomicrographs of *T. album* in Figure 10.1 was taken while the fungus was growing in a specially designed, thin, glass culture chamber (Cole and Kendrick 1968a). In A, the wall of the conidiogenous cell circumscribes an apical swelling which, 3/4 hour later (B), is delimited by a basal septum (arrow) and differentiates into the primary conidium. Thirty minutes later (C), the conidiogenous cell gives rise to a new growing point (arrow) located below and to one side of the first-formed, terminal conidium. During the next 3/4 hour, the new apex swells and develops into a second "terminal conidium" (D,E). A third growing point, appearing 3/4 hour later (F), gives rise to another extension of the conidiogenous cell which terminates with the formation of a third conidium at the new apex during the next 1-3/4 hours (G-I). At this point, the sympodial pattern of proliferation has become evident as the location of new growing points alternates from left to right at the base of successive terminal conidia. During the next 24 hours (J-R) at least six additional conidia are formed. In Q, the youngest of the acropetally produced conidia is labelled Y at the apex of the extended conidiogenous cell.

In our developmental study of *Curvularia inaequalis* (Shear) Boedijn (Kendrick and Cole 1968), we also described a process of sympodial proliferation of the conidiogenous cell. In Figure 10.2 A, the first-formed conidium of *C. inaequalis* has matured, the conidiogenous cell has developed a new subterminal growing point, has thereby elongated, and has just begun to produce a second conidium initial. The new conidium enlarges, becomes septate, and assumes the obovoid shape of a mature conidium over the next 1-1/4 hours (B-D). Three and three-quarters hours later, a slight protuberance appears just below and to the left of the apical conidium (arrow in F). During the next 30 minutes (G-H) it is apparent that this subterminal swelling is the initiation of a new period of proliferation of the conidiogenous cell which, 45 minutes later, terminates with the formation of a new conidium initial (I). This third conidium gradually expands during the next hour (J-K). Figure 10.2 L shows a conidiogenous cell which has produced nine successive apices: each has, in turn, ceased extension growth and developed a terminal conidium.

In both of these species, the conidia are produced acropetally at successive apices of sympodially proliferating conidiogenous cells. Therefore, according to Hughes's concept and Kendrick's terminology, we may refer to the asexual reproductive cells of *T. album* and *C. inaequalis* as sympodulae.

In our examination of the conidiogenous cell of *Deightoniella torulosa* (Syd.) Ellis, we found that considerable variation in the process of proliferation can occur (Figure 10.3). In A a conidiogenous cell has produced a sequence of three swellings, two of which each have a single conidium attached to them. A scar on the proximal swelling (arrow) indicates the point of attachment of a dehisced conidium. The conidia appear to have been produced alternately from opposite sides of a sympodially proliferating cell axis. In B (12-1/4 hours later) the conidiogenous cell proliferates through the lateral wall of the apical swelling. During the next 11-1/4 hours (C,D) a conidium initial develops at the new swollen apex and gradually matures into a septate, pigmented conidium. At the same time, a cylindrical branch emerges through the lateral wall of the third swelling below the apex (arrow in D), elongates, and forms a terminal swelling during the next 11 hours (E-G). A conidium, which unfortunately grew out of focus, formed at the apex of this new proliferation of the conidiogenous cell. In I, the branch to the left of the central axis of the conidiophore has produced a linear sequence of swellings. The terminal extension of this branch appears to have originated at the base of the penultimate swelling, growing out through its upper wall which may have been weakened by the detachment of a conidium. In contrast, a bifurcation has occurred at the base of this branch, presumably because of the strengthening effect which the undetached conidium had on the upper wall of the terminal swelling. From the above observations, it is clear that the conidiogenous cells of *D. torulosa* are not restricted to a sympodial pattern of proliferation.

Nelson (1966), while experimenting with crosses of two genetically differentiated strains of *Helminthosporium carbonum* Ullstrup, concluded that conidiophore morphology in this species is quite variable. His wild-type strain produced sympodially proliferating conidiogenous cells with a single conidium developing at each fertile node. The aberrant strains, however, produced irregularly branching conidiogenous cells, with long compartments constricted at the septa and conidia produced apically on the branches. From the composite data of all crosses studied, Nelson concluded that the differences in conidial arrangement between wild type and aberrant type strains are controlled by the interaction of only two genes.

At this stage I began to question the validity of the concept that developmental characters of the conidiogenous cell and not the conidium display the diagnostic features of Hughes's section II. Only confusion results when we attempt to classify *D. torulosa*, for example, in Hughes's system on the basis of the developmental characters of its conidiogenous cells. In fact, Hughes (1953) considered that this species should be included in his section III, composed of annellophore-forming genera such as *Scopulariopsis* Bainier. This decision probably resulted from the observation of percurrent proliferation as shown in one fertile branch of the conidiogenous cell in Figure 10.3 I. We examined the process of conidium formation in *D. torulosa* (Figure 10.4) and arrived at quite a different conclusion. At the apex of the conidiogenous cell in A, a slight protuberance is visible (arrow) which enlarges during the next 1/2 hour (B, C). The wall of this protuberance appears distinct from the wall of the swollen conidiogenous cell apex, and lateral expansion at the base of the bud seems to be restricted. These observations strongly suggest that a pore has developed in the primary wall releasing a cytoplasmic bud clad in a newly laid-down wall. As the bud rapidly increases in size over the next 1-1/4 hours (D, E), a ring of thickened wall material forms at its base (arrow in D). Clearly, a conidium is differentiating at the apex of the conidiogenous cell. During the next 3-3/4 hours (F, G), the initial elongates considerably while its girth increases very little. In H, about 12-3/4 hours after the appearance of a tiny protuberance on the swollen apex of the conidiogenous cell, a six-septate, obpyriform, light brown propagule has formed

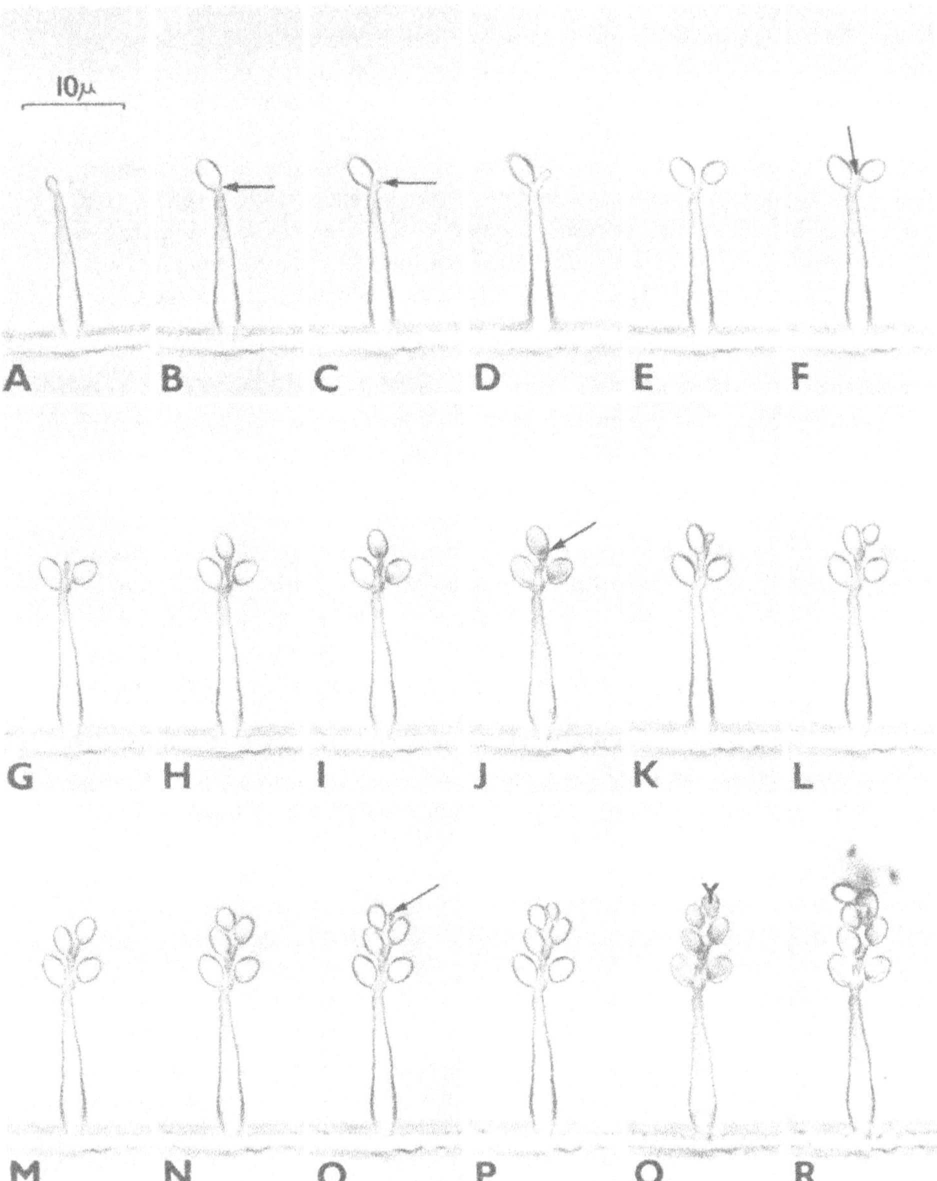


Figure 10.1 *Tritirachium album*, 35-mm time-lapse sequence of sympodula development. Arrows in B, C, F, J, and O indicate the location of new growing points. The letter Y in Q indicates the location of the youngest conidium at that particular stage of development. A-R at 0, ¾, 1¼, 1¾, 2, 2¾, 3¼, 4, 4½, 5¼, 5½, 5¾, 7¼, 9¾, 11½, 11¾, 12¼, and 28½ hours, respectively.

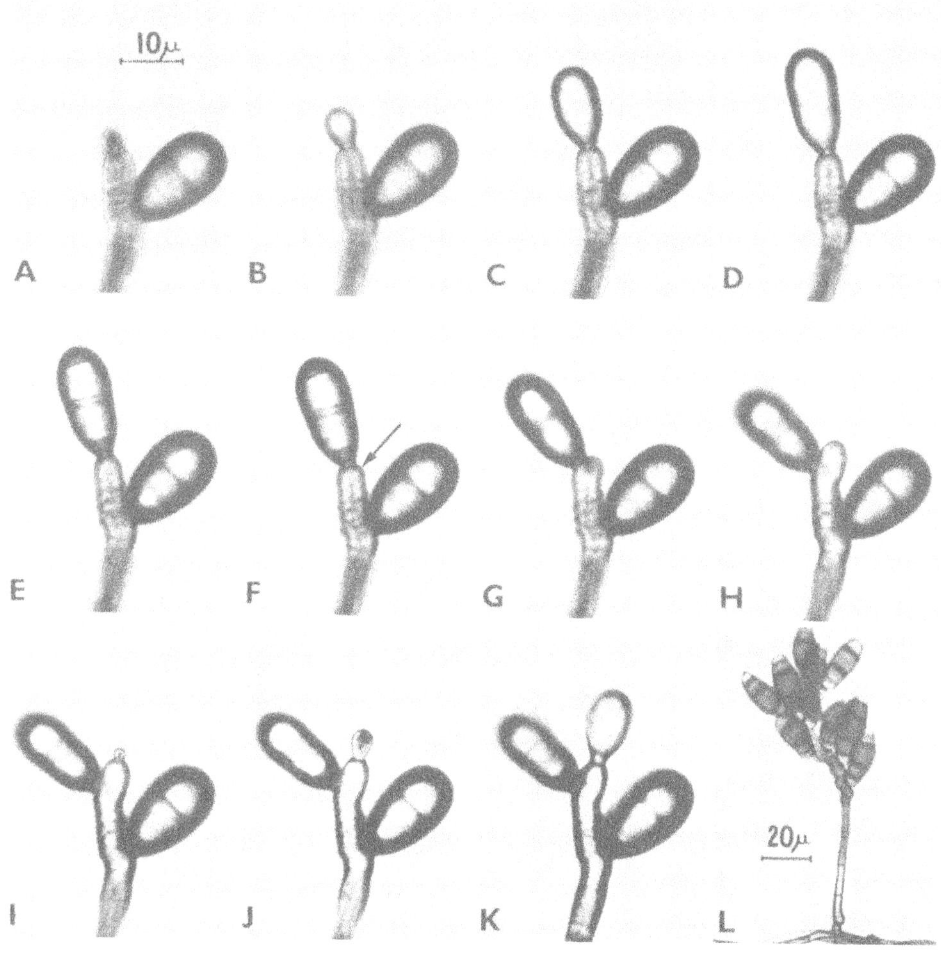


Figure 10.2 *Curvularia inaequalis*. A-K: 35-mm time-lapse sequence of sympodula development. Arrow in F indicates the location of a new growing point. L: A mature conidiogenous cell. A-L at 0, ¼, 1, 1¼, 4, 4¾, 5¼, 6¼, 7¼, 7¾, 8, and 8¾ hours, respectively.

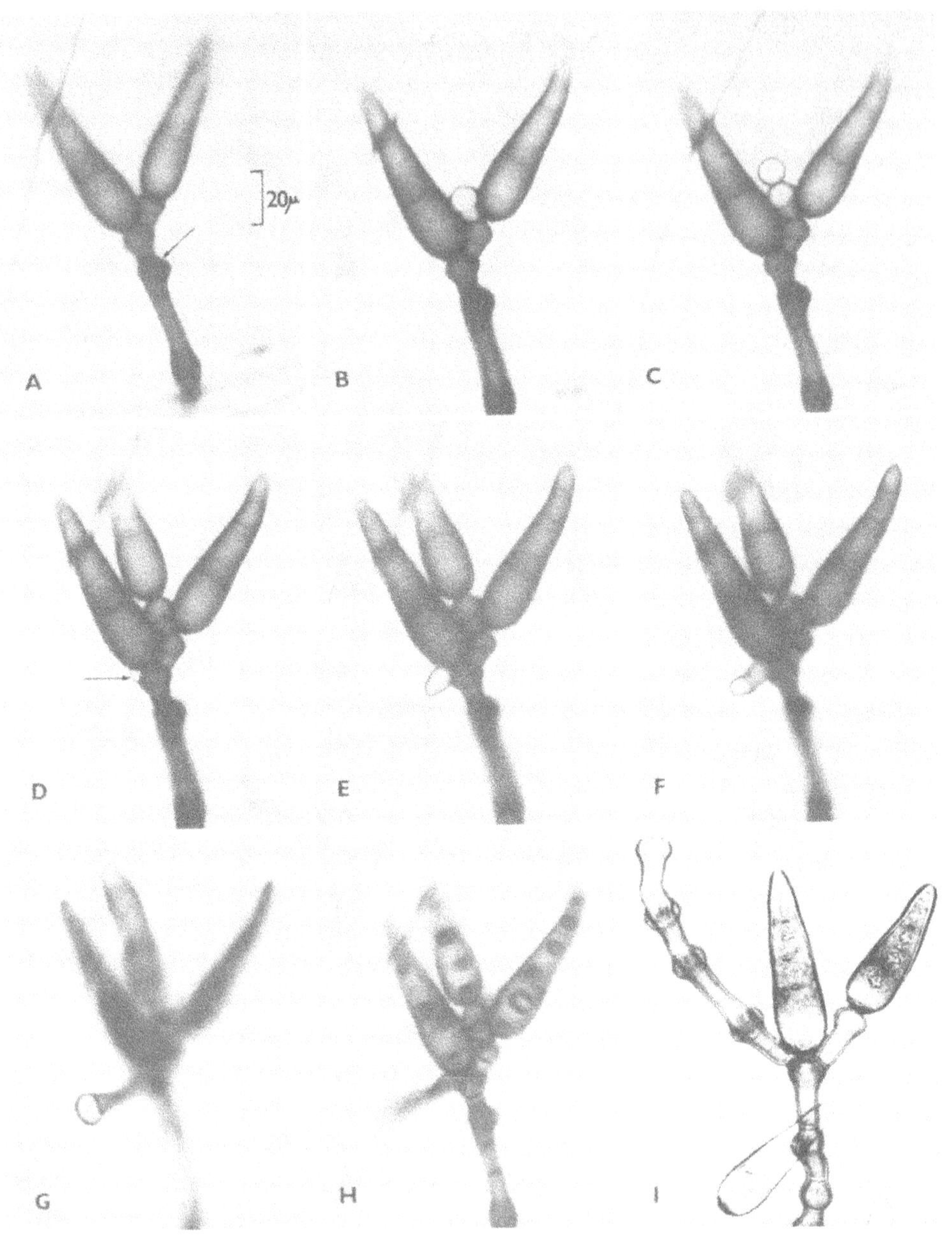


Figure 10.3 *Deightoniella torulosa*. A-H: 35-mm time-lapse sequence of conidiogenous cell proliferation. Arrow in A locates the point of previous attachment of a dehisced conidium. Arrow in D indicates a branch initial. A-H at 0, 12¼, 14¼, 23½, 25¾, 28½, 34½, and 53 hours, respectively. I: A mature conidiophore showing both percurrent and sympodial proliferation.

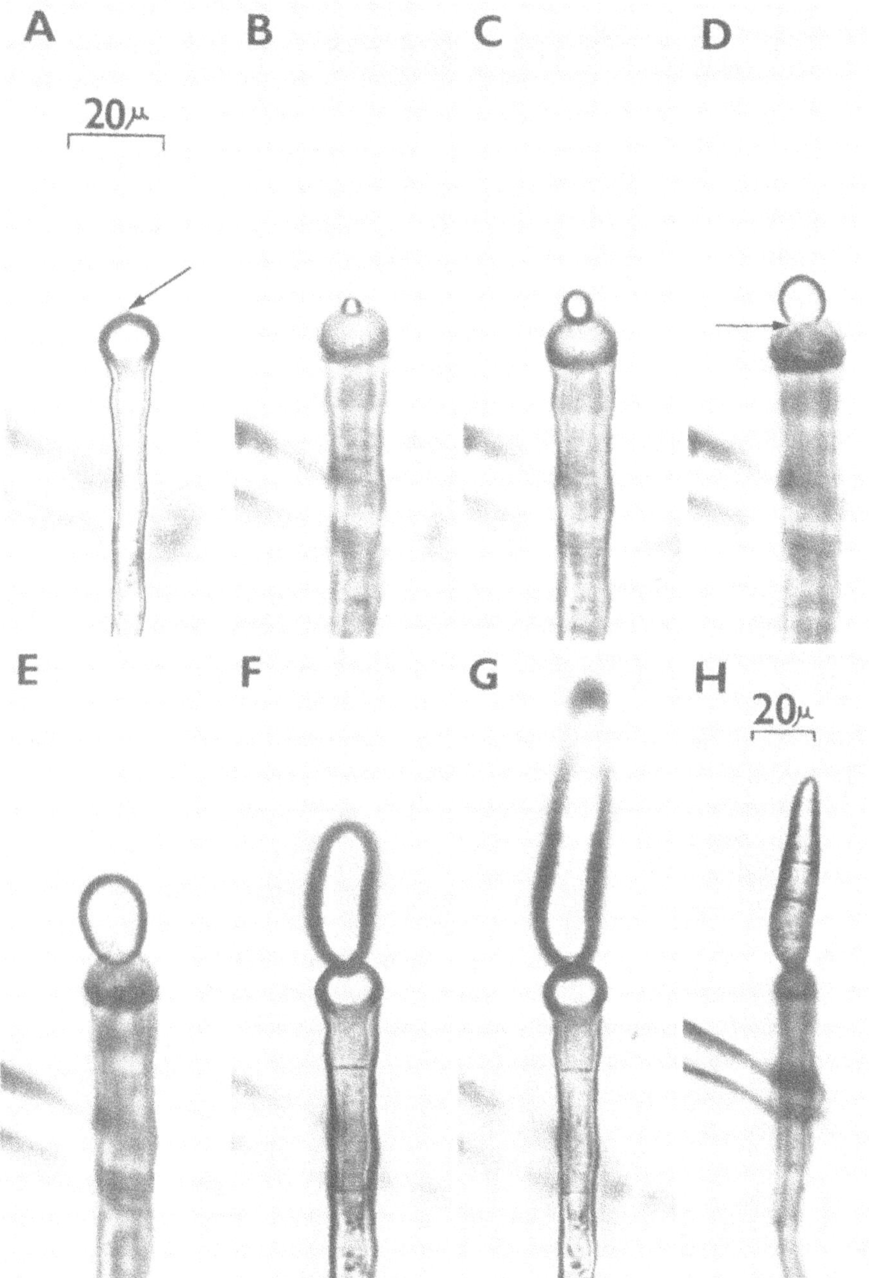


Figure 10.4 *Deightoniella torulosa*, 35-mm time-lapse sequence of poroconidium formation. Arrow in A indicates the apex of a conidium initial. Arrow in D indicates the ring of thickened wall material at the base of the developing conidium. A-H at 0, ¼, ½, 1, 1¾, 3, 5½, and 12¾ hours, respectively.

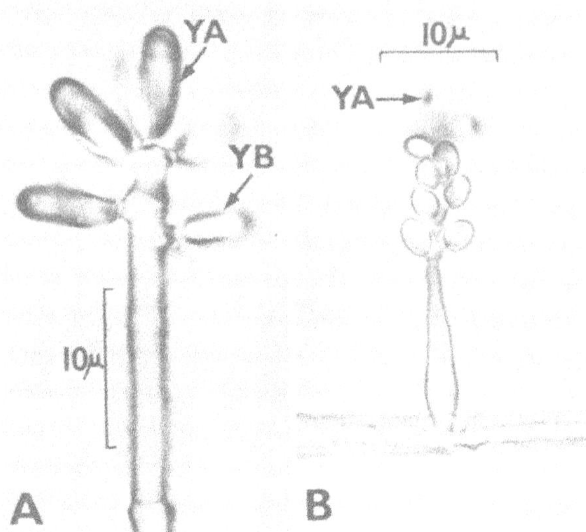


Figure 10.5 A, *Nodulisporium hinnuleum*; B, *Tritirachium album*. Conidiogenous cells which have produced a cluster of conidia at their respective apices. YA represents the youngest conidium at the apex of the conidiogenous cell, and YB represents the youngest conidium at the base of the conidiogenous cell.

which may now be referred to as a "poroconidium." The details of conidium formation in *D. torulosa* and *Curvularia inaequalis* (Figure 10.2) show striking similarities, and on the basis of *these* developmental characters the two fungi could be conveniently grouped together with other poroconidium-forming genera in Hughes's section VI (see Chapters 5 and 6).

It is often difficult to be sure whether a conidiophore or conidiogenous cell has proliferated sympodially when only the mature state is examined. This dilemma is exemplified by the conidiogenous cell of *Nodulisporium hinnuleum* (Preuss) Smith in Figure 10.5 A which has produced a cluster of conidia about its apex. If we eliminate the possibility that the conidia have arisen synchronously, we are left with two explanations of the process of conidium formation: an acropetal succession with the youngest conidium produced at the apex of the conidiogenous cell (YA), or a basipetal succession with the youngest conidium at the base (YB). If we choose the former and assume that the conidia develop at the apices of a succession of new growing points produced by a sympodially proliferating conidiogenous cell, then the developmental process essentially re-

Figure 10.6 *Nodulisporium hinnuleum*. A-G: 35-mm time-lapse sequence of conidium formation. Arrowheads indicate common reference points. Arrow in A indicates the denticle at the base of a mature conidium, and the arrow in C points to a young conidium initial. Arrow in F points to the youngest conidium, which is also labelled Y in G. A-G at 0, ¾, 1½, 2½, 2¾, 13½, and 14½ hours, respectively. H: Mature conidiogenous cells with exposed denticles (arrows).

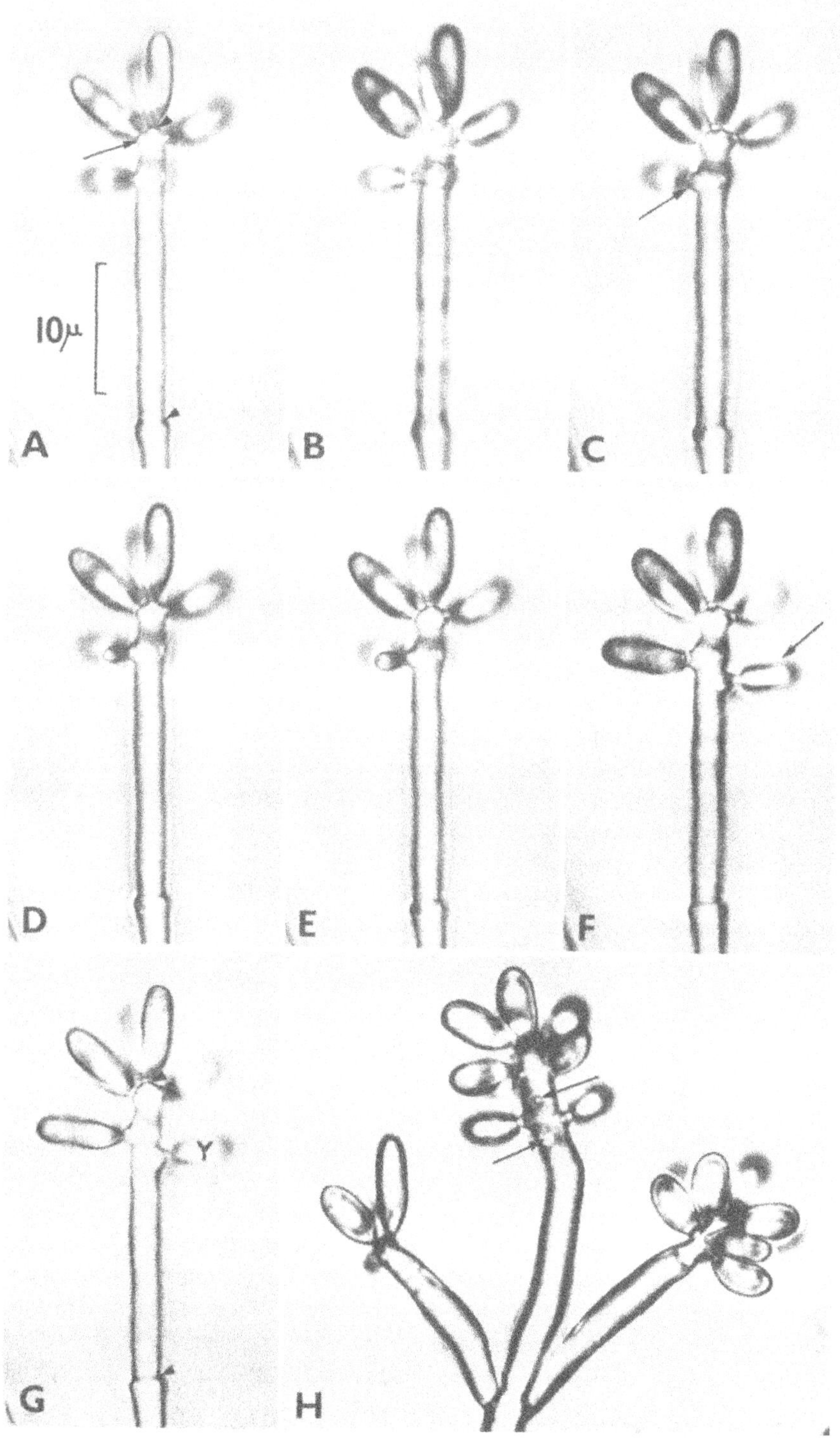
10μ
A
B
C
D
E
F
Y
G
H

sembles sympodula formation in *Tritirachium album* (Figure 10.1 and 10.5 B). In fact, the reproductive cells of *N. hinnuleum* have been interpreted as sympodulae giving rise to typical sympodioconidia by several authors (Hughes 1953; Tubaki 1958; Chesters and Greenhalgh 1964; Rogers 1966; Greenhalgh and Chesters 1968; Barron 1968). Our developmental study of this species, however, has yielded contradictory data (Figure 10.6). In A, seven conidia are clustered about the apex of a conidiogenous cell. Each has arisen from, and is attached at its base to a short, truncate, conical eruption of the outer wall (arrow). During the next 1-1/2 hours (B, C) a conidium initial appears (arrow in C) 4 μ *below* the apex of the conidiogenous cell. This conidium also remains attached at its base to a short denticle (D, E). Twelve hours after the formation of the conidium initial in C, a ninth conidium appears 6 μ below the conidiogenous cell apex (arrow in F). If we now compare the distance between the apex and the base of the conidiogenous cell (arrowheads) in A and G, representing a time interval of 14-1/2 hours during which two conidia develop, we find no change in length. The youngest conidium (Y in G) is *farthest from the conidiogenous cell apex*. In H, three conidiogenous cells are at different stages of development. The central one has matured and lost several of its conidia, leaving the denticles (arrows) clearly visible.

The conidia of *N. hinnuleum* are produced in a basipetal succession, clustering about the apex of the conidiogenous cell, which does not itself change in length during the process of conidium formation. I suggest that many other species producing alternately arranged conidia on conspicuous denticles may also possess previously unsuspected ontogenetic subtleties of the kind shown by *N. hinnuleum*, and that such fungi should, therefore, be subjected to developmental scrutiny. This caution applies especially to those species, cited by Hughes (1953), in which the conidiogenous cell "apex becomes swollen with the development of successive conidia."

If we are to establish conceptual stability in defining sympodioconidium formation, we must assign primary importance to the developmental characters of the conidium rather than the conidiogenous cell. In Hughes's original definition, he stated that the conidia are produced "*as blown-out ends singly* at the apex of the conidiophore" (the italics are mine). Two significant points made in this statement require further consideration. The term "blown-out ends" has also been applied to definitions of poroconidia, blastoconidia, and annelloconidia. As a result of our studies of conidium ontogeny, we suggest that greater specificity of terminology is required with regard to wall relations in defining these developmental processes. In many cases, this will necessitate an ultrastructural exploration of conidium formation. However, our light microscope study of *Gonatobotryum apiculatum* (Peck) Hughes (Kendrick, Cole, and Bhatt 1968) indicated that the formation of each primary blastoconidium initial from the swollen conidiogenous cell apex involves the laying down of a new secondary wall within the primary wall. The conidium initials appear to burst through the outer wall of the conidiogenous cell, leaving a conspicuous denticle at their point of rupture

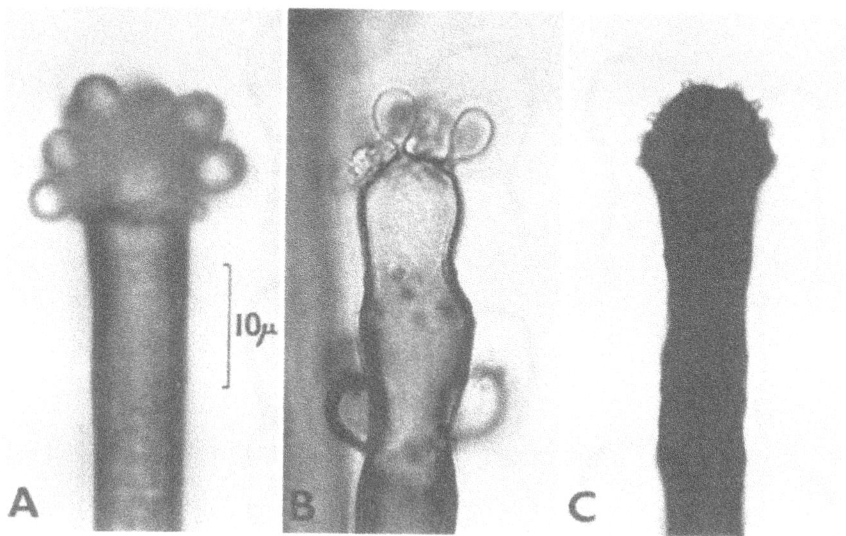


Figure 10.7 *Gonatobotryum apiculatum*: A, cluster of primary blastoconidia produced at the swollen conidiogenous cell apex; B, C, conidiogenous cells, fixed in 2.5 per cent glutaraldehyde followed by 1 per cent osmium tetroxide, showing denticles formed during conidium development.

(Figure 10.7 A-C). Scanning electron micrographs of *G. apiculatum* (Figure 19.2 A, B) support this interpretation.

During our study of annelloconidium formation in *Scopulariopsis brevicaulis* (Sacc.) Bain. (Cole and Kendrick 1968b), we found that the outer wall of the first conidium is continuous with the wall of the annellophore and that each subsequently formed conidium is clad in the outer wall of the proliferating conidiogenous cell (Figure 10.8). Results of our ultrastructural studies of conidium formation in this species (see Chapter 19) also support our light microscope interpretations. These same wall relations exist during conidium formation in *Tritirachium album* (Figure 10.9). However, ontogeny in *S. brevicaulis* and *T. album* can be distinguished on the basis of the kind of proliferation produced by their respective conidiogenous cells. Whereas successive proliferations in *S. brevicaulis* are percurrent, the conidiogenous cell of *T. album* proliferates sympodially from a point just below and to one side of the terminal conidium.

Hughes also emphasized, in his description of sympodioconidium formation, that each newly formed apex of the conidiogenous cell is terminated by a "single" conidium. Although the conidiogenous cells of *Cladosporium variabile* (Cooke) De Vries, *Codinaea fertilis* Hughes and Kendrick, and *Sympodiella* Kendrick proliferate sympodially (Figure 10.10), they would be excluded from Hughes's section II because each newly formed apex is terminated by *more than*

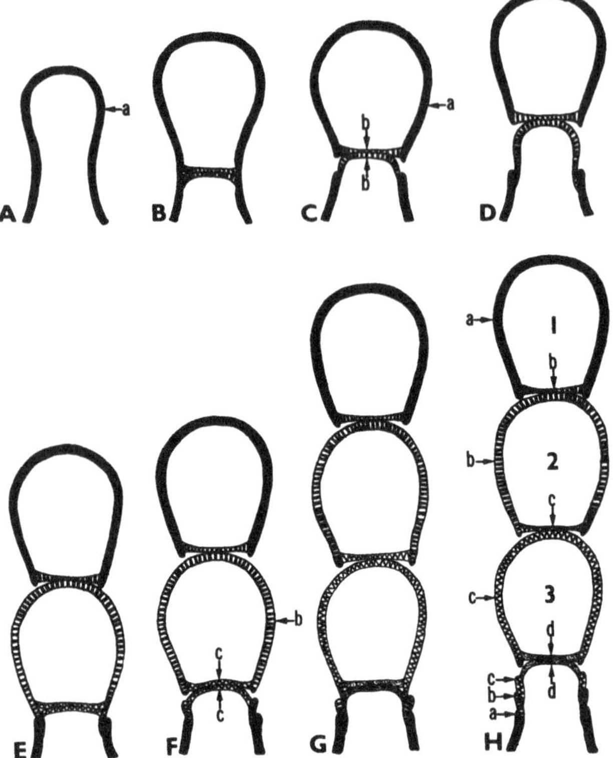


Figure 10.8 Diagrammatic representation of our light microscope interpretation of conidium formation in *Scopulariopsis brevicaulis*. Successive wall layers are differentiated by line hatching and are labelled a, b, c, and d, and conidia are labelled 1, 2, and 3 to indicate their order of formation.

*one conidium*. Tubaki (1963) united sympodula-forming hyphomycetes producing single conidia, or what he called "termino-radulaspores," with those producing chains of conidia, or "pleuro-radulaspores," in his classification. I disagree with Tubaki's proposal and would segregate these forms on the precept that characters of conidium formation, because of their greater stability, should have taxonomic priority over characters of conidiophore or conidiogenous cell development. In other words, the formation of acropetal chains of blastoconidia in *Cladosporium variabile* indicates close natural relationships to other members of *Cladosporium*, whereas its geniculate conidiophore is only an incidental variation of the simpler kind of conidiophore development in the genus. The characteristic process of phialoconidium formation in *Codinaea fertilis* and arthroconidium formation in *Sympodiella* would also persuade me to classify these fungi with other forms possessing similar kinds of conidium development.

In summary, sympodioconidium formation, as exemplified by *Tritirachium*

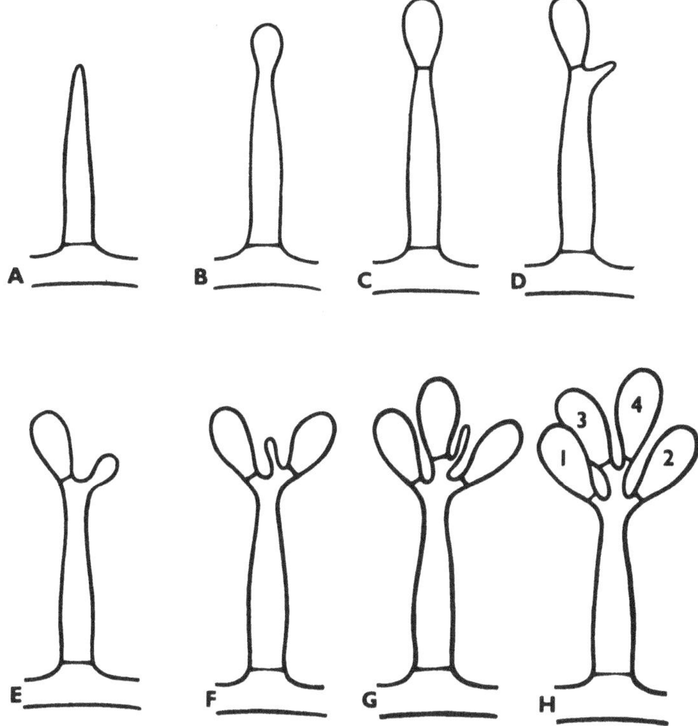


Figure 10.9 Diagrammatic representation of our light microscope interpretation of conidium formation in *Tritirachium album*. Conidia are labelled 1, 2, 3, and 4 to indicate their order of formation.

*album*, can be differentiated from annelloconidium, poroconidium, blastoconidium, phialoconidium, and arthroconidium formation on the basis of the following characters of both conidium and conidiogenous cell development:

1. A sympodioconidium is formed blastically, by a process of expansion and conversion of the conidiogenous cell apex which maintains the continuity of the outer wall until the conidium is delimited by a double septum at its base.
2. A single conidium terminates each new growing point of the sympodially proliferating conidiogenous cell.
3. Successive new growing points originate just below and to one side of the terminal conidium.

Perhaps the confusion which has arisen from the more general use of the terms sympodioconidium and sympodula will necessitate that new terms be proposed to represent this kind of developmental process. However, the terminology should be sufficiently descriptive to convey the three developmental characters outlined above. In the case of *T. album*, I suggest that we are examining simply the formation of blastoconidia from a specialized conidiogenous cell.

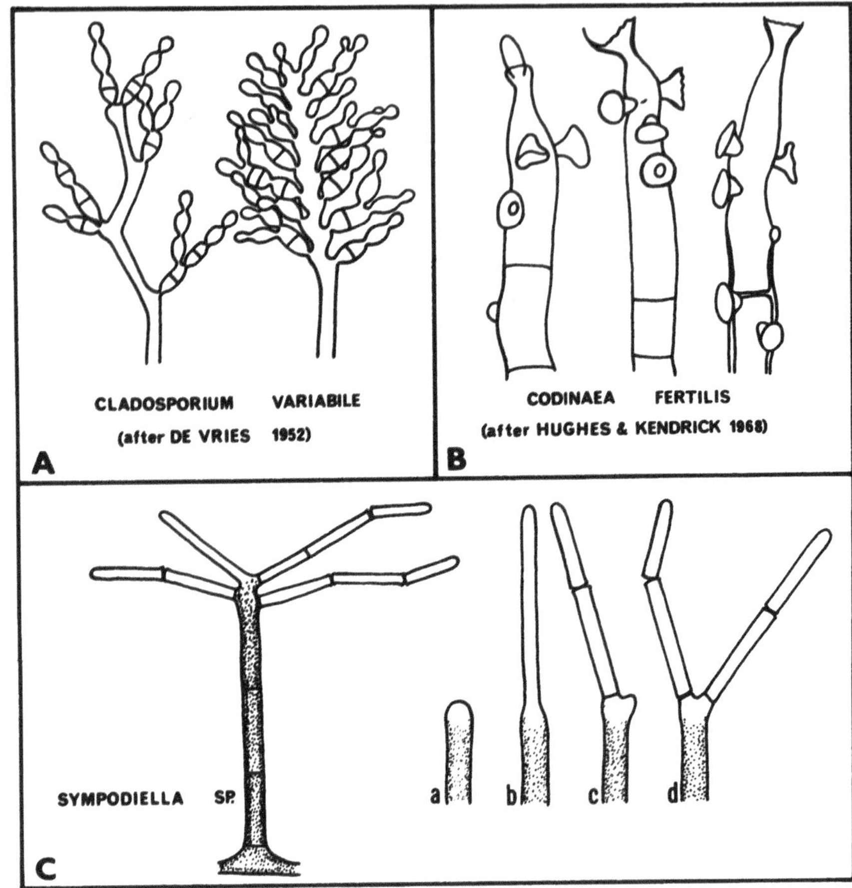


Figure 10.10 A: *Cladosporium variabile* (after De Vries 1952). B: *Codinaea fertilis* (after Hughes and Kendrick 1968). C: *Sympodiella* sp.; our interpretation of sympodial proliferation and arthroconidium formation is shown in a-d.

ACKNOWLEDGMENTS

The author thanks the National Research Council of Canada and the Ontario Department of University Affairs for financial support and is indebted to Dr W.B. Kendrick for valuable advice during the course of this research. Dr D.S. Meredith and Mr G.C. Bhatt supplied some of the fungal cultures.

REFERENCES

Barron, G.L. 1968. The genera of hyphomycetes from soil. Williams and Wilkins, Baltimore, Md.

Chesters, C.G.C. and Greenhalgh, G.N. 1964. *Geniculosporium serpens* gen. et sp. nov., the imperfect stage of *Hypoxylon serpens*. Trans. Brit. Mycol. Soc. 47: 393-401

Cole, G.T., and Kendrick, W. B. 1968a. A thin culture chamber for time-lapse photomicrography of fungi at high magnifications. Mycologia 60: 340-4

- 1968b. Conidium ontogeny in hyphomycetes. The annellophores of *Scopulariopsis brevicaulis.* Can. J. Botany 47: 925-9

Greenhalgh, G.N., and Chesters, C.G.C. 1968. Conidiophore morphology in some British members of the Xylariaceae. Trans. Brit. Mycol. Soc. 51: 57-82

Hughes, S.J. 1953. Conidiophores, conidia, and classification. Can. J. Botany 39: 577-659

Hughes, S.J., and Kendrick, W.B. 1968. New Zealand fungi. 12. *Menispora, Codinaea, Menisporopsis.* N.Z. J. Botany 6: 323-75

Kendrick, W.B. 1962. The *Leptographium* complex. *Verticicladiella* Hughes. Can. J. Botany 40: 771-7

Kendrick, W.B., Cole, G.T., and Bhatt, G.C. 1968. Conidium ontogeny in hyphomycetes. *Gonatobotryum apiculatum* and its botryose blastospores. Can. J. Botany 46: 591-6

Kendrick, W.B., and Cole, G.T. 1968. Conidium ontogeny in hyphomycetes. The sympodulae of *Beauveria* and *Curvularia.* Can. J. Botany 46: 1297-1301

Mason, E.W. 1933. Annotated account of fungi received at the Imperial Mycological Institute, List II (Fascicle 2). C.M.I. Mycol. Pap. 3

Nelson, R.R. 1966. The genetic control of conidial morphology and arrangement in *Cochliobolus carbonum.* Mycologia 58: 208-14

Rogers, J.D. 1966. Notes on the conidial stage of *Hypoxylon fuscum.* Mycologia 58: 459-64

Tubaki, K. 1958. Studies on Japanese hyphomycetes. V. Leaf and stem group with a discussion of the classification of Hyphomycetes and their perfect stages. J. Hattori Bot. Lab. 20: 142-244

- 1963. Taxonomic study of hyphomycetes. Ann. Rep. Inst. Fermentation, Osaka 1: 25-54

De Vries, G.A. 1952. Contribution to the knowledge of the genus *Cladosporium* Link ex Fr. Vitgeverij and Drukkerij. Hollandia Press, Baarn

Vuillemin, P. 1910. Les Conidiosporés. Bull. Soc. Sci. Nancy, Ser. III, 11(2): 129-72

## DISCUSSION

DR KENDRICK

Since some of my work has been discussed by Dr Cole, I would like to make a few introductory remarks. Our knowledge of the sympodial phenomenon has greatly increased since 1962, when I proposed the term sympodula. Dr Hughes has adopted the term, restricting it to conidiogenous cells which produce conidia blastically, and which produce only one conidium from each apex. *If* we retain the term sympodula, it should be retained in that sense - a sense already implicit in the original description, but which should perhaps be made explicit here at Kananaskis. Dr Cole and I recently raised the question of conidiogenous cells which proliferated sympodially but did not produce conidia blastically; we were asking this meeting to clarify the situation. This I hope it will now do.

DR COLE

If we retain the term sympodioconidium, may it not be linked to any conidiogenous cell proliferating sympodially, rather than being restricted, as Dr Hughes and I, at least, would restrict it, to the blastic type of conidium? Thus you would still have people saying that *Curvularia* produces "sympodioconidia." We have found sympodially proliferating conidiogenous cells that produce blastoconidia, phialoconidia, poroconidia, and arthroconidia. I feel that perhaps we should insert the word blastic, or whatever the conidium happens to be, to avoid all possibility of misunderstanding.

DR HENNEBERT

Yes, the way in which an individual conidium is formed is more important than the way in which the conidiogenous cell may subsequently proliferate. We shouldn't use terms like aleuriosympodioconidium. It is sufficient that the term describe the conidium itself; there should be a separate term for the conidiogenous cell.

DR S.J. HUGHES

When Dr Kendrick proposed the term sympodula, he had in mind *Verticicladiella,* whose conidiogenous cell increases in length as more conidia are formed. But there are conidiogenous cells in which the production of a succession of conidia is accompanied by a swelling of the apex of the conidiogenous cell, rather than an elongation. These are obviously closely related to the elongating kind, and are accordingly called sympodulae. In Phanerogams, where the term sympodial originated, there is always an increase in length with the production of successive apices. In the Fungi, then, we have changed the rules a little. We concentrate on the fact that there is a succession of subterminal apices, and a succession of single conidia. I'd like to see "sympodula" retained for single conidiogenous cells producing conidia blastically.

DR KENDRICK

At Waterloo we did some work on *Dactylaria* and found that sometimes the conidiogenous cell swells and has pegs on it, and sometimes it elongates and has pegs on it. We certainly didn't consider that this constituted a valid generic distinction.

DR S.J. HUGHES

In the definition of the "sympodula" perhaps what needs to be stressed is that the conidia are produced on successive new growing points; the elongation or swelling of the sympodula is secondary. The nuclear processes involved in such conidium development are little known except in some Aphyllophorales and Uredinales, and require investigation in ascomycetous Fungi Imperfecti.

DR MÜLLER

I have an example from the Discomycetes which may help to show that this sympodial growth habit is widespread. The conidial state of *Ascocorticium* (Figure 10.11) shows sympodial proliferation; not only this, but ascus formation involves sympodial proliferation.

DR KENDRICK

*Nodulisporium hinnuleum* has what might be called a static or reversed sympodula. Each new apex forms behind and to one side of the previous conidium, but there's *no* vegetative growth at all. It produces a conidium immediately.

DR CARMICHAEL

I have several cultures of Xylariaceae, *Hypoxylon* in particular, and here I find all three kinds: sympodial development of an axis with the terminal conidia

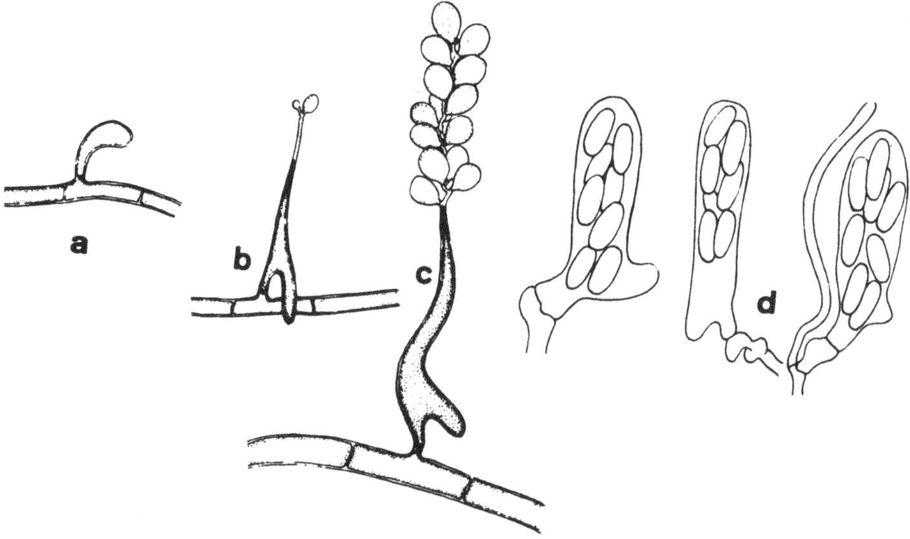


Figure 10.11 *Ascocorticium anomalum.* (a-c) *Rhinocladiella* conidial state: (a) reflexed conidiophore initial, (b) young conidiophore with a narrow conidiogenous extension, (c) older conidiophore with a narrow sympodially proliferated conidiogenous axis. (d) Asci with basal proliferations as in the pleurobasidia of Xenasmataceae. (X 1,000)

being youngest, inflated conidiogenous cells, and conidiogenous cells with the younger conidia lower down. We should consider the difference between the elongation of a conidiophore or vegetative hypha, and the specialized condition found in *Beauveria*, where there is a very fine, narrow meristematic structure - much finer than the vegetative hypha.

DR HENNEBERT

The phenomenon observed by Dr Kendrick and Dr Cole in *Nodulisporium hinnuleum* may be compared with one found in *Wardomyces,* where there is an inflated conidiophore that produces conidia first apically, then successively lower down on the swelling. Professor Mangenot has seen the conidiogenous cells of *Beauveria* growing immersed in agar, and failing to proliferate sympodially, but rather forming conidia somewhat in the manner of *Aureobasidium pullulans.* Here is another example of an induced change in the method of conidiogenesis.

DR COLE

With reference to Dr Hennebert's comment on *Wardomyces,* I'd like to quote from Dr Barron's book: "In ... *Wardomyces* ... the first spore is apical and terminal. A succession of spores is produced below and to one side of the apex. I ... consider [these] as botryose forms of the Aleuriosporae. The spores have fairly broad attachment to the sporogenous cell and do not secede readily. They do so by rupture of the sporogenous cell, part of which is left as a fringe at the base of the spore."

MR BHATT

I question the existence of valid differences between *Geniculosporium* and *Nodulisporium.*

DR ELLIS

*Geniculosporium* was erected for two reasons: (i) the conidiogenous cells in *Nodulisporium* are clearly polyblastic but not clearly sympodular, whereas those of *Geniculosporium* are obviously sympodular; (ii) the group of *Hypoxylons* to which *H. serpens* belongs is distinct from the others. This was a matter of the perfect states.

DR KENDRICK

But this is the kind of thing that happened in *Ceratocystis,* with the erection of *Grosmannia* and *Endoconiodiophora,* and it all collapsed.

DR CARMICHAEL

I agree with Dr Kendrick. I don't think that we need worry about coordinating the classifications of imperfect and perfect states. One need only consider *Thielaviopsis basicola* with its two vastly different conidial states: it will have to be classified simultaneously in two different sections of the Hyphomycetes.

DR KENDRICK

I'd like to suggest a comparison with the Zygomycetes, where zygospores are rather uniform but conidial states very diverse. It seems that much more varied selection pressures must have been brought to bear on those conidial states. Probably the same thing has happened to *Ceratocystis:* its conidial states have "speciated" much more than the perfect states.

Now I'd like to air a few thoughts about secondary plasticization of previously consolidated wall. I don't think the wall below the apical conidium can be thought of as an active apex; some consolidation will usually have occured. But, for some reason, in the sympodially proliferating conidiogenous cell this wall becomes plastic again, there is considerable intussusception of new wall material, and a new vegetative apex grows out. There is some similarity between that and the process of conidium formation in *Basipetospora*, *Trichothecium*, and *Cladobotryum*, where what one considers to be consolidated cell wall replasticizes and is involved in active growth. The difference between the two is chiefly one of the position of the proliferation. In the "sympodula" it is unilateral; in *Basipetospora* it forms a zone all round the cell. In the sympodula there can be a certain amount of vegetative extension growth before another conidium is formed; in *Basipetospora* the conidium forms immediately. In the sympodula the net result is an elongation or swelling of the cell; in *Basipetospora* the net result is a shortening of the cell. But there are similarities.

DR PIROZYNSKI

Dr Ellis (*C.M.I. Mycol. Pap.* 114 [1968], 27, 29) has recently illustrated *Spiropes dialii* and *S. melanoplaca* (Figure 10.12). They are congeneric with other fungi which have regular sympodial proliferations, but these two species frequently proliferate percurrently. Each conidium is formed apically. The darkly pigmented lower half of the double septum does not, as in the typical annellophore, become part of the integument of the next conidium. Dr Ellis's drawing

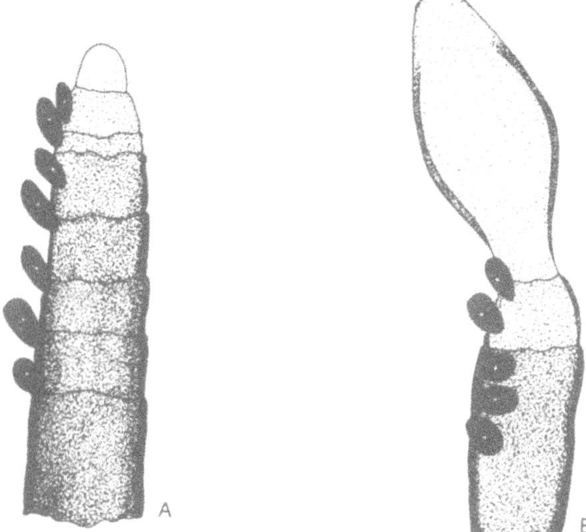


Figure 10.12 *Spiropes melanoplaca*, conidiogenous cells showing aberrant percurrent proliferations (after Ellis 1968); see text.

shows it as being flipped to one side by a newly erumpent apex which is clearly derived from an inner wall (Figure 10.12 A). The thickened, darkly pigmented septum is clearly not itself amenable to replasticization, but it has a peripheral line of weakness which makes it a logical exit for a proliferation. In these fungi the same conidiophore often apparently proliferates sympodially as well as percurrently (lower three scars in Figure 10.12 B).

MRS POLLACK

We must separate our thinking on conidiophores from our thinking about conidia. The situation is apparently more complex than Dr Hughes suspected in 1953. In his scheme you could apply the same, or a related, adjective to the conidiophore *and* the conidium. It is clear that we cannot do that any longer.

# 11
# Arthroconidia and Meristem Arthroconidia

BRYCE KENDRICK

Socrates, the gad-fly, began it all. "Define your terms," he told the youth of Athens. Unequivocal definitions are no less important to us today, and so I must begin by attempting to define "arthroconidium" and "meristem arthroconidium."

Going back to first principles, it seems to me that there are three ways in which a conidium can be formed: (i) virtually entirely de novo, by active meristematic growth; (ii) virtually entirely by conversion of pre-existing elements, involving no active growth; (iii) by a combination of the first two processes, a mixture of conversion and growth.

Arthrospores, as originally defined by Vuillemin (1910) and as redefined by Hughes (1953): "conidia which arise by the (basipetal) *fragmentation* of conidiophores of determinate length which do not possess a meristematic zone," would appear to fit nicely under heading (ii), "conversion."

Meristem arthrospores, as originally defined by Hughes (1953) and as redefined by Cole and Kendrick (1968): "conidia originating by the meristematic growth of the apical region of the conidiophore and its concurrent basipetal conversion into spores," would appear to fit nicely under heading (iii), "conversion and growth". But there are also hyphomycetes whose conidia, while arising by conversion of a determinate conidiophore with no meristematic zone, nevertheless apparently exhibit some growth; for example, *Amblyosporium*. I'll have more to say about these definitions later.

When I was attempting to subdivide the raw material of this conference into viable units, it seemed to me that arthroconidia and meristem arthroconidia had sufficient developmental similarity to warrant treating them in a single discussion. I no longer believe this, as you will see. But, as we have examined each of

Department of Biology, University of Waterloo, Waterloo, Ontario, Canada.

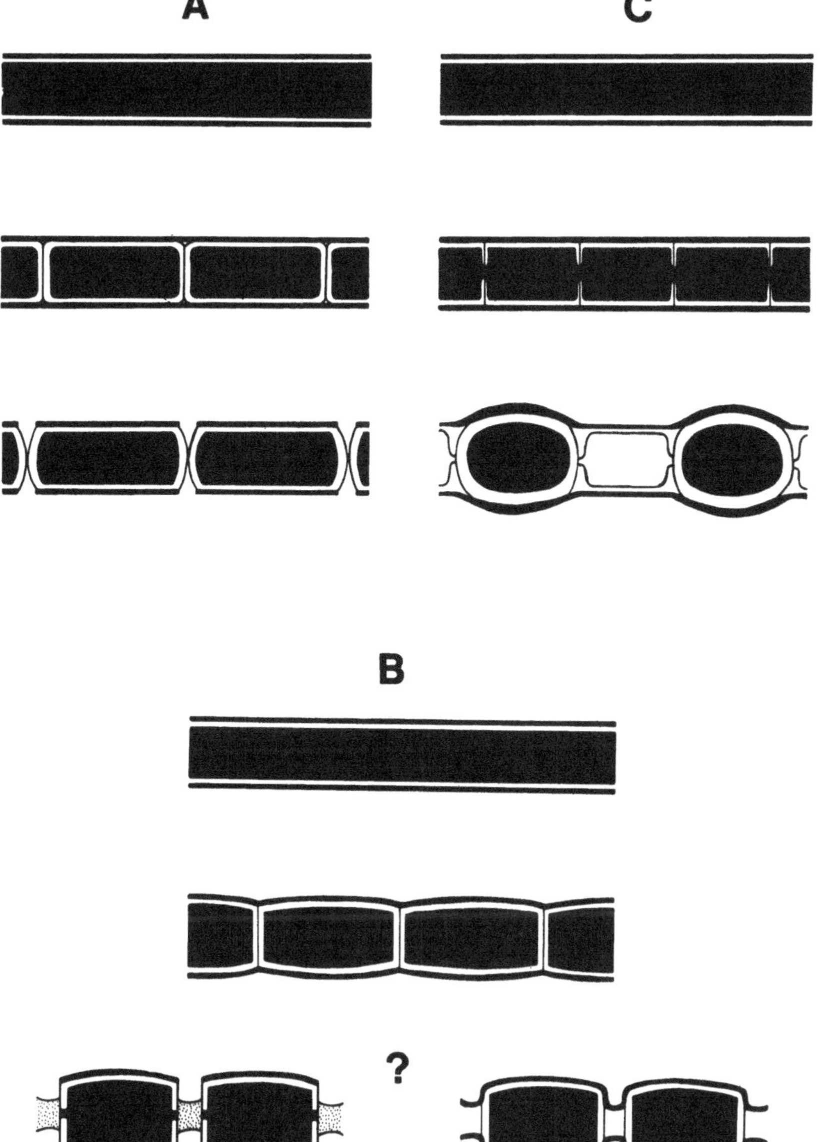


Figure 11.1 Diagrammatic representations of arthroconidium formation. Cell contents and outer wall are in black, inner wall and septa in white. A: arthroconidia type 1, as in *Geotrichum candidum*. B: arthroconidia type 2, as in *Oidiodendron truncatum*. The two lowest diagrams show alternative interpretations of the "connectives": that on the left suggests that the connective may be of mucilaginous material secreted through septal pores; that on the right suggests that the two halves of the double septum between contiguous conidia have separated, and that only collapsed outer wall now links the conidia. C: arthroconidia type 3, as in *Amblyosporium spongiosum*.

these two major ontogenetic groups using the technique of time-lapse photomicrography, we have become convinced that each requires further subdivision.

Now, please don't accuse me of being a hair-splitter. I am only carrying out the time-honoured processes of observation and analysis that must always precede synthesis. At present I can distinguish six kinds of arthroconidium and two kinds of meristem arthroconidium. It may be possible, even desirable, to reduce these numbers by amalgamation. It may, on the other hand, become necessary to recognize further subdivisions as we analyse spore development in more fungi. Let me delineate the various types before returning to a more general discussion.

ARTHROCONIDIA TYPE 1, as formed by *Geotrichum candidum* and the *Geotrichum* state of *Endomyces magnusii* (Figures 11.1 A, 11.2, 18.5, 19.2 C, D) (See Cole and Kendrick 1969)

When an ordinary vegetative hypha breaks under stress, it doesn't break at a septum, but rather at some point between septa. The septum is obviously a strengthening device, among other things. In *Geotrichum*, when a hypha, having ceased apical growth, is converted into arthroconidia, disarticulation occurs at a septum in every case. Clearly these septa are not ordinary cross-walls, and because of their apparent dual nature, I call them "double septa." Their ontogeny may be somewhat different from that of ordinary hyphal cross-walls, and is in dire need of detailed study at the ultrastructural level. At the visible level we have ascertained that in *Geotrichum* the order in which these double septa form is *not* basipetal, as has always been assumed in the literature, but is at first acropetal, and later more or less random. Soon after each double septum has been laid down, its two major components separate, as a circumscissile, centripetal split, originating at the outer wall, develops between them, and each, in its new role as the end wall of a conidium, becomes convex, heralding maturation and secession of the conidia. It is apparent, from the random order of septation, that, once linear growth of the fertile hypha has ceased, no part of the hypha subsequently delimited by double septa has any need of food substances translocated from below. It is quite clear that no significant translocation can take place along these chains of developing arthroconidia.

ARTHROCONIDIA TYPE 2, as formed by *Oidiodendron truncatum* (Figures 11.1 B, 11.3) (See Cole and Kendrick 1969)

The principal difference between these and type 1 lies in the fact that between contiguous conidia of *Oidiodendron* develops something called a "connective." We aren't sure what it is or how it forms. It may indeed be gelatinous as Barron (1962) suggested, but in this, as in other methods of conidium ontogeny, once the various stages have been characterized by time-lapse photomicrography, confirmation or refutation of the theories thus arrived at devolves upon the electron microscopist. Nevertheless, it is clear that in *Oidiodendron* the idea of "simple fragmentation" is an oversimplification, and that considerable ontogenetic subtlety remains to be resolved.

In *Oidiodendron*, though maturation of conidia is a more or less basipetal

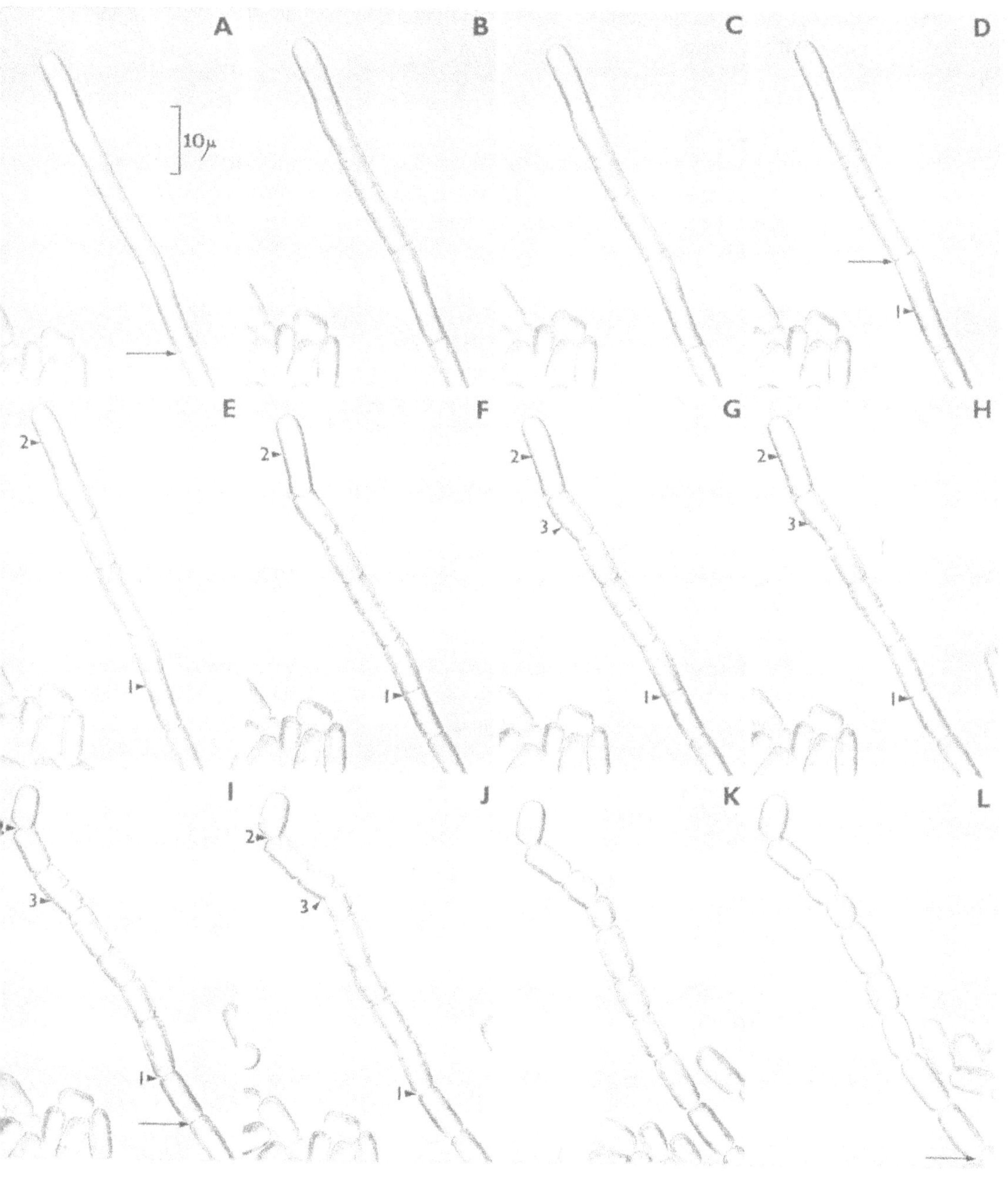


Figure 11.2 *Geotrichum candidum*, time-lapse sequence of arthroconidium formation. Note that the sequence is not regularly basipetal or acropetal. (From Cole and Kendrick 1969)

 process, the initial septation of the fertile hypha does not always follow a regular basipetal sequence.

ARTHROCONIDIA TYPE 3, as in *Amblyosporium spongiosum* (Figure 11.1 C) (See Pirozynski 1969)

Here only every other cell of the conidiogenous hyphae becomes a conidium, an empty cell remaining between adjacent conidia. There is obviously a migration of cytoplasm from the cells which become empty into those which become conidia, and this seems to me to constitute an important difference between this fungus and *Oidiodendron.*

At least partly because of this influx, the mature conidia of *Amblyosporium* are usually much wider than the hyphae which give rise to them. This swelling must involve a good deal of intussusception of new wall material, and is thus a form of "conversion and growth." Such conspicuous increase in size is unusual among arthroconidia and helps to differentiate type 3 from type 5 (q.v.).

ARTHROCONIDIA TYPE 4, "endoarthroconidia" as formed by *Thielaviopsis basicola* (Figure 11.4 A)

When the short, broad fertile hypha has ceased apical growth, and acropetal septation is complete, ultrastructural studies (Tsao and Tsao 1970) reveal that there is a middle lamella between the components of the double septa, which are initially pierced by a septal pore. Soon, however, the inner wall thickens considerably, obliterating the septal pores; each endoarthroconidium can now be seen to be completely enclosed in its own wall, and to be completely cytoplasmically isolated from its neighbours. Although septation is acropetal, subsequent maturation and secession do not appear to be similarly sequential. In the angle created by the intersection of the lateral and end wall of each cell is an annular electron-transparent wedge (shown by dotted lines in Figure 11.4 A). Eventually the original outer wall layer and the middle lamella both break down (in the soil, presumably as a result of attack by microbial enzymes) and the individual arthroconidia are freed. Subsequently, the cell wall splits along the "wedge" during germination. This phenomenon has led some mycologists to dub the dehiscent end wall an "operculum."

Tsao and Bricker (1970) consider that, because the development of what they call the "chlamydospores" of *T. basicola* is different from that shown by *T. paradoxa* (the type species of *Thielaviopsis*), the two fungi are not congeneric. In his forthcoming monographic study of *Chalara* and related genera Nag Raj comes to the conclusion (which I support) that the most important generic characters in *T. basicola* and *T. paradoxa* centre around their phialidic states, and he transfers both to *Chalara*, though maintaining them in different sections of the genus because of their chlamydospores.

ARTHROCONIDIA TYPE 5, "endoarthroconidia" as formed by *Sporendonema purpurascens* (Figure 11.4 B) (See Cole and Kendrick 1969)

The fertile hypha ceases apical growth, and a burst of septation ensues, which is

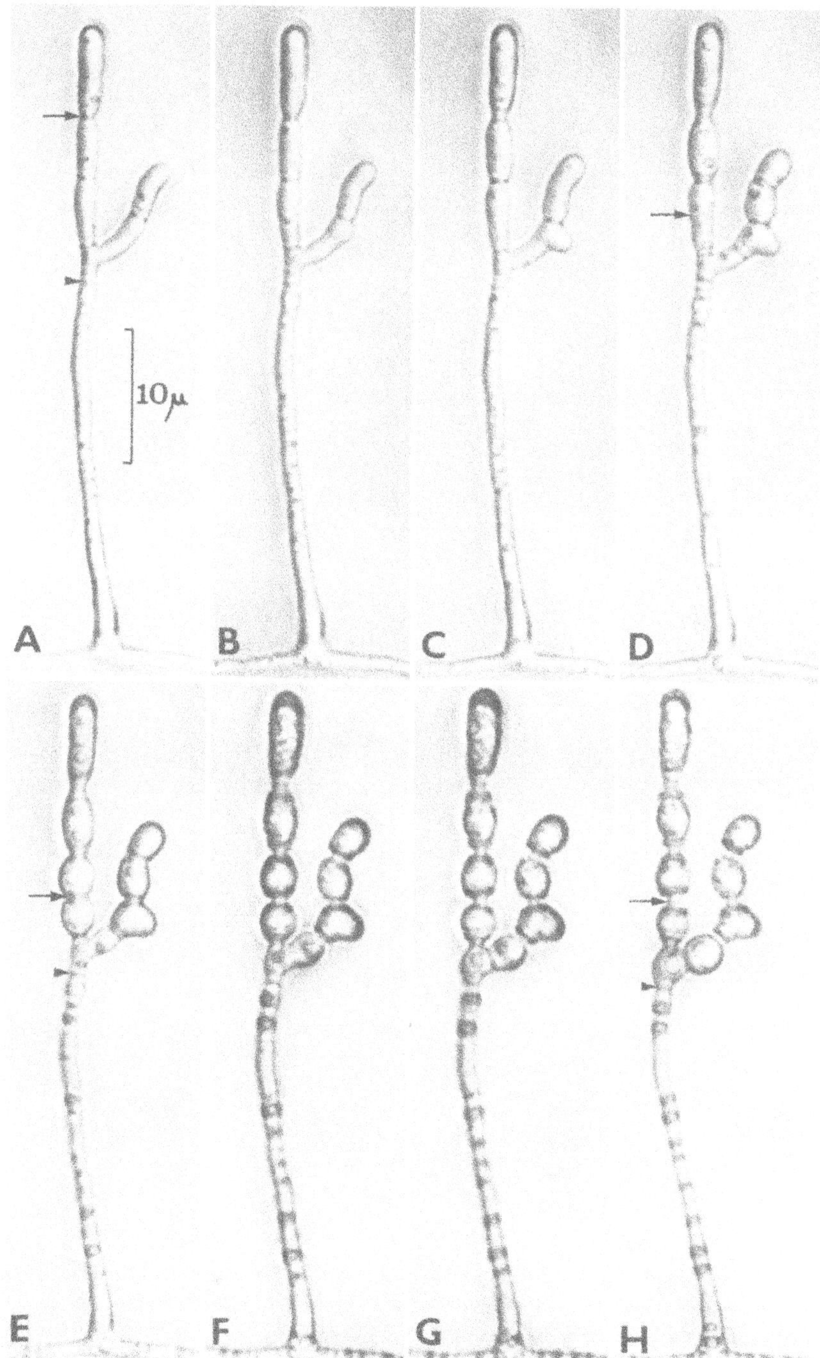


Figure 11.3 *Oidiodendron truncatum*, time-lapse sequence of arthroconidium formation. Note that in this species the sequence is not regularly basipetal. (From Cole and Kendrick 1969)

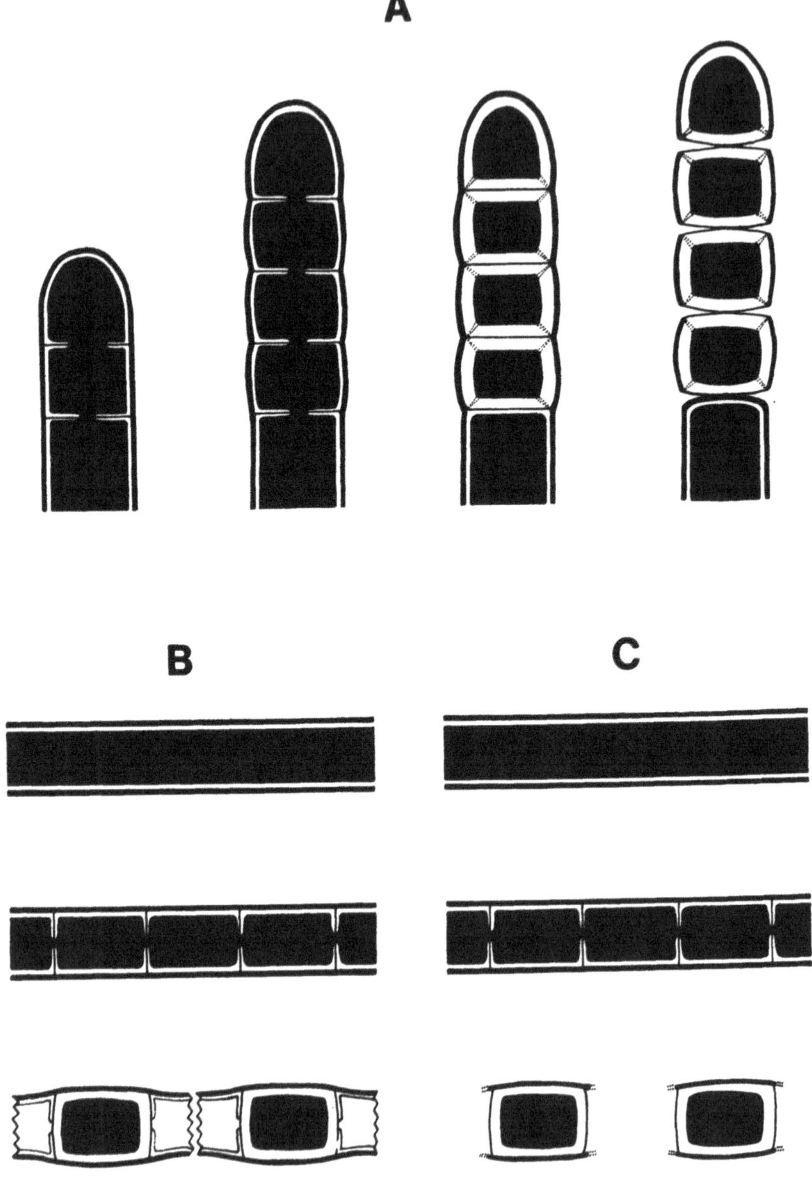


Figure 11.4 Diagrammatic representations of "endoarthroconidium" formation: A, arthroconidia type 4, as in *Thielaviopsis basicola*; B, arthroconidia type 5, as in *Sporendonema purpurascens*; C, arthroconidia type 6, as in *Coremiella ulmariae.*

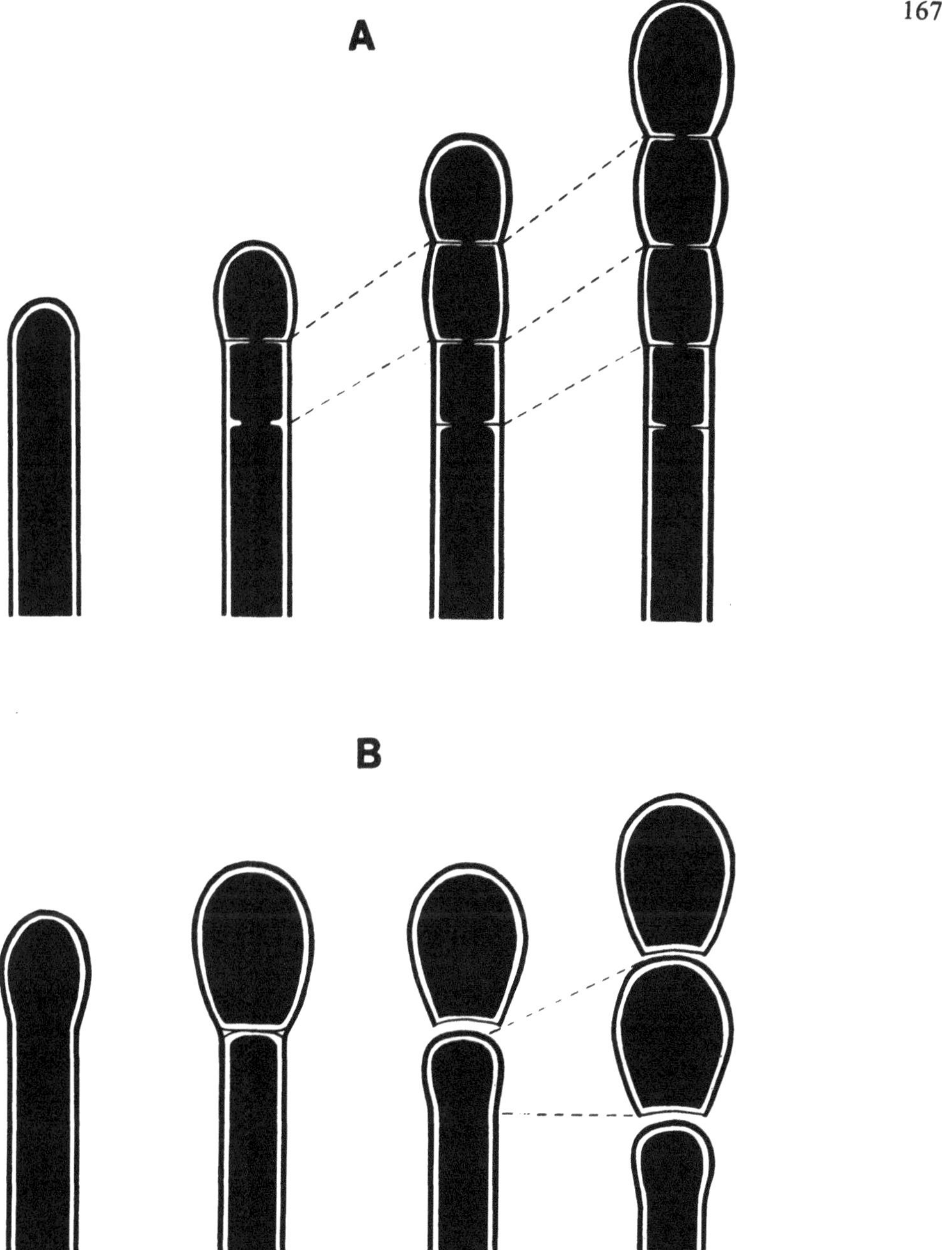


Figure 11.5 Diagrammatic representations of "meristem arthroconidium" formation: A, meristem arthroconidia type 1, as in conidial *Erysiphe polygoni*; B, meristem arthroconidia type 2, as in the *Basipetospora* state of *Monascus ruber*.

at first acropetal, later random. Following the completion of septation there is a retreat of cytoplasm from certain compartments and its concentration and condensation in adjacent compartments. This cytoplasmic migration apparently takes place through the septal pores. The outer wall of the empty compartments remains intact. Next, a complete new inner wall is laid down around the periphery of the full compartments, completing their conversion into endoarthroconidia. These conidia are eventually released by the rupture or breakdown of the outer wall of the empty compartments.

The septa in this fungus may not be double like those involved in the formation of other kinds of arthroconidium, although only ultrastructural studies could decide whether this is, in fact, the case.

Note the similarity, as regards cytoplasmic migration, between this fungus and *Amblyosporium* (type 3).

ARTHROCONIDIA TYPE 6, "endoarthroconidia" as formed by *Coremiella ulmariae* (Figures 11.4 C and 4.3 D)

This type resembles *Sporendonema*, except that the outer wall of the intervening empty cells autolyses. The mechanism involved is not understood.

Dr Tubaki in 1958 recognized some of the complexity of arthroconidial ontogeny when he subdivided Dr Hughes's section VII into section VIIA for those producing conidia "exogenously" and section VIIB for those producing conidia "endogenously." I have here found it necessary to further subdivide both of Dr Tubaki's subsections, VIIA into types 1, 2, and 3, and VIIB into types 4, 5, and 6. I hope that this is all the subdivision that will be necessary, though I am not overly optimistic since we have so far made time-lapse studies of the ontogeny of only a few arthroconidium-forming fungi.

This brings me to a consideration of meristem arthroconidia.

MERISTEM ARTHROCONIDIA TYPE 1, as formed by conidial *Hysterium insidens, Phragmotrichum chailletii,* and conidial *Erysiphe polygoni* (Figure 11.5 A)

Dr Hughes in 1953 introduced the term meristem arthrospore to describe the conidia of *Hysterium insidens, Erysiphe polygoni,* etc., and erected section V of his new classification to accommodate them. In these fungi there is presumably a basipetally maturing chain of cytoplasmically connected conidia. We could say that there appears to be a somewhat diffuse or extended meristematic zone which can sometimes make it difficult to tell just where the conidiophore ends and the chain of conidia begins. Although we have so far been unable to obtain time-lapse sequences of development in any of these fungi, it appears that extension growth in the apical region of the conidiophore probably keeps pace with its conversion into conidia, so that the conidiophore does not necessarily become shorter as conidium formation proceeds. Double septa are presumably laid down between conidium initials, and the septal pores must not be sealed off until just before secession from the apex of the chain. The conidia might con-

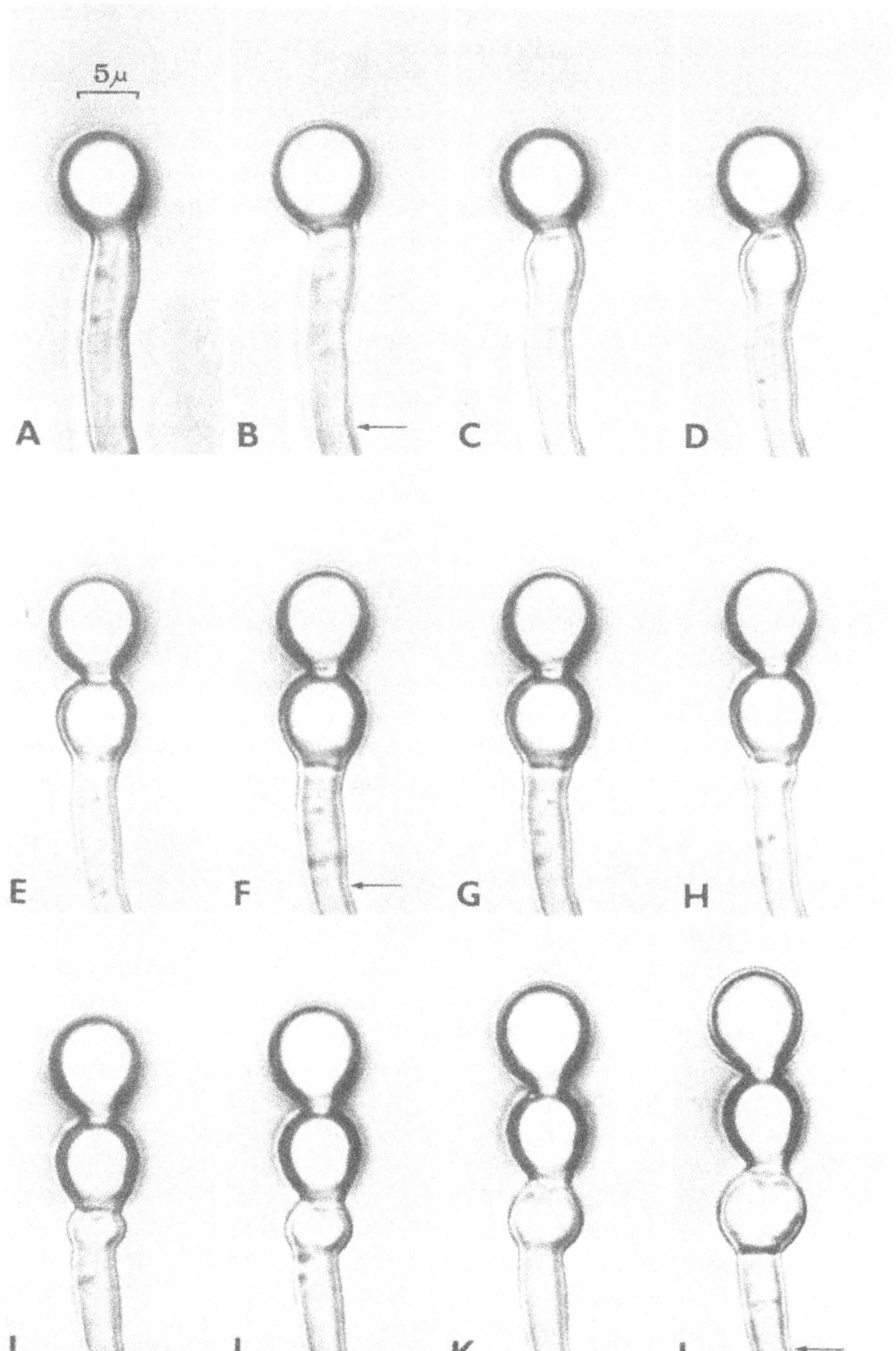


Figure 11.6 *Basipetospora* state of *Monascus ruber*, time-lapse sequence of conidium formation (From Cole and Kendrick 1968).

 ceivably be called simultaneous or concurrent meristem arthroconidia because the development of several proceeds at once.

MERISTEM ARTHROCONIDIA TYPE 2, as formed by *Trichothecium roseum, Cladobotryum variospermum*, and the *Basipetospora* state of *Monascus ruber* (Figures 11.5 B and 11.6)

Dr Cole and I have examined conidium formation of these three fungi by time-lapse photomicrography (Cole and Kendrick 1968; Kendrick and Cole 1969; Cole and Kendrick 1971). Each of them produces basipetal chains of conidia by a combination of growth and conversion. This would seem to qualify them for Hughes's section V. But there are three significant differences between these three fungi and those discussed by Hughes. First, these species have a meristematic zone that is restricted to the single conidium being formed at any given time. Second, there is, we believe, no cytoplasmic continuity between the conidia of a chain. Third, each conidium incorporates a segment of the fertile hypha and, since there is no meristematic activity other than that involved in the formation of each spore, the fertile hypha becomes progressively shorter. These differences, first observed in conidial *Monascus ruber* (Figure 11.6), persuaded us to subdivide Dr Hughes's section V into section VA (meristem arthroconidia type 1) for those fungi in which cytoplasmic continuity is maintained along a chain of developing conidia, and section VB (meristem arthroconidia type 2) for those fungi in which the cytoplasmic connection between fertile hypha and conidium is presumably severed before the initiation of the next conidium.

The conidia might conceivably be called sequential (or solitary) meristem arthroconidia because only one is being formed at any given time.

Now, having established and differentiated such groups of arthroconidia and meristem arthroconidia as we know to exist so far, it is appropriate to re-examine their various characteristics and see if it is possible to arrive at an over-all definition of these phenomena.

Conidia of the kinds I am considering here may be formed by conversion in the absence of a meristem (arthroconidia), or by the activities of a definite meristem which may involve some existing hyphal material (meristem arthroconidia). Conidia may be formed in a basipetal sequence (all meristem arthroconidia and some arthroconidia), or irregularly. Conidia may be almost entirely clad in original hyphal wall, or in original wall plus newly interpolated wall, or in an entirely new, secondary wall.

Formation of conidia may involve a large increase in wall area (generally more than five times; all meristem arthroconidia), a much smaller increase in wall area (less than twice; arthroconidia types 1, 2, and 4), or a marked decrease in final wall area (arthroconidia types 3, 5, and 6 because of their reduced number of functional cells). Cytoplasmic continuity may be maintained between members of a chain of developing conidia (meristem arthroconidia type 1) or it may not (meristem arthroconidia type 2 and most arthroconidia). The single or double nature of some of the septa involved is still unresolved and cannot at present be used for purposes of comparison.

What seemed at first sight a relatively simple group has exploded, at least to my eyes, into a fascinating diversity that makes over-all definition difficult if not impossible. I believe that meristem arthroconidia must be unequivocally separated from arthroconidia. In addition, it is becoming increasingly clear that the two kinds of meristem arthroconidium are themselves very different. The mechanism which Dr Cole and I have discovered in *Basipetospora, Trichothecium*, and *Cladobotryum* in some ways resembles that of Subramanian's "gangliospore," an apically produced blastoconidium whose formation involves a large portion of the apex of the hypha. The relationship of this new kind of "meristem arthroconidium" with the blastoconidium may be closer than with any arthroconidium. I must now ask the Conference as a whole to help resolve this situation. Perhaps the answers will be accessible only after many more time-lapse and ultrastructural studies have been made.



ACKNOWLEDGMENTS

I would like to thank the National Research Council of Canada for research support, and Dr Garry Cole for his skill and patience with the time-lapse photomicrography of some of the fungi discussed in this paper.

REFERENCES

Barron, G.L. 1962. New species and new records of *Oidiodendron*. Can. J. Botany 40: 589-607

Cole, G.T., and Kendrick, W.B. 1968. Conidium ontogeny in hyphomycetes. The imperfect state of *Monascus ruber* and its meristem arthrospores. Can. J. Botany 46: 987-92

- 1969. Conidium ontogeny in hyphomycetes. The arthrospores of *Oidiodendron* and *Geotrichum*, and the endoarthrospores of *Sporendonema*. Can. J. Botany 47: 1773-80

- 1971. Conidium ontogeny in hyphomycetes. Development and morphology of *Cladobotryum*. Can. J. Botany 49: 595-9

Hughes, S.J. 1953. Conidiophores, conidia, and classification. Can. J. Botany 31: 577-659

Kendrick, W.B. and Cole, G.T. 1969. Conidium ontogeny in hyphomycetes. *Trichothecium roseum* and its meristem arthrospores. Can. J. Botany 47: 345-50

Pirozynski, K.A. 1969. Reassessment of the genus *Amblyosporium*. Can. J. Botany 47: 325-34

Tsao, P.H., and Bricker, J.L. 1970. Acropetal development in the "chain" formation of chlamydospores of *Thielaviopsis basicola*. Mycologia 62: 960-6

Tsao, P.W., and Tsao, P.H. 1970. Electron microscopic observations on the spore wall and "operculum" formation in chlamydospores of *Thielaviopsis basicola*. Phytopathology 60: 613-16

Tubaki, K. 1958. Studies on Japanese hyphomycetes. V. Leaf and stem group with a discussion of the classification of hyphomycetes and their perfect stages. J. Hattori Bot. Lab. 20: 142-244

Vuillemin, P. 1910. Matériaux pour une classification rationelle des Fungi Imperfecti. C.R. Acad. Sci., Paris 150: 882-4

EDITOR

*The discussion of arthroconidia was prefaced by a viewing of sequences from the Cole-Kendrick time-lapse film "Conidium Ontogeny in Hyphomycetes" analysing conidiogenesis in* Geotrichum candidum, *the* Basipetospora *state of* Monascus ruber, *and* Cladobotryum variospermum.

DR CARROLL

I'd like to know what evidence you have that the cytoplasm in the emptying cells of *Sporendonema* is actually migrating. It seems more likely that it is degenerating or lysing.

DR KENDRICK

The cells which are becoming conidia tend to swell a little and their contents become denser and more granular. In addition, one of my graduate students, Miss Chang, has also seen nuclei migrating through the septal pores into cells which are differentiating as conidia.

DR CARMICHAEL

Under phase contrast, we can see that the central vacuole in the intervening cells enlarges until there is only a thin film of cytoplasm left on the walls. I'd suspect that at least one nucleus would also remain. But I don't think there's any doubt that there is migration and concentration of cytoplasm going on.

DR KENDRICK

The question is, how closely related would this kind be to the kind where the wall of the empty cell lyses? Probably very closely. How closely related would either of these be to *Amblyosporium*, where there is also an empty cell between neighbouring conidia? Probably rather closely.

DR CARROLL

The fact that the cells swell might simply indicate imbibition of water from the surroundings.

DR PIROZYNSKI

In *Amblyosporium spongiosum* the enlargement of conidia is very pronounced, and takes place after the head of developing spores has been lifted on a conidiophore up to 1 cm in length above the damp substrate. The conidial chains break up into a powdery mass. The fungus seems to be doing its very best to avoid moisture.

DR CARMICHAEL

As you look at more of these fungi, you'll likely decide that the distinction between endoarthroconidia and arthroconidia is a matter of degree rather than of kind; and also that the process of cell evacuation, and the size and occurrence of empty cells, are very irregular.

DR KENDRICK

But there are a lot of fungi in which it doesn't happen at all, so there is a definite dichotomy. However, it's quite possible that only two things may distinguish endoarthroconidia: (i) the laying down of a rather thick new wall all the way round the conidium; (ii) the fact that both layers of the original hyphal outer wall may hold conidia together - in ordinary arthroconidia only the very thin outer layer of the hyphal wall is left. 

DR S.J. HUGHES

There seems to be very little difference between intercalary chlamydospores and some of the arthrospores we've been discussing. It would perhaps be reasonable to call such intercalary chlamydospores arthroconidia, since they represent conversion of existing hyphal elements.

DR KENDRICK

If we delete the ideas of *basipetal* conversion and *total* conversion of a hypha, we could include these intercalary chlamydospores with the arthroconidia.

MRS POLLACK

Is anyone else uneasy about *Basipetospora* (conidial *Monascus*) being arthrosporic?

DR KENDRICK

Yes! Uneasy about *any* conidium of section VB really being an arthroconidium. I put them together with the arthroconidia perhaps because I was brain-washed by the name, and because there *is* some conversion. The most important feature in *Basipetospora, Trichothecium,* and *Cladobotryum* is growth - the reactivation-replasticization of existing wall and intussusception of new wall material; remember the figures I quoted for increase in wall area: six times in *Basipetospora*, ten times or more in *Cladobotryum*. The growth phenomenon is much more obvious than the conversion. Perhaps in our excitement at discovering the conversion phenomenon in these fungi, we overestimated its importance.

DR GOOS

Would you now call these blastic?

DR KENDRICK

Yes, *now* I would! They are actively "blowing out" on a wide front - a cylinder of the cell wall becomes involved. No basal septum is laid down until each conidium has more or less attained its final shape and size.

DR ELLIS

Certainly the fact that they are formed basipetally doesn't make them arthroconidia.

DR KENDRICK

What about the arthroconidia of *Amblyosporium*, in which there is also a considerable increase in the size of the cell?

DR ELLIS

But that is intercalary, in a previously delimited hypha.

DR KENDRICK

Whereas each conidium of *Basipetospora* is, in effect, apical. As soon as the organism has produced one, it forgets about it and gets on with the next.

DR ELLIS

In the meristem arthroconidia of section VA, the chain remains more or less cylindrical except for the apical conidium. The swelling occurs at a late stage. In section VB, the swelling occurs very early.

DR KENDRICK

You don't think VB should be kept in section V at all?

DR ELLIS

Why not take them out of section V and make a new, separate section for them called "meristem blastoconidia"?

DR SUBRAMANIAN

We are now rethinking all those terms. In section VB, there is a blastic type of development which we think is one of the basic types - whether it involves just a tiny spot at the tip of a cell, or a larger area, or the whole width of the apex of the cell. (See Figure 15.1)

DR CARMICHAEL

We are meeting the same problems as with the *Monilia* state of *Sclerotinia fructigena*: these are fission spores, and whether you call them blastoconidia or arthroconidia depends on how much they swell in relation to the size of the double septum where they split apart.

DR SUBRAMANIAN

If we stretch this idea, we may reach a point at which there are only two types: the phialides, and all the rest, which (since there is only one way of being liberated) are all "arthroconidia."

DR CARMICHAEL

Certainly you could think of these "arthroconidia" as being "blastoconidia" with a delayed dehiscence. This would enable you to distinguish between section VA, the kind that matures gradually with persistent translocation and delayed dehiscence (these would be arthroconidia), and section VB, where such a delay is absent (these would be blastoconidia). This would make the conidia of *Monilia* arthroconidia.

DR KENDRICK

But section VB does have a "delayed action" process. The hypha elongates first, then stops growing; a second growth phase precedes the much-delayed septation.

DR G.C. HUGHES

This is a recurring problem. We have blastoconidia and arthroconidia. "Blastoconidium" makes me think of how the spore forms, whereas "arthroconidium" makes me think about how it secedes. There is no method of actual formation - growth - that is "arthric." [*Laughter*] One of the basic things we have to decide is whether conidium formation or conidium separation is of primary importance.

DR PIROZYNSKI

Perhaps we should bring the conidiogenous cell into the discussion because, in the blastic type, the conidium arises as a new entity from a conidiogenous cell which remains unchanged (but may extend into an acropetal chain of conidia), or increases in length by sympodial or percurrent growth, whereas in the arthric type a preformed conidiogenous cell is transformed into conidia. We can distinguish between them, and for practical reasons they might as well be used as characters.

DR KENDRICK

Dr Cole and I have already found one or two previously unknown variations on the themes so well delineated by Dr Hughes. I don't think Dr Hughes felt, in 1953 when he wrote his paper, that he'd seen everything; and I don't think we have, even now. But while some of the differences I have pointed out between the kinds of arthroconidium may be minimal, and I was really pointing them out in an analytical spirit rather than a schismatic one, I still think we may find further valid variations on this theme.

EDITOR

*Further discussion of the various kinds of arthroconidia, and the conclusions we reached concerning terminology for them, may be found in Chapters 15 and 16.*

*We had now, after two days, had 18 hours of "on the record" group sessions, and many more hours of "off the record" dialogues. The next day, Sunday, was supposed to be our mid-conference day of rest and recreation. The sightseeing bus generously provided by the University of Calgary through Dr Parkinson rolled up to the lodge and soon we were headed for the magnificent Banff-Jasper Highway. Most of us had been co-opted onto one or other of the committees established to consider issues raised by keynote addresses and subsequent discussions, and to present reports and recommendations to the conference as a whole during a final "Terminology" session. As we rolled on through some of the finest mountain scenery in the world, the bus passengers, having grouped themselves in these committees, engaged in animated conversation about the Hyphomycetes. Stops at Banff Springs Hotel, Lake Louise, and Peyto Lake provided only temporary distractions. A trip by snowmobile on the Columbia Icefields was a highlight for our visitors; and on our way back to Kananaskis we were again temporarily diverted from the Fungi Imperfecti by the sight of enormous numbers of agarics -* Coprinus comatus *- growing along the verge of the highway. We stopped and collected a quantity of these for subsequent consumption and, having also seen other game in the form of bear, moose, and wapiti, we eventually arrived back at Kananaskis well satisfied with our day.*

*Committee meetings, lubricated by countless cups of coffee, continued well into the night. The* Coprinus comatus *accompanied steak for Monday dinner, our conference banquet. It is a pleasure to record that those who drank red wine with their* Coprinus *apparently suffered no ill-effects from the combination; unfortunately this experiment lacked controls!*

# 12
# The Basauxic Conidiophore

K. TUBAKI

The term basauxic conidiophore was introduced by Dr S.J. Hughes in 1953 for fungi which were characterized by basal, as opposed to apical, elongation of their conidiophores. These fungi comprised his section VIII. The basauxic conidiophore arises from a barrel-shaped, flask-shaped, subspherical, ampulliform or broadly clavate "conidiophore mother cell" which itself develops from a cell of a superficial or erumpent mycelial mat. Conidia are either borne solitarily at the apex, or apically and irregularly on short stalks along the increasing length of the conidiophore, or apically and in whorls between thickened septa of the conidiophore. The conidia arise as blown-out ends, and the oldest conidia are usually found towards the apex and the youngest towards the elongating base of the conidiophore. The conidiophore does not grow at its apex.

Conidiophores having a basal meristem are found in *Arthrinium, Endocalyx, Dictyoarthrinium, Cordella, Spegazzinia, Isthmospora*, and *Graphiola*. For taxonomic treatments of these genera the reader is referred to Hughes (1952, 1953a, 1953b), Damon (1953), Ellis (1965), Fischer (1883, 1922), Rao and Rao (1964) and Sydow and Butler (1907).

Figure 12.1 shows "mother cells" of *Arthrinium phaeospermum*, *A. japonicum*, and *Isthmospora trichophila* (= *Spegazzinia trichophila*). Those shown in A, B, and C were stained with acid-fuchsin. It is fairly difficult to find these mother cells in culture, but easier when the fungi are grown on sterilized leaves.

As far as I know, the only members of this group to have been connected to perfect fungi are *Arthrinium* states of *Apiospora* and *Pseudoguignardia* species.

I believe that the behaviour of the conidiophore is of importance as a diagnostic criterion in the higher taxa of the Hyphomycetes because it involves the behaviour of the growing points, which in turn implicates the active, multiplying

Institute for Fermentation, Osaka, Japan.

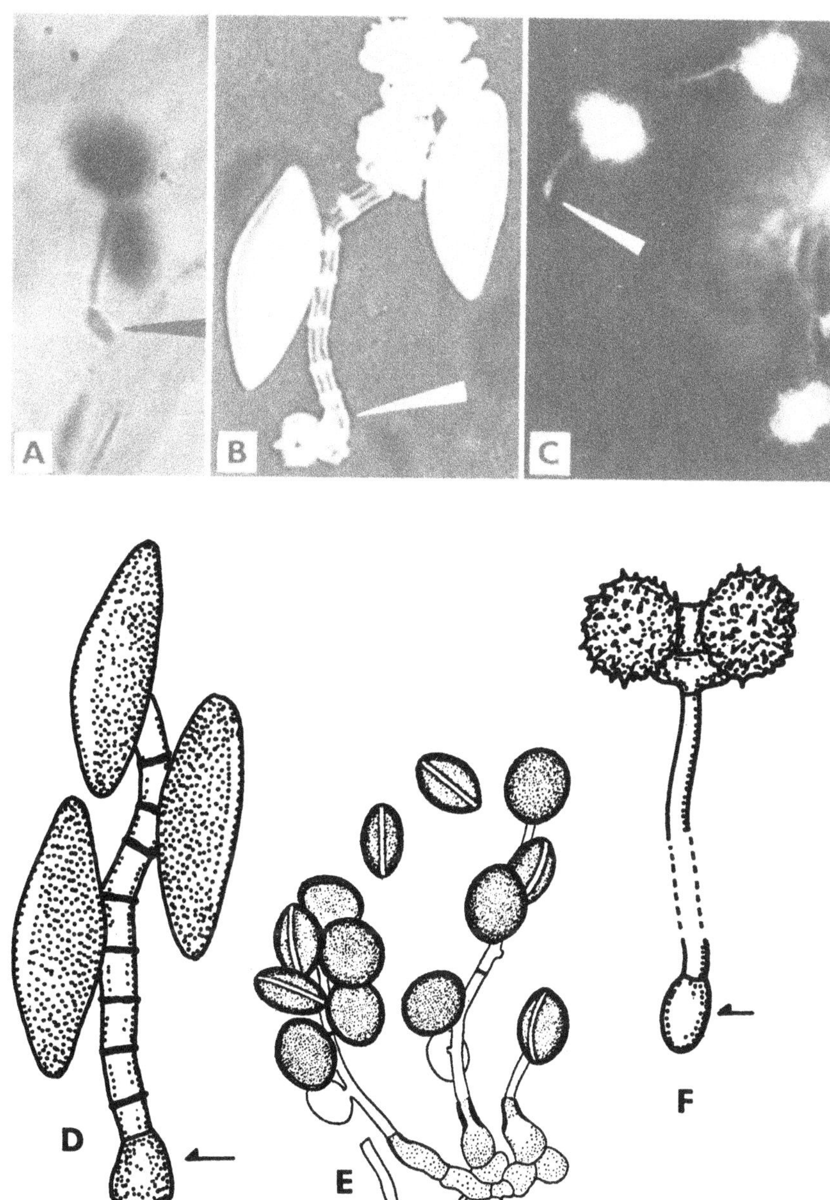


Figure 12.1 Examples of basauxic conidiophores: A, E, *Arthrinium phaeospermum*; B, D, *Arthrinium japonicum*; C, F, *Isthmospora trichophila*. Arrows indicate "mother cells." A, E, ×1,000; B, C, ×400.

 nucleus or nuclei in the "mother cell." However, it should be remembered that the conidia developing from basauxic conidiophores are not all of the same kind. They are sometimes dehiscent blastoconidia, sometimes indehiscent "chlamydospores," and so on. Therefore, although the characteristics of the basauxic conidiophores are indeed unique, conidia developing from such conidiophores should still be referred to their own ontogenetic groups.

If, as it would appear, our approach to hyphomycete taxonomy is principally concerned with the ontogeny of the conidia themselves, then the problem of the basauxic conidiophore might be considered a separate and, for the time being, peripheral problem. Actually, the basauxic phenomenon has usually been inferred rather than observed. More data, derived perhaps from time-lapse photomicrographic studies of the kind carried out by Kendrick and Cole, are needed.

I thank Dr M.B. Ellis and the Commonwealth Mycological Institute for permission to reproduce Figure 12.1 E.

## REFERENCES

Barron, G.L. 1968. The genera of hyphomycetes from soil. Williams and Wilkins, Baltimore, Md.

Brierley, W.B. 1918. The microconidia of *Botrytis cinerea*. Kew Bull. Misc. Inf. 4: 129-46

Damon, S.C. 1953. Notes on the hyphomycetous genera *Spegazzinia* Sacc. and *Isthmospora* Stevens. Bull. Torrey Bot. Club 80: 155-65

Ellis, M.B. 1963. Dematiaceous hyphomycetes. V. C.M.I. Mycol. Pap. 93

- 1965. Dematiaceous hyphomycetes. VI. C.M.I. Mycol. Pap. 103

Fischer, E. 1883. Beitrag zur Kenntnis der Gattung *Graphiola*. Bot. Zeit. 41: 745-801

- 1922. Weitere Beiträge zur Kenntnis der Gattung *Graphiola*. Ann. Mycol. Notitiam Sci. Mycol. Univ. 20: 228-37

Goos, R.D. 1969. Conidium ontogeny in *Cacumisporium*. Mycologia 61: 52-6

Hughes, S.J. 1952. Fungi from the Gold Coast. I. C.M.I. Mycol. Pap. 48

- 1953a. Fungi from the Gold Coast. II. C.M.I. Mycol. Pap. 50

- 1953b. Conidiophores, conidia, and classification. Can. J. Botany 31: 577-659

- 1969. *Strigopodia*. Can. J. Botany 46: 1099-1107

Hughes, S.J., and Kendrick, W.B. 1968. New Zealand fungi. 12. *Menispora, Codinaea, Menisporopsis*. N.Z. J. Botany 6: 322-75

Mason, E.W. 1933. Annotated account of fungi received at the Imperial Mycological Institute. List II (Fascicle 2). C.M.I. Mycol. Pap. 3

Rao, P.R., and Rao, D. 1964. Some allied Dematiaceae-Dictyosporae from India. Mycopath. Mycol. Appl. 23: 23-8

Subramanian, C.V. 1965. Spore types in the classification of the Hyphomycetes. Mycopath. Mycol. Appl. 26: 373-84

Sydow, H.P., and Butler, E.J. 1907. Fungi Indiae orientalis, pars. II. Ann. Mycol. Notitiam Sci. Mycol. Univ. 5: 489-90

## DISCUSSION

### DR PIROZYNSKI

An obvious omission from the table I presented a few days ago (Table 3.1) is

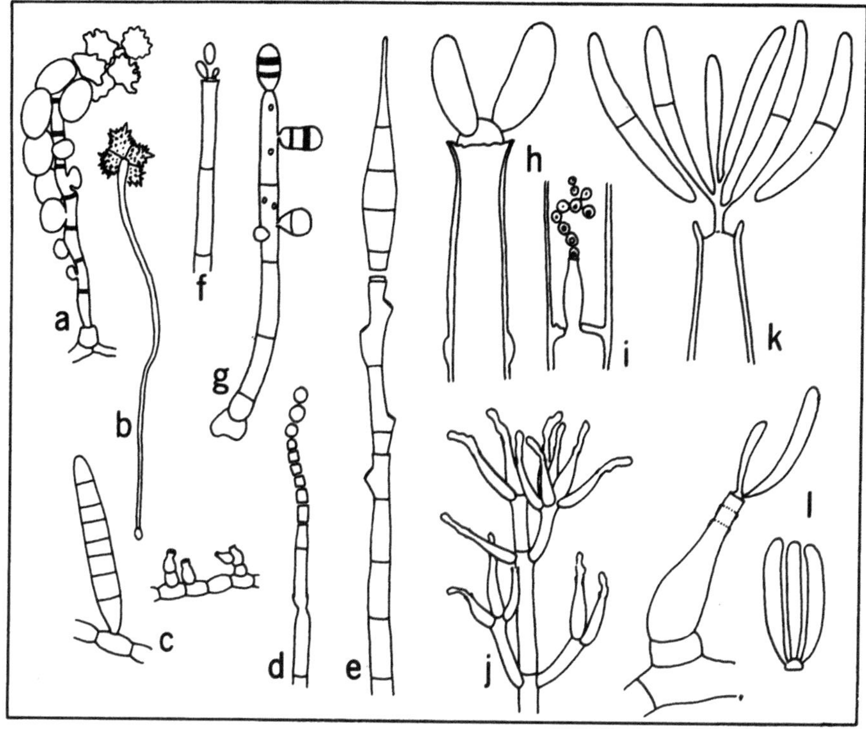


Figure 12.2 a, *Arthrinium lobatum*; b, *Spegazzinia tessarthra*; c, *Strigopodia resinae*; d, *Wallemia ichthyophaga*; e, *Eriomycopsis* sp.; f, *Bisporomyces* (*Chloridium*) *chlamydosporis*; g, *Spadicoides obovata*; h, *Cacumisporium capitulatum* (redrawn from Goos 1969); i, *Botrytis cinerea* (redrawn from Brierley, in Mason, 1933); j, *Dactylium dendroides*; k, *Codinaea setosa*; l, *Gyrothrix* sp.

Hughes's section VIII, grouping species with basauxic conidiophores. At this point I would like to return to "conidiophores" and discuss these species, as well as some other fungi which produce more than one type of conidiogenous cells in one colony or on one conidiophore. In the fungi now classified in the basauxic section, e.g. *Arthrinium lobatum* (Figure 12.2 a) and *Spegazzinia tessartha* (Figure 12.2 b), the spore-bearing hyphae or filaments arise from minute conidiophore mother cells, and can, according to Subramanian (1965), "theoretically bear some of the different spore types based on ontogeny." The examples known, as he pointed out, are blastic: blastoconidia in *Arthrinium* and aleuriospores in *Spegazzinia.*

Many fungi produce more than one type of conidiogenous cell in the course of development or life history, or under certain abnormal conditions (e.g., in culture): aleuriospores and phialoconidia, blastoconidia and arthroconidia, etc. Excellent examples of such pleomorphism are to be found in sooty moulds such as *Strigopodia resinae* (Hughes 1969) (Figure 12.2 c). Single conidiophores bearing different types of conidiogenous cells have also been recorded. Brierley

 (1918) found phialides within supporting hyphae of *Botrytis cinerea* (Figure 12.2 i). The conidiophore of *Spadicoides obovata* (Figure 12.2 g) produces an apical blastoconidium and lateral poroconidia (Hughes 1953b; Ellis 1963). In some *Eriomycopsis* spp. (Figure 12.2 e) conidiogenous cells of a single conidiophore appear to proliferate both sympodially and percurrently. These seemingly confusing examples can, however, be more easily accounted for if one considers the conidiophore of *Spadicoides* or *Eriomycopsis* to be a hypha comparable to that of *Strigopodia*, but which assumed an ascending habit, and partly or wholly eliminated sterile intercalary cells.

More difficult to accommodate are conidiophores in which the conidiogenous cell itself produces conidia by different mechanisms. One of the species showing, according to Barron (1968), "anomalous spore formation" is *Wallemia ichthyophaga* (Figure 12.2 d) in which "the phialide acts as a meristem from which the sporogenous hypha continues to grow while the arthrospores develop above." In *Bisporomyces* (*Chloridium*) *chlamydosporis* (Figure 12.2 f), several conidia may develop almost simultaneously, apparently in a sympodial manner, from a single phialide. Judging by a photomicrograph in Hughes and Kendrick (1968), a very similar case is presented by *Codinaea setosa* (Figure 12.2 k), and possibly also by *Dactylium dendroides* (Figure 12.2 j). In *Cacumisporium capitulatum* (Figure 12.2 h), Goos (1969) demonstrated not only that conidia arise sympodially from the open, phialide-like tip of a conidiogenous cell, but that the conidiogenous cell itself can also proliferate percurrently to produce a new crop of conidia at a higher level. I think the same type of conidiogenous cell is formed by members of *Circinotrichum* and *Gyrothrix* (Figure 12.2 l), but the minute size of the relevant structures makes accurate assessment difficult. These cells proliferate percurrently leaving behind a series of annellations, but the apex, which is easily dislodged, bears conidia sympodially.

There is a definite analogy, and probably homology, between these fungi and the founder members of section VIII. The difference seems to lie only in the degree of development of the "conidiogenous mother cell" in relation to the "conidium-bearing filament." Thus the concept of the basauxic conidiophore could be extended to include representatives of spore types other than blastoconidia: sympodioconidia in *Chloridium, Codinaea, Cacumisporium, Circinotrichum*, and *Gyrothrix*, and arthroconidia in *Wallemia*.

*Codinaea setosa* is, however, perfectly congeneric with other species of *Codinaea*, and unless it is found to have a perfect state taxonomically far removed from *Chaetosphaeria*, it would be very difficult to segregate it into a different genus.

DR ELLIS

Recently I have worked on *Spegazzinia lobulata* (Figure 12.3). My preparations show that the conidium has a very thick wall and, at the base, a marked frill. The tip of the conidiophore mother cell also exhibits a marked frill, exactly matching that on the conidium. The outer thick wall of the conidium corresponds to the outer wall of the mother cell. The thin-walled column which raises the conidium above the mother cell arises inside the mother cell's frill, and can be seen inside

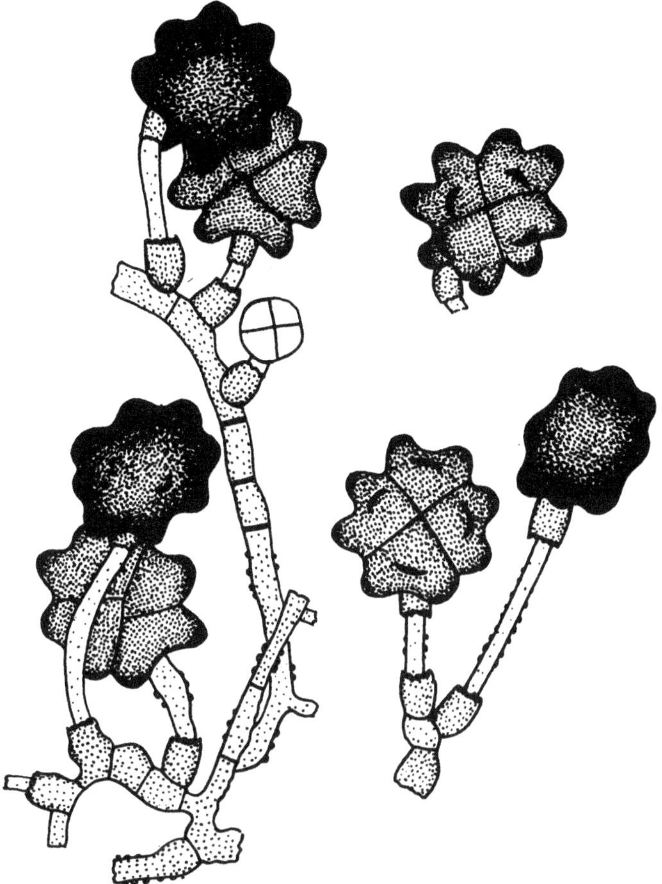

Figure 12.3 *Spegazzinia lobulata.*

the frill at the base of the conidium. It undergoes very rapid extension growth, and I have seen all stages - from the conidium still attached to the mother cell to the conidium considerably elevated. We don't know whether the wall of the columnar extension is really the inner wall of the mother cell, or whether it is a new wall not corresponding in any way to the wall of the mother cell. We may be looking at an extension of the phialide concept, in which only the first conidium actually differentiates.

*Spegazzinia lobulata* gives a very clear look at a phenomenon that is also probably exhibited by *Cacumisporium*. The column in *Spegazzinia* corresponds to the thin-walled protrusion in *Cacumisporium*. Mr Mason, Dr Hughes, and I have long considered this a very curious phenomenon. In all species of *Arthrinium* the apical conidium corresponds to the "aleuriospore." All other conidia in this fungus, produced laterally, derive their walls from that of the column - their wall does not correspond to that of the first conidium.

DR KENDRICK

An extension of the principle that the first conidium often differs considerably from the rest. I have only one question about your comparison of *Spegazzinia* and *Cacumisporium*, and it concerns the location of the meristematic region. Surely in *Spegazzinia* it's at the bottom of the column, but perhaps in *Cacumisporium* it is at the top.

DR ELLIS

That is so. The similarity was only in the nature of the column and its wall. The location of the meristem must be a separate consideration.

DR KENDRICK

Your discussion this morning reinforces the point that our emphasis may be shifting from the location of the meristem to the nature of the walls.

DR ELLIS

I think Dr Pirozynski is right: there is some correspondence between the column in *Spegazzinia* and the extension in *Cacumisporium*; and, as Dr Kendrick has pointed out, the main difference lies in the location of the meristem. Dr Tubaki published a good photograph of *Chloridium viride*, the type species of *Chloridium*, in which the protrusion of the column above the thick outer wall is clearly shown.

DR KENDRICK

In the phialide committee discussions we were trying to define the collarette as being the distance from the point at which the outer wall ruptures to the meristem. If the meristem is above the collarette, the distance is negative, and this way of defining the collarette wouldn't work. We have seen in *Chalara* that the meristem can be very deeply situated, in *Phialophora* that the meristem can be near the open end of the cell, and in *Penicillium* that it can be right at the tip. I don't see why it shouldn't go a bit farther and stick out.

DR MÜLLER

As an outsider (at least as far as hyphomycetes are concerned), I am always looking for generalizations. Is it not true that a change in the kind of conidium ontogeny is due to a change in the position of the meristematic zone? This change in position may not be restricted to the conidiogenous cell, but may involve the conidium. We know that some ascospores germinate to produce conidia; is it not possible that the first conidium may not fulfil its dispersal function, but germinate at once to produce another conidium? Perhaps the extension seen in one species of *Codinaea* is the first conidium.

MRS POLLACK

I have tried to work out a logical sequence, based on the location of the meristem after the production of the first spore: (a) meristem remains at the conidiophore apex; (b) meristem moves into the first conidium, which then proliferates to produce another conidium; (c) meristem remains at base of conidiophore (basauxic); (d) meristem splits up - conidiophore and conidium can both proliferate. And there are other possible groups, like arthroconidia.

What Dr Müller said about the first conidium in *Codinaea* is very interesting because exactly the reverse happens in *Arthrinium*. The first conidium in this species is non-functional as a unit of propagation. Because of this, the meristem must seek another exit; in this case, basal proliferation. 

DR KENDRICK

Why is the first conidium non-functional?

MRS POLLACK

In *Arthrinium japonicum*, which I have studied (though not developmentally, because I couldn't get it to grow), the first conidia form a canopy, clearly either for protection or for moisture retention. There appears to be no cytoplasm in them at maturity.

DR ELLIS

In some species the first conidium is functional, but in many the potential is lost. Sometimes as many as twenty conidia are non-functional. The conidiogenous locus must move down towards the base.

DR KENDRICK

From general evolutionary principles this isn't very surprising; one of the major tricks of evolution is a change in function of a pre-existing structure. Here we have something that originally evolved as a dispersal unit, but has now assumed a protective role.

DR ELLIS

In *Spegazzinia*, of course, the apical conidium is the only one there is, so it must be functional; it usually germinates through the abscission scar.

DR KENDRICK

Why does *Spegazzinia* push its one conidium up on such a long column? Perhaps a relevant question, but one we probably can't answer at present. In studies of birds or mammals or insects, one can usually decide upon a function for each part of the beast, but in hyphomycetes, one frequently has no idea why things are as they are. It is difficult to assign something a level of taxonomic significance when one does not know what it's there for. It is very frustrating and very challenging.

# 13
# Imperfect-Perfect Connections in Ascomycetes

E. MÜLLER

To the beginner it seems extraordinary that during the development of a fungus several completely different fructifications may be formed. When Tulasne (1851) published his first observation on this subject it was difficult for mycologists to accept this fact. But soon others confirmed his observations, so that de Bary (1866) had no hesitation about including this concept in his textbook on mycology. The introduction of pure culture techniques made it possible to prove the existence of such links between perfect and imperfect fungi, and since then hundreds of sexual and asexual states of fungi have been connected. But the number of proved connections is still small compared with the number of sexual states without any known asexual fructification and the number of Fungi Imperfecti not linked to any perfect state.

Within the Oomycetes, Trichomycetes, and Zygomycetes there is more or less uniform development and morphology of sexual states; the taxonomy of these groups is, therefore, based mainly on the asexual states, which show more morphological diversity. In the Basidiomycetes, and particularly in the Ascomycetes, not only the asexual but also the sexual states are extremely variable in both morphology and development. These groups are logically classified according to their sexual fructifications; any asexual states occurring within their life histories are treated as additional characteristics.

Unfortunately we are still forced to maintain two different systems within higher fungi: one for perfect states, including Ascomycetes and Basidiomycetes, which may be arranged according to their phylogeny; and a second for the asexual states, including imperfect states of Ascomycetes and Basidiomycetes as well as fungi which never produce asci or basidia. It is not possible to arrange these conidial forms in a natural system; any classification is no more than a "form" system which is not related to the phylogeny of these fungi.

Institut für Spezielle Botanik, Eidg. Technische Hochschule, Zurich, Switzerland.

Although the problems involved are known to every mycologist, only a few attempts have been made to find general rules for linking the two systems. There are several aspects to be considered, first the one which centres on taxonomy: will it be possible to connect the present phylogenetically oriented system of Ascomycetes and Basidiomycetes to the form system of Fungi Imperfecti? It was Tubaki (1958) who first tried to find an answer. He brought together numerous proved connections between perfect and imperfect states and made an effort to arrange them. In his arrangement he divided the Ascomycetes into some (then generally accepted) orders such as Endomycetales, Taphrinales, Eurotiales, Pezizales, Erysiphales, Sphaeriales, Hypocreales, Helotiales, Hysteriales, Microthyriales, and Dothideales, and the imperfect forms were classified according to Hughes (1953b). 

Since Tubaki's publication, our knowledge has increased and concepts of the taxonomy of the Ascomycetes and of imperfect forms have changed considerably. But neither of the two systems has yet become stable.

The basis for our present reflection is a simplified ascomycete system similar to the one used by Gäumann (1964) or Müller and Loeffler (1968). It considers ascus morphology as well as the morphology of fruiting structures, and is based on observations first expressed by Luttrell (1951). We have omitted some small orders, and in the Bitunicatae we do not differentiate between the Pseudosphaeriales (Pleosporales), Dothiorales (Hysteriales), Dothideales, and Hemisphaeriales because there is no general agreement at present. The imperfect states are arranged according to the modified system of Hughes (1953b).

The older systems of Fungi Imperfecti were based on the morphology of fruiting structures and so these fungi have been divided into Sphaeropsidales (including Discellaceae) with pycnidia, Melanconiales with acervuli, and Hyphomycetes without strictly united conidiogenous cells or well-defined fruiting structures. Within the groups with prototunicate asci (such as Eurotiales and Microascales) we have only Hyphomycetes, whereas in the eutunicate group of ascomycetes we find Hyphomycetes, Melanconiales, and Sphaeropsidales. Unfortunately, studies on conidium formation are much rarer for Melanconiales and Sphaeropsidales than for Hyphomycetes. We are therefore not able to include many of these in our considerations.

From Table 13.1 we see that some of the different kinds of conidium development are represented throughout the ascomycete system whereas others are limited to, or even typical of, one or a few orders of ascomycetes.

Phialoconidia seem to be most widespread and are represented in the following orders:

Eurotiales: e.g. *Byssochlamys, Eurotium, Emericella, Sartorya, Talaromyces, Eupenicillium, Levispora, Cephalotheca, Melanospora.*
Microascales: e.g. *Ceratocystis.*
Sphaeriales: e.g. *Chaetosphaeria, Chaetosphaerella* (Sphaeriaceae), *Nectria, Hypocrea, Calonectria, Gibberella, Neocosmospora, Thyronectria* (Hypocreaceae), *Bombardia, Lasiosphaeria, Coniochaeta, Neurospora, Sordaria* (Sordariaceae).
Diaporthales: e.g. *Melanochaeta, Gaumannomyces.*
Xylariales: e.g. *Griphosphaeria.*

TABLE 13.1. The ascomycete system connected to the conidial states. The conidial states are represented by numbers which correspond to their conidium development according to Hughes (1953b). IA: Blastoconidia [acropetal chains] (*Haplographium* and *Hyaloscypha* in the Pezizales, Hughes 1953b; *Cladosporium* and *Mycosphaerella* in the Bitunicatae, von Arx 1950). IB: Botryoblastoconidia [synchronous] (*Oedocephalum* and *Ascophanus* in the Pezizales, Tubaki 1958). II: Sympodioconidia [blastoconidia produced on a sympodially proliferating conidiogenous cell] (*Fusicladium* and *Venturia* in the Bitunicatae, Hughes 1953a; *Verticicladium* and *Desmazieriella* in the Pezizales, Hughes 1953b). IIIA: Annelloconidia [blastoconidia produced on a percurrently proliferating conidiogenous cell] (*Scopulariopsis* and *Microascus* in the Microascales, Morton and Smith 1963; *Spilocaea* and *Venturia* in the Bitunicatae, Hughes 1953a). IIIB: "Chlamydospores" (*Trichophyton* and *Nannizzia* in the Eurotiales, Stockdale 1963). IV: Phialoconidia (*Penicillium* and *Talaromyces* in the Eurotiales, Benjamin 1955; *Trichoderma* and *Hypocrea* in the Sphaeriales, Brefeld 1891; *Paecilomyces* and *Cordyceps* in the Clavicipitales, Brown and Smith 1957). VA: Meristem arthroconidia (*Oidium* and *Erysiphe* in the Erysiphales, Hughes 1953b). VB: Retrogressive blastoconidia (*Basipetospora* and *Monascus* in the Eurotiales, Cole and Kendrick 1968). VI: Tretoconidia (*Stemphylium* and *Pleospora* in the Bitunicatae, Wiltshire 1938). VII: Arthroconidia (*Oidiodendron* and *Arachniotus*, Müller and Pacha-Aue 1969).

Euascomycetidae

| | Unitunicatae | | |
|---|---|---|---|
| | Operculatae | Inoperculatae | Bitunicatae |
| Eutunicatae | | Diaporthales<br>IIIA, IV | |
| | Erysiphales<br>VA | Clavicipitales<br>II, IV, VB, VII | |
| | | Sphaeriales<br>IA, II, IV, VII | IA, II, IIIA, IIIB |
| | | Xylariales<br>II, IIIA, IV | IV, VA, VI |
| | Pezizales<br>IB, II | Helotiales<br>IA, IB, IV | |
| | | Phacidiales<br>IIIB | |
| Prototunicate | Microascales<br>IIIA, IV | | Meliolales |
| | | Eurotiales<br>II, IIIA, IIIB,<br>IV, VII | |

Clavicipitales: e.g. *Claviceps, Cordyceps, Hypomyces, Pyxidiophora.*
Helotiales: e.g. *Pyrenopeziza, Mollisia, Godronia, Streptotinia.*
Bitunicatae: e.g. *Ophiocapnodium.*

Phialoconidia have not been reported, and are quite probably lacking, in the Erysiphales, Pezizales, and Phacidiales: they seem to be rare in the Bitunicatae (only represented in the Capnodiaceae; Hughes 1967). The phialide type of conidium development is most frequent in the Eurotiales (*Penicillium, Aspergillus*, and related genera), Sphaeriales (*Cephalosporium*), and Helotiales. In some cases, however, it is not clear whether phialoconidia are functioning as normal conidia or only as spermatia; in the Laboulbeniales and among basidiomycetes in the Uredinales, phialides produce only spermatia as far as we know.

Although in all these cases conidia are produced in the same characteristic manner, there are considerable differences in their morphology, in their conidiogenous cells, and in the supporting structures. It is therefore possible to divide them into well-defined genera and species of Fungi Imperfecti, and there are many examples of the fact that similar conidial structures belong to related perfect states. Some of these have been known and used for taxonomical arrangements for a long time. The *Aspergillus nidulans* group (Raper and Fennell 1965), which differs from other groups of aspergilli by some minor but well-marked characteristics, belongs - as far as perfect states are known - strictly to the genus *Emericella* (Benjamin 1955) defined by red or violet ascospores. Another recent example of such a strict coordination of imperfect and perfect forms is found in the *Chaetosphaeria* complex.

The genus *Chaetosphaeria* Tul. (Sphaeriales, Sphaeriaceae) was misinterpreted by Fuckel (1869/70) and by many others after him, e.g. Winter (1887). Only Booth (1957, 1958) accepted the original concept of Tulasne and Tulasne (1863). He compared the known species carefully and had to remove *Sphaeria phaeostroma* Dur. et Mont. and similar species from the genus. Unfortunately *Sphaeria phaeostroma* had until then been regarded as the type species of *Chaetosphaeria* Tul. Booth (1958) included *Sphaeria phaeostroma* in *Thaxteria* Speg.; however, it is quite different from *Thaxteria didyma* Speg. (the type species, which we have recently examined), and so has to be put in a new genus, *Chaetosphaerella* (Booth and Müller, in press). Recently more species have been added to *Chaetosphaeria* by Müller and von Arx (1962), Hughes (1965), and Hughes and Kendrick (1967). Other somewhat similar fungi included in *Chaetosphaeria* by Hughes (1966) differ considerably and therefore belong in the new genus *Melanochaeta* Müller, Harr et Sulmont (Müller, Harr and Sulmont 1969) which is related to *Melogramma* Fr. (Diaporthales). This arrangement based on the morphology of the perfect state is supported by the conidial states (see Table 13.2).

The phialoconidia of *Chaetosphaeria* species belong to different form genera according to their differences in phialide arrangement, phialide morphology, and conidium morphology (Figure 13.1 a-h). *Sporoschisma* - also a phialidic conidial genus - differs by its deep-seated conidiogenous locus (Figure 13.1 k-m). The conidial states of *Chaetosphaerella,* the third segregate of *Chaetosphaeria,* are mainly blastoconidia, but at least one of the two species forms an additional phialide state (Hughes and Hennebert 1963) (Figure 13.1 i, j).

Annelloconidia (blastoconidia produced on a percurrently proliferating conidiogenous cell) are also widespread in the ascomycete system:

Eurotiales: e.g. *Arachniotus* (Müller and Pacha-Aue 1969).
Microascales: e.g. *Microascus* (Barron, Cain, and Gilman 1961; Morton and Smith 1963), *Ceratocystis* (Tubaki 1958).
Xylariales: e.g. *Broomella* (Shoemaker and Müller 1963), *Clathridium* (Shoemaker and Müller 1964).
Diaporthales: e.g. *Melanconis* (Hughes 1953b).
Bitunicatae: e.g. *Venturia* (Hughes 1953a), *Otthia* (*Stigmina,* unpublished).

TABLE 13.2

| Perfect | Imperfect | Conidium development | References |
|---|---|---|---|
| *Chaetosphaeria* | | | |
| *innumera* Tul. (type) | *Chloridium* | Phialoconidia | Booth 1957; Hughes 1965 |
| *myriocarpa* (Fr.) Booth | *Chloridium* | Phialoconidia | Booth 1957; Hughes 1965 |
| *cupulifera* (Berk. et Br.) Sacc. | *Catenularia* | Phialoconidia | Booth 1958; Hughes 1965 |
| *novae-zelandiae* Hughes et Shoem. | *Catenularia* | Phialoconidia | Hughes 1965 |
| *pomiformis* (Fr. ex Fr.) E. Müller | *Gliobotrys* | Phialoconidia | Booth 1957 |
| *callimorpha* (Mont.) Sacc. | *Codinaea* | Phialoconidia | Hughes and Kendrick 1967 |
| *dingleyae* Hughes, Kendrick et Shoem. | *Codinaea* | Phialoconidia | Hughes and Kendrick 1967 |
| *pulchriseta* Hughes, Kendrick et Shoem. | *Codinaea* | Phialoconidia | Hughes and Kendrick 1967 |
| *talbotii* Hughes, Kendrick et Shoem. | *Codinaea* | Phialoconidia | Hughes and Kendrick 1967 |
| *brevispora* Shoem. | *Zanclospora* | Phialoconidia | Hughes and Kendrick 1967 |
| *pulviscula* (Curr.) Booth | *Menispora* | Phialoconidia | Hughes and Kendrick 1963 |
| *bramleyi* Booth | | Phialoconidia | Booth 1958 |
| *Melanochaeta* | | | |
| *hemipsila* (Berk. et Br.) Müller et al. | *Sporoschisma* | Phialoconidia | Hughes 1966; Müller, Harr, and Sulmont 1969 |
| *actearoae* (Hughes) Müller et al. | *Sporoschisma* | Phialoconidia | Hughes 1966 |
| *Chaetosphaerella* | | | |
| *phaeostroma* (Dur. et Mont.) Booth et Muller | *Oedemium* | Blastoconidia | Hughes 1953b; Hughes and Hennebert 1963 |
| | *Phialocephala* | Phialoconidia | |
| *fusca* (Fuckel) Booth | *Oedemium* | Blastoconidia | Booth 1958; Hughes and Hennebert 1963 |

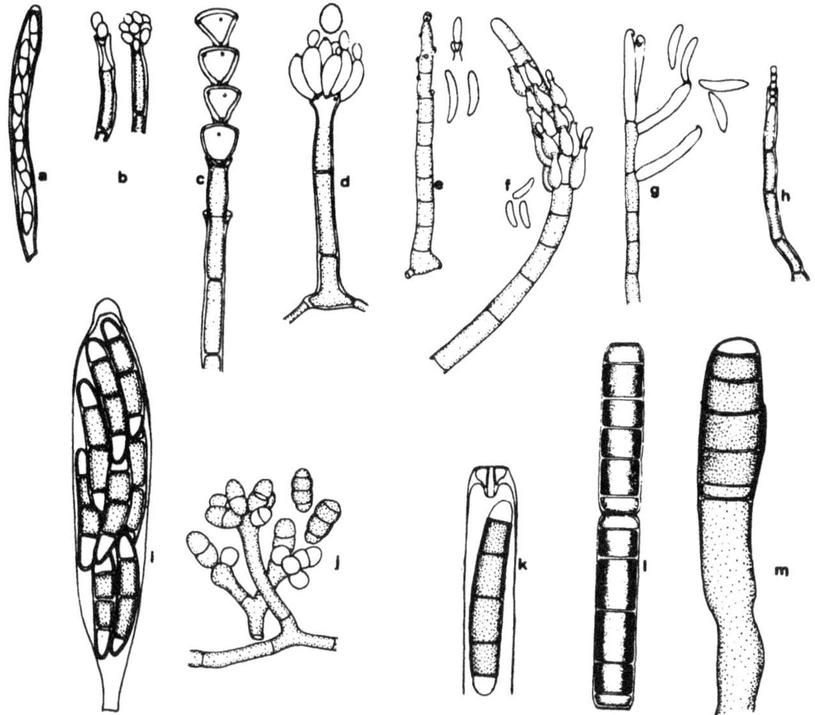


Figure 13.1 a, *Chaetosphaeria innumera*, ascus after Booth 1957; b, *Chloridium* conidial state, after Booth 1957; c, *Catenularia* conidial state of *Chaetosphaeria novae-zelandiae* Hughes et Shoem., after Hughes 1965; d, *Gliobotrys* conidial state of *Chaetosphaeria pomiformis* (Pers. ex Fr.) Müller, after Booth 1957; e, *Codinaea* conidial state of *Chaetosphaeria callimorpha* (Mont.) Sacc., after Hughes and Kendrick 1967; f, *Zanclospora* conidial state of *Chaetosphaeria brevispora* Shoem., after Hughes and Kendrick 1965; g, *Menispora* conidial state of *Chaetosphaeria pulviscula* (Curr.) Booth, after Booth 1957; h, conidial state of *Chaetosphaeria bramleyi* Booth, after Booth 1958; i, *Chaetosphaerella phaeostroma* (Dur. et Mont.) Booth et Müller, ascus; j, *Oedemium* conidial state; k, ascus apex with one ascospore of *Melanochaeta hemipsila* (Berk. et Br.) Müller, Harr et Sulmont; l, two conidia of the *Sporoschisma* conidial state; m, phialide with one young conidium of the *Sporoschisma* conidial state.

The formation of annelloconidia is typical for *Microascus* (*Scopulariopsis*, Morton and Smith 1963) and a small group of Xylariales investigated by Shoemaker and Müller (1963, 1964, 1965). In other cases this kind of conidium occurs occasionally in addition to other types of conidia.

Another relatively widespread type is represented by sympodioconidia (blastoconidia produced by sympodially proliferating conidiogenous cells):

TABLE 13.3

| Species | Macroconidial form | Microconidial form | References |
|---|---|---|---|
| *Leptosphaeria* | | | |
| *acuta* (Fuckel) Karst. | | *Phoma* (*Leptophoma*) | Müller and Tomasevic 1957 |
| *maculans* (Desm.) Ces. et de Not. | | *Phoma* (*Plenodomus*) | Müller 1953; Boerema and von Kesteren 1964 |
| *ogilviensis* (Berk. et Br.) Ces. et de Not. | *Camarosporium* | *Phoma* (*Leptophoma*) | Müller and Tomasevic 1957 (sub. *maculans*) |
| *millefolii* (Fuckel) Niessl | *Camarosporium* | *Phoma* (*Leptophoma*) | Müller and Tomasevic 1957 |
| *macrospora* (Fuckel) Thuem. | *Rhabdospora* | | Müller 1953 |
| *anemones* Hollos | *Rhabdospora* | | Müller 1950 |
| *pratensis* Sacc. et Briard | *Stagonospora* | | Jones and Weimer 1938 |
| *polygonati* Müller et Tomasevic | *Hendersonia* | | Müller and Tomasevic 1957 |
| *bellynckii* (West.) Auersw. | | | Müller and Tomasevic 1957 |
| *fallaciosa* Berl. | | | Müller 1962 |

TABLE 13.4

| Species | Hosts | Acospore morphology | Conidial state | References |
|---|---|---|---|---|
| *Venturia* | | | | |
| *inaequalis* (Cooke) Winter | *Pyrus malus*<br>*Sorbus* spp. | Upper cell short and thicker than the longer basal cell | *Spilocaea* = blastoconidia produced on percurrently proliferating conidiogenous cell (annellide) | Hughes 1953a |
| *crataegi* Aderh. | *Crataegus* spp. | Same | *Fusicladium* = blastoconidia propuced on sympodially proliferating conidiogenous cell | Aderhold 1903 |
| *pirina* Aderh. | *Pyrus communis* | Upper cell longer than the basal cell | *Fusicladium* | Aderhold 1903 |

Eurotiales: e.g. *Pseudeurotium* (Stolk 1955), *Ascotricha* (Boulanger 1897). 
Microascales: e.g. *Ceratocystis* (Tubaki 1958).
Pezizales: e.g. *Desmazierella* (Hughes 1953b).
Helotiales: e.g. *Ascocorticium* (Oberwinkler, Casagrande, and Müller 1967).
Sphaeriales: e.g. *Chaetosphaerella* (Hughes 1953b).
Xylariales: e.g. *Daldinia, Hypoxylon, Rosellinia* (Tubaki 1958).
Clavicipitales (including Ostropales): e.g. *Cordyceps* (Kobayashi 1941), *Acrospermum* (Webster 1956).
Bitunicatae: e.g. *Mycosphaerella* (Klebahn 1918), *Tubeufia* (Webster 1951).

Blastoconidia arising on sympodially proliferating conidiogenous cells are produced typically in the Xylariales (e.g. *Hypoxylon* with *Nodulisporium*) and the Bitunicatae (e.g. *Mycosphaerella* with *Cercospora, Cercosporella, Ovularia, Ramularia,* and *Heterosporium*; Tubaki 1958). In the Bitunicatae there are unusual examples of this kind of blastoconidium. Linder (1929) proved that *Helicoma curtisii* Berk., which has helicoid blastoconidia, belongs to "*Lasiosphaeria pezicula*," which is the type of *Thaxteriella* Petr. (Petrak 1953). Webster (1951) found helicoid blastoconidia (*Helicosporium phragmitis* v. Höhn.) in *Tubeufia helicomyces* v. Höhn., and Hughes (1953b) found them for *Ophionectria cerea* (Berk. et Curt.) Ell; which also belongs in the genus *Tubeufia* (Booth 1959). All three proved connections are to closely related species of the Bitunicatae.

Tretoconidia (poroconidia) are typical for Bitunicatae, e.g. *Pleospora* (*Alternaria*), *Pyrenophora* (*Drechslera*), *Cochliobolus* (*Bipolaris*) (Tubaki 1958). A similar strict relationship occurs in the Erysiphales, which have only meristem thalloconidia.

Imperfect-perfect connections are not always so simple as the ones mentioned above. Certain ascomycete genera have several different conidial states. In *Leptosphaeria* Ces. et de Not. (in the strict sense given by Holm 1957) macroconidial and microconidial forms may occur (see Table 13.3). It seems that all these conidial states have blastoconidia formed solitarily or on a sympodially proliferating conidiogenous cell. Unfortunately there is no information available on conidium development. Differences in the form of the conidia are indicated by the different generic names of Fungi Imperfecti mentioned.

In *Venturia* de Not. (in the sense of Korf 1956 and Müller and von Arx 1962) most of the species have no conidial state. In some species, however, annelloconidia or sympodially produced blastoconidia are formed, and within certain groups (e.g. the apple scab group) the species are even characterized by their conidial states (see Table 13.4).

In Meliolales no conidial states are known.

A second aspect is often neglected in considering imperfect-perfect connections: when and why does a fungus fruit asexually, and when and why does it fruit sexually? Ascomycetes may have a strict rhythm. Under natural conditions *Venturia inaequalis* (Cooke) Winter produces its asexual state (*Spilocaea pomi* Fr.) on living apple leaves; the sexual state develops in the spring on dead leaves. In the laboratory, however, it is possible to obtain the two states at any time on moist

TABLE 13.5. Factors influencing reproduction of *Talaromyces helicus*

| Factor | Asexual reproduction, *Penicillium* | Sexual reproduction, cleistothecia |
|---|---|---|
| $(NO_3)^-$ nutrition | Favoured | Suppressed |
| $(NH_4)^{2+}$ nutrition | Suppressed | Favoured |
| Light | Favoured | Suppressed |
| C:N ration | | |
| >240 and <10 | Suppressed | Inhibited |
| >10 to <240 | Favoured | Favoured |

TABLE 13.6. Nutrition and fructification of *Talaromyces helicus*. Basic medium composed of 1,000 ml $H_2O$ (dist.), 20 g agar (Difco), 1 g $KH_2PO_4$, 0.4 g $Na_2HPO_4$, 0.5 g $MgSO_4$, 0.01 g $FeSO_4$, 0.01 g $MnSO_4$, 0.01 g $ZnSO_4$, and different concentrations of glucose (indicated by the amount of C), $KNO_3$, and $(NH_4)_2HPO_4$ (indicated by the amount of N). The organism was incubated in constant artificial daylight (L), or dark (D), for 28 days at 27°C. A = asexual sporulation (*Penicillium* state); S = sexual sporulation (cleistothecia). Roman letters indicate scanty to weak fructification, bold letters indicate dense fructification over the whole agar surface. and italic letters indicate a moderate amount of fructification.

| | N (g/l) | | | | | | | | | | | |
|---|---|---|---|---|---|---|---|---|---|---|---|---|
| Glucose | 2.24 | | 0.56 | | 0.14 | | 0.035 | | 0.0088 | | 0.0022 | |
| C (g/l) | L | D | L | D | L | D | L | D | L | D | L | D |
| $KNO_3$ | | | | | | | | | | | | |
| 33.2 | *A* | *A* | *A* | *A* | *A* | *A S* | A | A | A | A | A | A |
| 8.3 | *A* | *A* | A | A | A S | A S | A S | A S | *A* | *A* | A | A |
| 2.075 | *A* | *A* | *A* | *A* | A | A S | A S | A S | A | A | A | A |
| 0.52 | A | A | A | A | A | *A* | A | *A* | A | A *S* | A | A |
| 0.13 | A | A | A | A | A | A | A | A | A | A | A | A S |
| $(NH_4)_2HPO_4$ | | | | | | | | | | | | |
| 33.2 | S | S | *S* | S | | A | A | A | A | A | A | A |
| 8.3 | A | A | A S | A S | A S | A S | A | A S | A | A | A | A |
| 2.07 | A | A | *A* | *A* | *A* | *A* S | *A* S | *A* S | A | A | A | A |
| 0.52 | A | A | A | A | A | A | *A* | A S | A S | A S | A | A |
| 0.13 | A | A | A | A | A | A | A | A | A | A S | A | A S |

leaves or on humid artificial media kept at relatively low temperatures. It is therefore evident that fructification may be induced and directed by edaphic factors in addition to genetic ones.

At present our knowledge of factors influencing fructification is still poor. According to Turian (1966, 1969), who has summarized the available information, it is not possible to make general rules. Factors such as nutrition (e.g., carbon source, nitrogen source, phosphate concentration), temperature, pH, and light may affect sporulation. All these factors direct metabolism, and sporulation is only the response of the organism to a certain pathway in its metabolism. Turian (1960) has shown for *Neurospora* that conidium differentiation is connected with a lesion in the Krebs cycle, while protoperithecial morphogenesis (the first step of sexual reproduction) apparently requires full functioning of the Krebs cycle.

We have investigated the behaviour of *Talaromyces helicus* (Raper et Fennell) Benjam. which produces a *Penicillium* conidial state and yellow cleistothecia. Asexual and sexual reproduction depend on the presence or absence of certain ex-

TABLE 13.7. Factors influencing reproduction of *Talaromyces emersonii* 

| Factor | Asexual reproduction, *Penicillium* | Sexual reproduction, cleistothecia |
|---|---|---|
| $(NO_3)^-$ nutrition | Favoured | Favoured |
| $(NH_4)^{2+}$ nutrition | Suppressed | Suppressed |
| Light | No influence | No influence |
| $(PO_4)^{3+}$, 1 g/1,000 ml | Inhibited | Suppressed |
| Temperature | | |
| 37°C | Slightly suppressed | Favoured |
| 45°C | Favoured | Suppressed |
| 53°C | Suppressed | Inhibited |

TABLE 13.8. Nutrition and reproduction of *Talaromyces emersonii.* Basic medium composed of 1,000 ml $H_2O$ (dist.), 20 g agar (Difco), 20 g glucose, 1 g $KH_2PO_4$, 0.4 g $Na_2HPO_4$, 0.5g $MgSO_4$, 0.01 g $FeSO_4$, 0.01 g $MnSO_4$, 0.01 g $ZnSO_4$, and different concentrations of nitrogen sources indicated by the concentration of pure nitrogen. G = vegetative growth, S = sexual reproduction, AS = asexual reproduction.

| N conc. | $KNO_3$ | | | $(NH_4)_2HPO_4$ | | | $(NH_4)_2SO_4$ | | |
|---|---|---|---|---|---|---|---|---|---|
| (g/l) | G | S | AS | G | S | AS | G | S | AS |
| 11.2 | – | – | – | – | – | – | – | – | – |
| 8.4 | + | + | + | – | – | – | – | – | – |
| 5.6 | + | + | + | – | – | – | – | – | – |
| 4.5 | + | + | + | – | – | – | – | – | – |
| 3.4 | + | + | + | – | – | – | + | + | + |
| 2.8 | + | + | + | – | – | – | + | + | + |
| 2.2 | + | + | + | + | + | – | + | + | + |
| 1.7 | + | + | + | + | + | – | + | + | + |
| 1.1 | + | + | + | + | + | – | + | + | + |
| 0.6 | + | + | + | + | + | – | + | + | + |
| 0.3 | + | + | + | + | + | – | + | + | + |

ternal factors, some of which are represented in Table 13.5. In addition to the factors mentioned, temperature (above 30°C there is effective suppression of both types of sporulation) and the concentration of both the carbon and nitrogen source may influence the amount of sporulation. Conidium and ascospore production are best with a glucose concentration of 5-20 g and a potassium nitrate concentration of 0.25-1 g in 1,000 ml medium (equivalent to 0.15-0.7 g ammonium phosphate). These factors are also indicated in Table 13.6.

The thermophilic *Talaromyces emersonii* Stolk behaves differently in many respects (see Table 13.7). Light has no influence on its fructification, but temperature has a much greater influence than in the example mentioned above. In this species the most interesting effect comes from the phosphate in comparatively high concentrations (more than 1 g phosphate in 1,000 ml medium) as indicated in Table 13.8.

The example of these closely related ascomycetes shows the complexity of the fructification problem. It seems that we are still unable to generalize on this subject.

A third aspect concerning imperfect and perfect states of ascomycetes is their

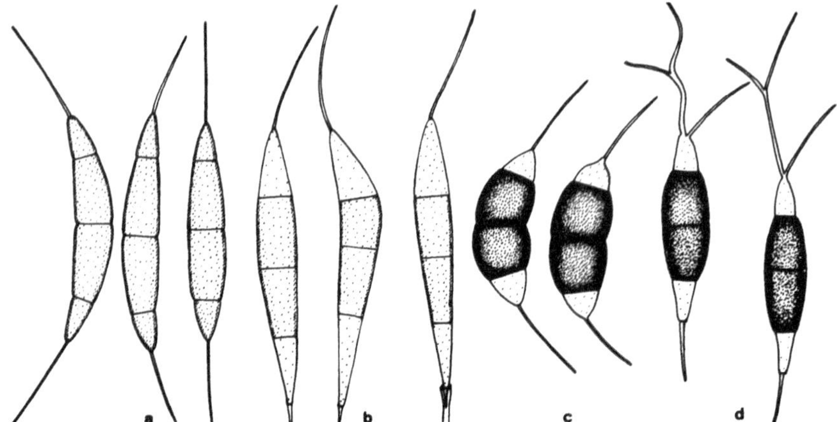


Figure 13.2 Ascospores and conidia (*Pestalotia*) of *Broomella* species: a and b, ascospores and conidia, respectively, of *Broomella vitalbae* (Berk. et Br.) Sacc.; c and d, ascospores and conidia, respectively, of *Broomella acuta* Shoem. et Müller.

degree of morphological similarity. Of course there are many cases in which such large differences exist that their genetic connections are hard to accept: e.g., the long filiform ascospores of *Cochliobolus* and the dark, fusoid conidia of *Drechslera*; or the globose or ellipsoidal, coloured conidia of *Coniothyrium* and the cylindric, mostly only slightly coloured ascospores of the *Paraphaeosphaeria* perfect states (Webster 1955; Hedjaroud 1969).

There are, however, many examples where there is a very close morphological resemblance between perfect and imperfect fructifications. Ascostromata of *Leptosphaeria macrospora* (Fuckel) Thuem. and pycnidia of its conidial state *Rhabdospora bernardiana* Sacc. occur mixed on the same stems of their composite hosts. Since these have the same form and size, they are very hard to distinguish macroscopically. Microscopically the peridial walls have the same texture, the openings of the two types of fruit bodies are similar, and their juvenile stages even have the same appearance. Similarly, *Heterosphaeria* species (Helotiales) have the same kind of apothecial fruit bodies in their sexual as in their asexual state (*Heteropatella*), and in *Phaeosphaeria typharum* (Desm.) Holm, both states may develop within the same fruit body (Parguey-Leduc 1963).

Similarities between conidia and ascospores are more common. When the morphology of the two spore kinds is simple this is perhaps to be expected, but if we consider the highly complicated structures of both ascospores and conidia of *Broomella* species, their similarity is quite impressive (Shoemaker and Müller 1963). Figure 13.2 shows ascospores and conidia of two *Broomella* species with their characteristic coloration and their appendages. *Chaetosphaerella phaeostroma* and *Melanochaeta hemipsila* (Figure 13.1) are other examples in which ascospores and conidia are similarly coloured.

At present our information on connections between imperfect and perfect states is still poor. We are not yet able to find general rules to help us understand the complicated life cycles of these fungi or the reasons why they sometimes form sexual fructifications and at other times asexual ones. Further investigations on such aspects as genetics, morphogenesis (and its relationship to metabolic pathways), and definite proof of connections between many more sexual and asexual fructifications are urgently required. 

## ACKNOWLEDGMENT

I would like to thank Mrs Judith Fineran for her help with the English of this paper.

## REFERENCES

Aderhold, R. 1903. Kann das *Fusicladium* von *Crataegus* und von *Sorbus* auf den Apfelbaum übergehen? Abh. Biol. Abt. Land-und Forstwirtschaft. Kais. Gesundheitsamt. 3: 436-9

von Arx, J.A. 1950. Ueber die Ascusform von *Cladosporium herbarum* (Pers.) Link. Sydowia (Ann. Mycol. II) 4: 320-4

Barron, G.L. Cain, R.F., and Gilman, J.C. 1961. The genus *Microascus*. Can. J. Botany 39: 1609-31

Benjamin, C.R. 1955. Ascocarps of *Aspergillus* and *Penicillium*. Mycologia 47: 669-87

Boerema, G.H., and Von Kesteren, H.A. 1964. The nomenclature of two fungi parasitizing *Brassica*. Persoonia 3: 17-28

Booth, C. 1957. Studies of Pyrenomycetes. I. Four species of *Chaetosphaeria*, two with *Catenularia* conidia. II. *Melanopsamma pomiformis* and its *Stachybotrys* conidia. C.M.I. Mycol. Pap. 68

- 1958. The genera *Chaetosphaeria* and *Thaxteria* in Britain. The Naturalist 1958: 83-90

- 1959. Studies of Pyrenomycetes. IV. *Nectria*. C.M.I. Mycol. Pap. 73

Booth, C., and Müller, E. The generic position of *Sphaeria phaeostroma* Dur. et. Mont. (in press)

Boulanger, E. 1897. Sur une forme imparfaite dans le genre *Chaetomium*. Rev. Gén. Botan. 9: 17-26

Brefeld, O. 1891. Untersuchungen aus dem Gesamtgebiet der Mykologie. 10: 157-378

Brown, A.H.S., and Smith, G. 1957. The genus *Paecilomyces* Bainier and its perfect stage *Byssochlamys* Westl. Trans. Brit. Mycol. Soc. 40: 17-89

Cole, G.T., and Kendrick, W.B. 1968. Conidium ontogeny in hyphomycetes. The imperfect state of *Monascus ruber* and its meristem arthrospores. Can. J. Botany 46: 987-92

De Bary, A. 1866. Morphologie und Physiologie der Pilze, Flechten und Myxomyceten. Engelmann, Leipzig

Fuckel, L. 1869/70. Symbolae mycologicae. Jahrb. Nassau. Ver. Naturkunde 23-24

Gäumann, E. 1964. Die Pilze. Birkhäuser, Basel

Hedjaroud, G.A. 1969. Études taxonomiques sur les *Phaeosphaeria* Miyake et leurs formes voisines (Ascomycetes). Sydowia (Ann. Mycol. II) 22: 57-107

Holm, L. 1957. Études taxonomiques sur les Pléosporacées. Symb. Bot. Upsal. XIV (3)

Hughes, S.J. 1953a. Some foliicolous hyphomycetes. Can. J. Botany 31: 560-76

- 1953b. Conidiophores, conidia, and classification. Can. J. Botany 31: 577-659
- 1965. New Zealand fungi. 3. *Catenularia* Grove. N.Z. J. Botany 3: 136-50
- 1966. New Zealand fungi. 6. *Sporoschisma* Berk. et Br. N.Z. J. Botany 4: 77-85
- 1967. New Zealand fungi. 9. *Ophiocapnocoma* with *Hormiokrypsis* and *Capnophialophora* states. N.Z. J. Botany 5: 117-33
Hughes, S.J., and Hennebert, G.L. 1963. Microfungi. X. *Oedemium, Dimera, Diplosporium, Gongylocladium,* and *Cladotrichum.* Can. J. Botany 41: 773-809
Hughes, S.J., and Kendrick W.B. 1963. Microfungi. IX. *Menispora* Persoon. Can. J. Botany 41: 693-718
- 1965. New Zealand fungi. 4. *Zanclospora* gen. nov. N.Z. J. Botany 3: 151-8
- 1967. New Zealand fungi. 12. *Menispora, Codinaea, Menisporopsis.* N.Z. J. Botany 6: 323-75
Jones, F.R., and Weimer, J.L. 1938. *Stagonospora* leaf spot and root rot of forage legumes. J. Agric. Res. 57: 791-812
Klebahn, H. 1918. Haupt- und Nebenfruchtformen der Ascomyceten. Bornträger, Leipzig
Kobayashi, Y. 1941. The genus *Cordyceps* and its allies. Sci. Rep. Tokyo Bunrika Daigaku, Sect. (b) 5: 53-260
Korf, R.P. 1956. Nomenclatural notes. I. Misuse of neotypes for *Venturia* and *Phaeosphaerella.* Mycologia 48: 591-5
Linder, D.H. 1929. A monograph of the helicosporous Fungi Imperfecti. Ann. Missouri Bot. Gard. 16: 227-388
Luttrell, E.S. 1951. The system of Pyrenomycetes. Univ. Missouri Studies 24 (3)
Morton, F.J., and Smith, G. 1963. The genera *Scopulariopsis* Bainier, *Microascus* Zukal and *Doratomyces* Corda. C.M.I. Mycol. Pap. 86
Müller, E. 1950. Die schweizerischen Arten der Gattung *Leptosphaeria* und ihrer Verwandten. Sydowia (Ann. Mycol. II) 4: 185-319
- 1953. Kulturversuche mit Ascomyceten I. Sydowia (Ann. Mycol. II) 7: 325-34
- 1962. Kulturversuche mit Ascomyceten IV. Sydowia (Ann. Mycol. II) 16: 115-20
Müller, E., and Von Arx, J.A. 1962. Die Gattungen der didymosporen Pyrenomyceten. Beitr. Krypt. Fl. Schweiz. II (2)
Müller, E., Harr, J., and Sulmont, P. 1969. Deux ascomycètes dont le stade conidien présent des conidies phaeophragmiées endogènes. Revue de Mycol. 33: 369-78
Müller, E., and Loeffler, W. 1968. Mykologie. Thieme-Verlag, Stuttgart
Müller, E., and Pacha-Aue, R. 1969. Untersuchungen an drei Arten von *Arachniotus* Schroeter. Nova Hedwigia 15: 551-8
Müller, E., and Tomasevic, M. 1957. Kulturversuche mit einigen Arten der Gattung *Leptosphaeria* Ces. et de Not. Phytopath. Z. 29: 287-94
Oberwinkler, F., Casagrande, F., and Müller, E. 1967. Ueber *Ascocorticium anomalum* (Ell. et Harkn.) Earl. Nova Hedwigia 14: 283-9
Parguey-Leduc, A. 1963. Sur les fructifications mixtes à la fois pycnidienne et périthèciales, du *Leptosphaeria typharum* (Desm.) Karst. C.R. Acad. Sci., Paris 256: 4275-7
Petrak, F. 1953. Ein Beitrag zur Pilzflora Floridas. Sydowia (Ann. Mycol. II) 7: 103-20
Raper, K.B., and Fennell, D.I. 1965. The genus *Aspergillus.* Williams and Wilkins, Baltimore, Md.
Shoemaker, R.A., and Müller, E. 1963. Generic correlations and concepts: *Broomella* and *Pestalotia.* Can. J. Botany 41: 1235-43
- 1964. Generic correlations and concepts: *Clathridium* (= *Griphosphaeria*) and *Seimatosporium* (=*Sporocadus*). Can. J. Botany 42: 403-9
- 1965. Types of pyrenomycete genera. *Hymenopleella* and *Lepteutypa.* Can. J. Botany 43: 1457-60

Stockdale, P.M. 1963. The *Microsporum gypseum* complex (*Nannizzia incurvata* Stockd. *N. gypsea* (Nann.) comb. nov. *N. fulva* sp. nov.). Sabouraudia 3: 114-26 

Stolk, A.C. 1955. The genera *Anixiopsis* Hansen and *Pseudeurotium* van Beyma. Ant. van Leeuwenhoek 21: 65-79

Tubaki, K. 1958. Studies on the Japanese hyphomycetes. V. J. Hattori Bot. Lab. 20: 142-244

Tulasne, L.R. 1851. Note sur l'appareil reproducteur dans les lichens et les champignons (1ère partie). C.R. Acad. Sci., Paris 32: 470-5

Tulasne, L.R., and Tulasne, C. 1863. Selecta Fungorum Carpologia II. Paris

Turian, G. 1960. Déficiences du métabolisme oxydatif et différenciation sexuelle chez *Allomyces* et *Neurospora*. Pathol. Microbiol. 23: 687-99

- 1966. Morphogenesis in Ascomycetes. *In* Ainsworth, G.C., and Sussman, A.S. (ed.) The Fungi. Vol. II. Academic Press, New York and London. pp. 339-85

- 1969. Différenciation fongique. Masson, Paris

Webster, J. 1951. Graminicolous Pyrenomycetes. I. The conidial stage of *Tubeufia helicomyces*. Trans. Brit. Mycol. Soc. 34: 318-21

- 1955. Graminicolous Pyrenomycetes. V. Trans. Brit. Mycol. Soc. 38: 347-65

- 1956. Conidia of *Acrospermum compressum* and *A. gramineum*. Trans. Brit. Mycol. Soc. 39: 361-6

Wiltshire, S.P. 1938. The original and modern conceptions of *Stemphylium*. Trans. Brit. Mycol. Soc. 21: 211-39

Winter, G. 1887. Die Pilze. *In* Rabenhorst, L. Kryptogamenflora von Deutschland, Oesterreich und der Schweiz. 2nd ed. Vol. I (2)


## DISCUSSION

DR TUBAKI

Comparing notes with Dr Müller, I find that we have reached nearly the same conclusions. I have examined the conidial states of many ascomycetes, and am still gathering data. I believe that characteristics reflecting those of the ascomycetous parents should be discernible in the imperfect states. If the classification of the perfect states reveals their true phylogeny, some parallel phylogenetic relationships may be found among these imperfect states. What characteristics might show this? Perhaps those of conidium ontogeny. In the classical system ascomycete imperfect states are divided among innumerable form genera. But in the modern system, which is based on conidium ontogeny, they show a fairly definite trend. In Eurotiales, 15 imperfect states of 34 genera belong to section IVA, 13 belong to section III, and 2 have annellations. Xylariales have conidial states belonging to section II, and Erysiphales always have section V states. Pezizales have conidial states in section IB and section II, and 33 imperfect states of 15 genera of Hypocreales belong to section IV. Tretic conidia are restricted to Pseudosphaeriales, as indicated by Dr Müller.

I am most interested in the imperfect states of *Hypomyces*. They belong to *Cladobotryum, Mycogone, Sepedonium, Trichothecium*, and others. These genera all produce conidia by the method of my section IX (section VB, according to the recent precise observations of Dr Kendrick and Dr Cole). Other members of the Hypocreales have conidial states in section IV. From this evidence, I think it is reasonable to remove *Hypomyces* from the Hypocreales.

I'm also interested in yeast-like conidial states. The genus *Graphiola* is well known in the Tropics, but its taxonomic affinities are not clear. Some people put it in the Ustilaginales, others in the Fungi Imperfecti. We have obtained a *Rhodotorula* as the imperfect state of *Graphiola phoenicis* in Japan. Red yeast-like states occur in some members of the Tremellales, *Ustilago violacea* and *Rhodosporidium toruloides*. This may lend support to the idea that *Graphiola* has heterobasidiomycetous affinities. But we also have red yeast states for *Taphrina* and *Protomyces* in the lower ascomycetes.

I'd also like to focus some attention on the problem of conidial states and spermatial states. Some "phialoconidia" may act as spermatia, as in the Uredinales and some ascomycetes. Where we discover the sexual role of these spores, we call them spermatia; where we don't know, we call them microconidia or phialoconidia. They are found in *Sclerotinia, Lambertella*, and *Martinia* of the Discomycetes and *Neurospora, Sordaria*, and *Laboulbenia* of the Pyrenomycetes. The sexual function of these spores in certain discomycetes has been clearly demonstrated. We have at present in culture many *Phialophora*-like imperfect states, in some of which the "conidia" are suspected of having a sexual function. Sometimes if we add some water to a month-old culture and crush the surface of the colony, perfect states (often resembling *Coryne sarcoides*) develop two months later. Shall we call these states conidial states if a sexual function is proved? Does the term conidia include both asexual spores and spermatia? Although it is not a serious obstacle to practical classification, I wish to show that a problem exists.

MR BHATT

A note for soil mycologists: instead of throwing your plates away after a couple of weeks, if you keep them 4-6 weeks longer, perfect states sometimes develop. I have found many interesting ascomycetes in soil samples by this simple expedient.

DR S.J. HUGHES

I'd like to thank Dr Müller for his stimulating address. For some years people have been trying to turn the mucronate hyphopodium of *Meliola* into a phialide. It is quite similar in appearance to the so-called antheridia of the Laboulbeniales and to the phialides of some sooty moulds, but I don't think anyone has seen a phialoconidium associated with it. So far as I know, no one has suggested a function for it. Has Dr Müller looked for phialoconidia within the mucronate hyphopodium?

DR MÜLLER

I have had occasion to examine not only recent *Meliola* collections, but also 40-million-year-old specimens. I could show you slides in which you would not be able to distinguish the fossils from the recent. There is no sign of any conidia. If they ever were phialides, they have certainly lost their function. Of course these capitate hyphopodia are very different from those hyphopodia from which haustoria arise and penetrate the leaf.

DR S.J. HUGHES

Dr Müller, you mentioned that you know of no phialides in the Pseudosphaeriales. *Helminthosporium sorokinianum*, which has a pseudosphaeriaceous perfect state, produces pycnidia - but whether these contain phialides or annellophores hasn't been shown yet. There are also sooty moulds with the pseudosphaeriaceous type of development which have tretic conidia and phialides. 

DR MÜLLER

Most of the sooty moulds constitute a natural group which belongs in the Pseudosphaeriales in the Bitunicatae.

DR S.J. HUGHES

There are two kinds of development in the sooty moulds, but it's not appropriate to go into that at present.

DR MÜLLER

You also mentioned the *Cephalosporium* species produced by different genera throughout your classification. This is where our classification is falling down. We cannot differentiate the phialide of one group of fungi from that of another, but it may be possible to do so after Dr Carroll and Dr G.C. Hughes have sliced them.

DR S.J. HUGHES

Dr Müller mentioned that the Prototunicatae produce only hyphomycetous imperfect states. These are primitive ascomycetes, and perhaps their substratum will not allow production of anything more complicated than simple conidiophores.

MME NICOT

Tropical seed-borne fungi produce large and beautiful synnemata - *Penicilliopsis*, for example.

DR KENDRICK

Is Dr Hughes suggesting that the Hyphomycetes are more primitive than pycnidial or acervular fungi?

DR CARMICHAEL

*Trichophyton* produces pycnidia - at least, they are as good pycnidia as the gymnoascaceous ascocarps are cleistothecia! Then there is *Onygena*, which produces very complex structures that contain the ascocarps even though it is off the bottom of Dr Müller's scale.

DR MÜLLER

In the Gymnoascaceae, I wouldn't say that even the perfect state has good fruit bodies, because the peridium consists only of a network. No further opening is needed, because the spores can easily escape through the network.

DR KENDRICK

In these "pycnidia" of *Trichophyton*, do the conidiogenous cells form a hymenial layer, or are they randomly arranged?

DR CARMICHAEL

Randomly arranged.

DR S.J. HUGHES

One of the big problems facing anyone who wishes to enumerate perfect-imperfect connections is that neither the perfect state nor the imperfect state may be correctly classified. We must check and double check. Different species of *Melanomma* produce pycnidia (with presumed phialides) and *Pleurophragmium*, respectively. There is no guarantee that all the *Melanomma* species are really congeneric, or that the *Pleurophragmium* is a *Pleurophragmium*. But if there is agreement in the connections, it is not as suspicious as if there isn't.

DR KENDRICK

May I raise once more the spectre of *Ceratocystis*, which the ascomycete people assure us, on ontogenetic grounds, is a good genus, yet produces a welter of imperfect states.

DR GOOS

*Mycosphaerella* is another ascomycete genus with diverse conidial states.

DR S.J. HUGHES

But in *Mycosphaerella* the macroconidial state is almost always a blastic sympodial conidium. In *Limacinia* phialides are almost always present. There are also usually tretic and sympodial conidia. In some species the tretic conidia are absent, in some the sympodial are absent. I still feel the parent species is a *Limacinia*, even if one or other of the conidial states has "dropped out."

DR KENDRICK

Yes, as the main propagative or dispersal phase of the organism, I expect the conidial states have been subjected to much more selection pressure than the perfect states. Dr Müller mentioned that different species of *Venturia* have percurrently and sympodially proliferating conidiogenous cells. I wonder if we could use that as an argument for linking these two methods of proliferation more closely? These particular states of *Venturia* are very similar in most other respects, differing chiefly in the way in which the succession of conidia is produced.

DR MÜLLER

Within *Venturia* there are species which never produce any conidial state. One species on *Amelanchier* has ascospores just like those of *V. inaequalis*, but never produces a conidial state. Unfortunately we have no *Venturia* which produces both conidial states.

DR TUBAKI

I have made single ascospore cultures from many ascomycetes which have been reported to have conidial states, and many of them have developed no conidial state. I suspect the original reports of conidial states were reports of contaminants!

DR KENDRICK

There's a high probability that some of the connections reported in the literature are not correct, and also that, as Dr Hughes says, either or both of the names given are not correct. So we must be very circumspect, perhaps to some extent basing our evaluations of the connections on the reputations of the people who made them.

DR MÜLLER

In 1968, it was stated that the conidial state of *Gymnoascus* was *Histoplasma*. Three months later it was demonstrated that this wasn't true: the *Gymnoascus* was a contaminant.

DR PIROZYNSKI

May I also point out that some mycologists have a greatly developed imagination. If one sees a structure such as a mucronate hyphopodium, which looks like a phialide, it is very tempting to illustrate a few round objects discreetly nearby. [*Laughter*] This has been done on at least two occasions. The fungi seem to be asking for this sort of misinterpretation. [*Laughter*]

DR KENDRICK
What about the practice of making connections by observing conidiophores growing on, say, a perithecium? How reliable can that be?
DR MÜLLER
We all know that there are many fungi which parasitize other fungi, so we must always be careful. Sometimes mere observations can be valid, but it is very difficult to tell when an organism is parasitic, and when it is really the conidial state. It's always better to culture both organisms.
DR S.J. HUGHES
It's worth mentioning the possibility. It may give someone a lead for subsequent cultural work.

# 14
# Pleomorphism in Fungi Imperfecti

G. L. HENNEBERT

Pleomorphism in fungi was first demonstrated in 1851 by L.R. Tulasne. Pleomorphic fungi were those which, simultaneously or successively, exhibited more than one reproductive state, one perfect and the other(s) imperfect. He expressed it thus: "These little fungi [Pyrenomycetes and Coniomycetes] are not, as generally believed, autonomous growths, and do not represent by themselves a complete plant species ... [but] the dissociated parts of a species composed of several elements."

Tulasne saw two problems which still preoccupy us today. One is "in bringing together the elements of a single fungus species when they will be found unassociated in nature." This difficulty has been only partially overcome by the introduction of pure culture techniques, and the affinities of many Fungi Imperfecti are still doubtful or unknown.

The second problem Tulasne recognized was that "the multiplicity of reproductive organs in all these fungi forces us to create new terms to distinguish these organs, one from the other" - particularly in spore terminology. He proposed reserving the term spores for propagules of perfect states and distinguishing in the Imperfecti several types among the conidia of Fries (stylospores, spermatia, conidia [restricted to "gemmes"]) except when it is impossible to define the exact nature of the asexual propagule.

Tulasne was also aware of the problems raised by the nomenclature of composite species. He said that it is necessary to look for the perfect state of Sphaeropsideae and Cytosporeae among the ascosporic Sphaeriaceae, and that these imperfect forms must some day be submerged in the Ascomycetes. He thus established a first principle of pleomorphic nomenclature, the *precedence* of the perfect state name over imperfect state names.

Laboratory of Systematic and Applied Mycology, University of Louvain, Heverlee, Belgium.

These two problems, the taxonomy and the nomenclature of pleomorphic fungi, are the focus of this chapter. 

## HISTORICAL COMMENTS ON PLEOMORPHISM AND POLYMORPHISM

De Bary (1854) supported Tulasne's thesis of pleomorphism in the Fungi and demonstrated the existence of a perfect state of *Aspergillus glaucus.* The Tulasne brothers (1861-1865) gave expanded illustrations of the phenomenon in "Carpologia Fungorum," showing in particular the connection of conidial and chlamydosporic states of the same ascosporic species. Brefeld (1893) must be credited with providing additional evidence through his pure culture techniques, especially in the connection of yeast states with smut fungi.

The potentialities of all these discoveries overly excited the imagination of some botanists. The lack of adequate microscopic equipment, the incompleteness of data on morphology and physiology of fungi and on methods of pure culture, the break in the classic concept of autonomous species caused by the demonstration of pleomorphism, and finally the then still current theory of spontaneous generation, all were circumstances contributing to misinterpretations of the theory of pleomorphism.

From pleomorphism of reproductive organs indicated by Tulasne and by de Bary, one came to the *polymorphism* of species and of genera, a polymorphism quite reminiscent of Darwin's theory of evolution of species. Thus Spring, Bail, Hoffmann, von Hallier, and also Carnoy produced pure fantasies. They held that *Mycoderma* produced a *Mucor*, which then metamorphosed into a *Penicillium*, and they claimed that under adequate culture conditions it would be possible to go back to one from the other.

Fortunately, Gilkinet (1875) and de Bary (1887) reviewed the theory of "polymorphism" and refuted it. Tulasne and de Bary adopted the term pleomorphism (or pleomorphy) to designate the diversity of reproductive forms in fungi. Delpino (1887) introduced the term pleont for the separate states of these pleomorphic fungi, but this term has seen little use. Gilkinet (1875) and Beauverie (1899), however, used only the term polymorphism (or polymorphy), which is still used in the French language for the phenomenon.

In the study of dermatophytes, Sabouraud (1900) applied the term pleomorphism to the degeneration of fungal colonies maintained in culture. This degeneration is characterized by the progressive disappearance of the typical aspect of the fungus and the change to a sterile mat of hyphae. This transformation is so fundamental that a return to the original aspect is impossible, and the culture is said to be "pleomorphosed." Grigoraki (1925), Langeron and Talice (1930), and Langeron and Milochevitch (1937) established that this "pleomorphism" is only a laboratory artefact caused by too high a concentration of simple carbohydrates (glucose, maltose) in the culture medium. Thus two different concepts of pleomorphism (excluding the aberration of the "polymorphists") developed: the pleomorphism of Tulasne and de Bary, and pleomorphism sensu Sabouraud. These two concepts are accepted separately by Dodge (1935) in his book *Medical Mycology*, as, respec-

Figure 14.1 A, *Doratomyces stemonitis*, chlamydospores of the *Echinobotryum* state and annellidic conidia (MUCL 7541); B, *Humicola grisea* Traaen, chlamydospores and phialidic conidia (Type, CBS 119.14 = MUCL 8008).

tively, "polymorphism" and "pleomorphism." However, Snell (1936), Gäumann (1952), and Snell and Dick (1957) synonymized the two terms, and defined them in the original sense of Tulasne. Ainsworth (1961) did likewise, but also indicated that pleomorphism is still used by dermatologists to describe degeneration of dermatophytes in culture. 

El-Ani (1968) objected to the use, by fungal geneticists, of "pleomorphism" to describe morphologically distinct mutants. He suggested that the term should be rejected. Savile (1969) agreed with El-Ani's first stricture, but insisted that a term cannot be abandoned simply because it is misused. Savile quoted the clear definition given in the shorter Oxford English Dictionary: "*Pleomorphic*, having more than one form: (a) *Biol.* exhibiting different forms at different stages of the life-history." Savile pointed out that although pleomorphism has been widely adopted by mycologists, polymorphism is generally used to describe the condition of insects with several adult forms, such as the castes of social insects. It would seem appropriate that this useful distinction between the names be maintained.

## THE SIGNIFICANCE OF PLEOMORPHISM IN TAXONOMY

The proven connection between the perfect and imperfect states of many fungi (Müller, Chapter 13) stimulates us to look for affinities among the many others for which connections have not yet been established. But pleomorphism also exists in Fungi Imperfecti for which no perfect state is known and may permit one to recognize natural affinities among species.

At first, *Wardomyces columbinus, W. dimerus*, and *W. ovalis* (Gams 1968) appeared to us as close to *Mammaria* as to *Echinobotryum* in their chlamydosporic states. In their annellidic states these *Wardomyces* species appeared more closely related to *Echinobotryum*, a suggestion which was strengthened by the discovery of a phialidic state of *Mammaria echinobotryoides* (Hennebert 1968). By analogy we should search for, or attempt to induce, the annellidic state of other species of *Wardomyces* now known only by their chlamydospores. Are, for example, the annellidic *Scopulariopsis humicola* (Barron 1966) and the chlamydosporic *Wardomyces inflatus* (Hennebert 1968) states of the same fungus, and what is their relationship to the ascosporic genus *Microascus*? The phialide state of *Mammaria echinobotryoides* is a *Phialophora*, apparently most closely related to those species of *Phialophora* with chlamydospores (*P. mutabilis* and others).

Orthotropic, hyaline phialides are known in *Humicola fuscoatra* and *H. grisea* (Traaen 1914) (Figure 14.1 B) and in *Trichocladium canadense* (Hughes 1959), and are quite similar to those in certain species of *Chaetomium*. Phialides of the same kind have now been observed by Domsch and Gams (1970) and by myself in *Trichocladium asperum* Harz (Figure 14.2 A), and more recently by me in *Thermomyces verrucosus* Pugh et al. (Figure 14.2 B). This similar association of states in *Humicola, Trichocladium*, and *Thermomyces* may lead us to use roughness of spores as a generic character, although in *Trichocladium* rough and smooth spored species are classified together at present. I propose to include the species of *Thermomyces* in the older form genus *Humicola*, even though some species, includ-

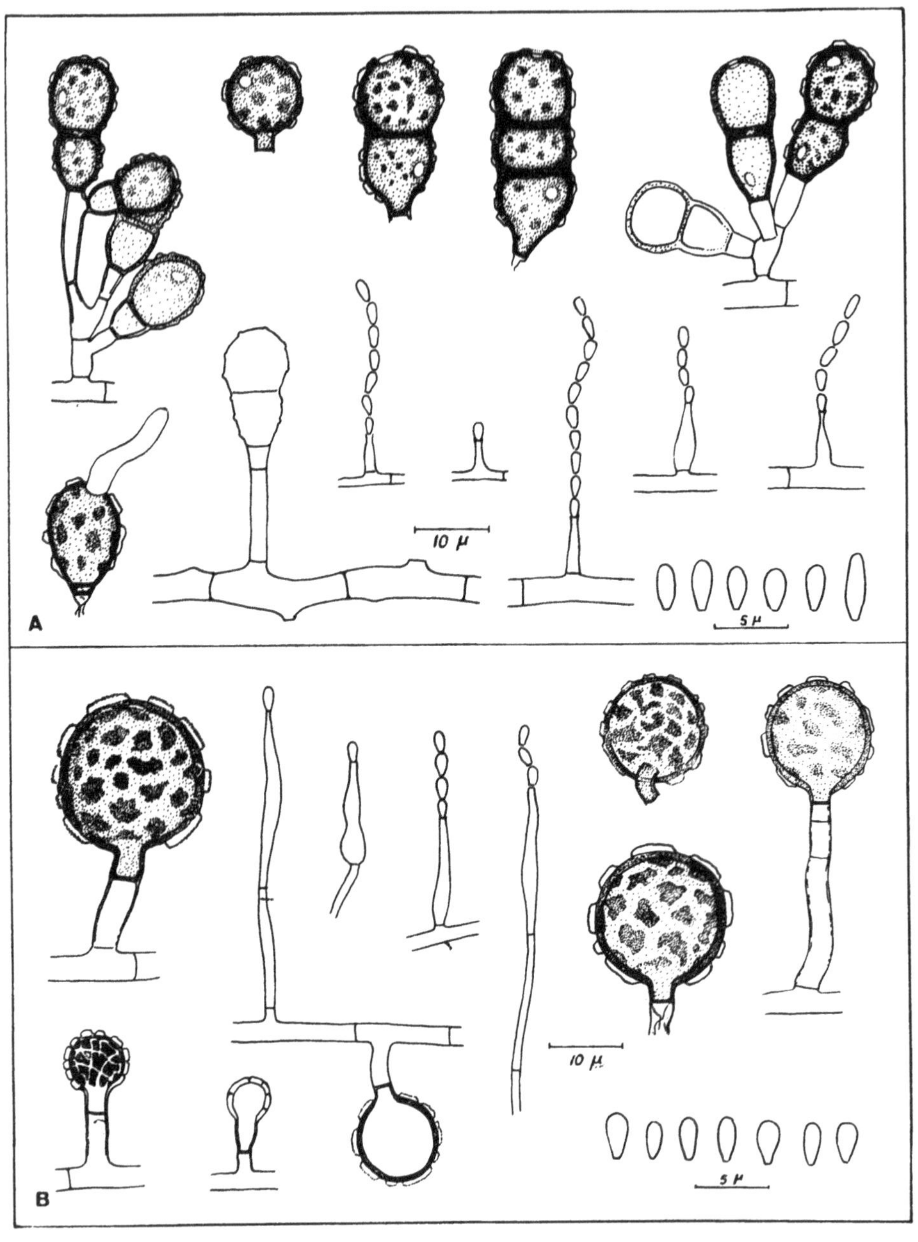


Figure 14.2 A, *Trichocladium asperum* Harz, chlamydospores and phialidic conidia (MUCL 872); B, *Thermomyces verrucosus* Pugh et al., chlamydospores and phialidic conidia (Type, CBS 116.64 = MUCL 8370).

ing the type species of *Thermomyces*, are only known in the chlamydosporic state. *Humicola*, with one-celled aleuriospores lacking germ pores, is, however, to be kept distinct from *Trichocladium* with multicellular spores possessing germ pores. Thus I hope to approach a more natural classification. 

Pleomorphism in *Trichosporonoides oedocephalis* (Haskins and Spencer 1967) illustrates anew, but now as a fact, the "metamorphism" from mould to yeast or from yeast to mould which the "polymorphists" of the past believed to be the rule. *T. oedocephalis* has yeast forms of the *Candida* and of the *Trichosporon* type, and a filamentous form of the *Oedocephalum* type. We already know of the yeast forms of more highly evolved ascomycetes, for example the connection between *Rhodotorula* and *Taphrina*. We have also known of the yeast forms of basidiomycetes since the time of Brefeld (1893), in *Ustilago, Tolyposporium, Anthrocoidea, Tilletia, Urocystis*, etc., and more recently, for example, *Cryptococcus* and *Schizosaccharomyces* forms of *Tremella, Naematelia*, and *Holtermannia* (Kobayashi and Tubaki 1965). The imperfect yeasts are too often considered separately from the Fungi Imperfecti because of differences in methods of identification. Unquestionably they merit more attention. The study of their morphology and sporogenesis may reveal, as in *Trichosporon oedocephalis*, other types of sporulation and perhaps new affinities among the imperfects.

## PLEOMORPHISM AND ITS TAXONOMIC AND NOMENCLATURAL IMPLICATIONS FOR FUNGI IMPERFECTI

### *On the Nomenclature of the Imperfects*

Before the discovery of pleomorphism in fungi, all fungus forms, including conidial, were considered as representing autonomous species and classified as such. Thus the classes Coniomycetes and Hyphomycetes took their place among other classes, Gasteromycetes and Hymenomycetes (Fries 1812-32). But, following Tulasne's discovery of the connection between conidial and ascosporic forms, the fungi classified as Coniomycetes, Hyphomycetes, and even Mycelia Sterilia were separated from the Ascomycetes and Basidiomycetes and grouped together in the class Deuteromycetes (Saccardo 1877, 1882). These fungi were then only considered as "parts," "stages," "forms," or "states" of sexually reproducing, pleomorphic fungi and their nomenclature as a potential duplicate of that of the perfect fungi.

However, the necessity of naming fungi known only in their asexual states forces us to treat their names as those of "imperfect species," grouped in genera, even families and orders. So the same principles as are applied in the classification and nomenclature of perfect fungi have been invoked. However, the Code of Nomenclature does clearly distinguish between the perfect fungi (Ascomycetes and Basidiomycetes) and the imperfects (Art. 59). On the one hand it allows the name of the perfect state to be the name of all states of any one species in ascomycetes or basidiomycetes, and thus to supersede the names given to particular imperfect states of the same species. On the other hand it rules that "in the case of imperfect states, the name refers only to the state represented by its type." This rule might imply that each state separately could be referred to by a name, as presumably representing a distinct species. In that case, emendation, synonymy, and transference of taxa or names referring to distinct states of the same fungus are hardly conceivable. The Code explicitly forbids the transfer of an epithet from the imperfect nomenclature to the perfect one. Further, for practical reasons, the successive Codes have provided for the maintenance or creation of binomials for asexual states in parallel with the names of the corresponding perfect states. Indeed, the provisions of the Code "shall not be construed as preventing the use of names of imperfect states ... [and] when not already available, specific or infraspecific names for imperfect states may be proposed at the time of publication of the name for a perfect state or later ..." (Art. 59, Edinburgh Code, 1966).

In mentioning the "imperfect state," the Code refers to the entire asexual portion or state of sexually reproducing, pleomorphic fungi (as in "imperfect states of pleomorphic fungi," Art. 59), the sexual phase of the fungus being denoted as the "perfect state." However, the term "Fungi Imperfecti" in the sense of the Code (Art. 7) includes both fungi known only in their imperfect phase and other imperfect species which are indistinguishable from states of known perfect species ("collective," "aggregate," or "complex" states or species).

But the Code goes no further. It does not distinguish between imperfect fungi known only by *one* state from those showing, simultaneously or successively, in close association or not, *several* states of reproduction or of vegetative growth organically connected. To distinguish these alternatives, I designate as monomorphic those imperfect fungi with one recognized state, and as pleomorphic those with multiple asexual states. The distinction between monomorphic and pleomorphic imperfects implies a clear definition of the "state" in such a way that its recognition as a morphological entity is possible. But we know that the concept of a fungal state is far from explicit, if it even exists other than conceptually. Surely, with the continuing refinement of characters, what is now thought to be a single state will be shown to be divisible into different elements. Conversely, evidence of a continuous association of what were thought to be separate states may force us to consider those "states" as no more than conjugated and corroborant characters of a single living species.

The Code of Nomenclature bases names on their nomenclatural types; a specific name on what is represented in its type specimen. But we do not know exactly what is supposed to be represented in the type specimen, except that, in Art. 59, the Code says that it is a "state." Either "state" is understood as the whole imperfect or asexual phase of the fungus, no matter how many types of spores or vegetative bodies are distinguished; or "state" is considered as a particular asexual spore or mycelial form, even if there are several such in the specimen. According to the definition chosen for "state," the names will refer to quite different things.

The rule in Art. 59 can thus be interpreted in divergent ways. Which interpretation should be adopted depends on the nature of the object being named in the imperfects (i.e. botanical taxa or anatomical categories) and the comparability of such nomenclature with the botanical nomenclature of the perfect fungi. The question is especially acute in the case of the pleomorphic imperfect fungi.

We shall consider these two interpretations and note their effects on the nomenclature of the Fungi Imperfecti at the specific and the generic levels. 

*Two Opposing Views on the Species Taxon in Fungi Imperfecti*

Mason (1937, pp. 71-5) clearly delineated the two opposite positions in the interpretation of the Code for the nomenclature of Fungi Imperfecti. I shall not quote from him here, but invite the reader to consult his text in full. However, Mason did not carry the logic behind each position to its conclusion, nor did he come to an explicit decision as to which was correct.*

The first position is founded on a literal interpretation - apparently the legal one - of the Code: names may be given to imperfect *states* of pleomorphic fungi, as well as to those of imperfect fungi, but these names refer only to the states represented in their types. Thus a monomorphic imperfect fungus will possess a single binomial, whereas a pleomorphic imperfect fungus will possess as many binomials as the states that the morphologist wishes to distinguish. Further, a pleomorphic perfect species will have, in addition to the perfect state name, which is the only legitimate name for the whole species, as many other binomials as there are recognized asexual states.

This literal interpretation is supported by various arguments. The first of these is that the distinct states of the same fungus frequently appear separately, both in culture and in nature, and it is desirable to name them as they are seen. Another argument resides in those pleomorphic imperfect fungi which have one state in common with several distinct Fungi Imperfecti (collective states, of which the individuals, although they belong to distinct fungi, are hardly distinguishable, and hence are still grouped together under a single state binomial). This case is common enough among Fungi Imperfecti to argue for denotation, state by state, of the Imperfecti.

The binomials, in such a systematics of Imperfecti, are thus reduced to mere monomorphic state names, names of anatomical value only; reference is made to different states of the same fungus. The concept of the state, which is here considered as a species on its own, cannot therefore be emended to include the characters of another morphologically distinct state of the same fungus (otherwise we are naming fungi, and not states of fungi). Since all of these names, without being synonyms, only designate form species or particular aspects of the fungus, the imperfect fungus species in its entire pleomorphic being can have no proper botanical name!

If then it was desired to name the composite pleomorphic species without emending one of the available state names, the only procedure would be to give it a new specific epithet to accommodate that pleomorphy. This method, seen from a Linnaean point of view, would be inadmissible under the Code because such an

*From his previous and later papers it is clear that Mason adopted the second of them, the one which should be adopted in general. Further, his discussion remaining at the specific level, he did not consider the choice of the generic names to be used in imperfect binomials, except that he also applied generic names for the denotation of a particular state of an imperfect pleomorphic species, using the expression "[form-generic name] state of [the binomial of the imperfect species]."

epithet would automatically be superfluous if its type included the type of an earlier name, or synonyms of these earlier names.

Such a situation shocks the botanist, without doubt. Therefore are we not right to ask whether a refined approach might not be preferable to a literal reading of the rule?

The second position, without doubt more sane, is based on the general botanical principle that for each species only one name can be correct (1966 Code, Principle IV), with a corollary that a name based on any state of the species may be the correct one for the species. As Mason (1937, p. 81) succinctly put it: "It is now, I think, generally agreed that seven binomials for one species is too many."

This same botanical principle implies a further corollary principle: "a binomial given to any state of reproduction or vegetative growth of a species, perfect or imperfect, is in fact given to the species itself in any of its potentialities." Such a corollary would allow synonymy between perfect and imperfect names, and would force the choice of the oldest epithet for the whole species. When applied in the same way to genera, that corollary would justify Raper's (1957) concept of a sexual genus *Penicillium* including both sexual and asexual species, and would justify the actual position of the Code in reference to the Phycomycetes. But in the case of the Ascomycetes and the Basidiomycetes, the concept of precedence of the perfect state over the imperfect, already enunciated by Tulasne in 1851, has been a part of the Botanical Code ever since the 1910 Brussels Congress. In pleomorphic fungi the binomial applied to the perfect state, even if younger, takes precedence for the composite species.

At the Stockholm Congress in 1950, however, a major concession was made to applied botanists who had refused, in the 40 years of its codification, to follow the one-species-binomial rule for pleomorphic fungi. The new concept adopted at Stockholm is that, although the binomial applied to the perfect state includes all imperfect states, botanists are authorized to use the names of the imperfect states in works referring to such states.

The Code has failed to concern itself with the correct name for a pleomorphic species of Fungi Imperfecti unconnected with an ascomycete or basidiomycete. Since most mycologists today, following Mason's (1937) lead, may freely synonymize and transfer epithets among the various states of the imperfect phase, it seems important to refine the Code, and to interpret it correctly, so that the corollary principle would now read: "Only a name given to the perfect state of a species can be the name of the species itself in any or all of its potentialities, sexual or asexual; the name given to any state of an imperfect species or of the asexual phase of a perfect species can be the name of the whole imperfect phase of that species."

This corollary principle has a highly significant facet: since all epithets applied to a pleomorphic imperfect species are nomenclaturally equal, priority (in some form) will dictate the choice of epithet to apply in the one correct binomial of a composite species; since all binomials necessarily encompass the whole species, each anatomical state remains unnamed, and cannot be named under the Code. This is, of course, the opposite situation from that of the first position proposed above. The practical necessity of having names for imperfect states prompts the proposal below for naming the different anatomical states of a pleomorphic imperfect species in a new, conventional system.

On the basis of these principles, the designation of a monomorphic imperfect species causes no problems. But in a pleomorphic imperfect species, where the different, but now connected, states have been previously named as separate species, more than one name will compete in naming the composite species. The correct epithet must logically be the oldest one (all later epithets being taxonomic synonyms) and must be combined with the appropriate generic name on the basis of the congenericity of the species with the type species.

Just as a perfect fungus species may be incompletely known in its asexual or vegetative phase and later prove to be pleomorphic, so originally monomorphic imperfect species may become known as pleomorphic. The adoption of the oldest epithet, no matter what state is represented in its type, implies an emendation by amplification of the diagnostic characters, but without excluding the type of the epithet adopted.

Taxonomic synonymy between names whose types represent different states of the same species is not quite like the usual taxonomic synonymy established by morphological identity of type specimens of names. The same kind of taxonomic synonymy occurs between the names of perfect and imperfect states of fungi, with the difference that the Code excludes the names of the imperfect states from considerations of priority, but does not restrict this (if the present interpretation is correct) among the Imperfecti themselves.

We can now see that a species of imperfect fungi is not a mere morphological state of a species, but a real fungus species known only in its asexual phase. The species in the Fungi Imperfecti should be called imperfect species. The term form species, avoided in the Code, but frequently encountered in the literature, should perhaps merely cover a single morphological state of the species, or should perhaps be abandoned.

### *The Form Genus in Fungi and in the Botanical Code*

Although the term form genus is used regularly by mycologists for denoting a genus of imperfect fungi (Ainsworth 1961), it is somewhat surprising that the Code avoids the term in reference to fungi and applies it only to fossils. This is doubly strange in that the term was apparently first proposed for use in the Code by Atkinson (1909), who defined it for the "*fungi imperfecti* dont la forme parfaite n'est pas à présent connue." Though many of Atkinson's proposals to the Brussels Congress of 1910 were incorporated into the Code, the term form genus was not adopted at that time. Finally the Code did accept the term (1952, 1956: Art. PB 1): "A *form-genus* is one that is maintained for classifying fossil specimens that lack diagnostic characteristics indicative of natural affinity but which for practical reasons need to be provided with binary names. Form genera are artificial in varying degree ... [and] have been recognized as pertaining to a *special morphological category* since 1828 (Adolphe Brongniart)." In the more recent Codes (1961, 1966), the special appendix for fossil plants disappears, and form genera are discussed in Arts. 3, 7, and 59, but always only in reference to fossils, not to fungi.

The palaeobotanical and mycological concepts of a form genus are very close in that both have a strong anatomical *essence*, and are created as artificial categories parallel to "real" genera in each group, and must not be referred to "real" families, since they may contain species belonging to different "real" families. The "original meaning" of a fossil form genus was to be maintained, "subsequent alteration of the diagnostic characters of form genera" not being desirable (1952 Code, Art. PB 6B).

Quite clearly, a special anatomical meaning is attached to the form genus of fossils, just as it is to the form genus of imperfect fungi; a wholly different *anatomical classification* appears to apply to the genera.

Most of the genera of the imperfects were originally described as monomorphic, based on one state, and including monomorphic species. For these, the nomenclature in either an anatomical system or a botanical one will be the same.

But in the case of the pleomorphic imperfect fungi, how must we treat genera whose names are typified by species based on *different states*? Are these genera, like their type (form) species, to have their diagnostic characters amplified so as to include all states (their names thus being synonymized) or not?

The Code does not answer these particular questions other than by (1) the well-known rule (Art. 59) "in the case of the imperfect states ["particular states" or "asexual phases"?], the name [generic, or specific, or generic and specific name?] refers *only* to the state represented in its type"; (2) by its acceptance of form genera, and (3) by its silence about so-called form species.

## *Three Choices of Generic Concept as the Basis of Distinct Nomenclatures for Fungi Imperfecti*

In our attempt to segregate a workable system of nomenclature of the imperfects, in which the fungal species as well as their different states can be unequivocally named, it is time to examine the different concepts of the generic taxon and to determine the nomenclatural consequences of their use.

From possible interpretations of the Code, and from common practice, three concepts of the genus suggest themselves for the imperfects: (i) the "form genus," an anatomically defined group of monomorphic fungal states of form species; (ii) what I shall call an "imperfect genus," a taxon including real - though imperfect - fungus species, congeneric in any of their characters or states; (iii) what could be imperfect and form genus combined, let us say an "anatomical imperfect genus," i.e. a taxon of botanical value, for grouping real fungus species, but also of strict anatomical meaning which, as in form genera, is monomorphic and restricted to the one particular state serving as common denominator among the included species.

According to the option taken, the nomenclatural type of the generic name will differ. The type (i) of the name of the form genus will be a form species, (ii) of the name of the imperfect genus will be an imperfect species, and (iii) of the name of the so-called anatomical imperfect genus an imperfect species in only one of its states. 

The Code provides that the nomenclatural type of the genus is a species, but fails to define the exact nature of the nomenclatural type of the form genus. Is it a "state" (Art. 59), which can mean either a form species or a state of an imperfect species, or is it simply an imperfect species?

On the basis of these generic concepts three distinct systems of the Fungi Imperfecti can be obtained.

1. The Anatomical System

In a true anatomical system, the form genus groups together morphologically similar states of fungi, each state being considered as representing a "species," the form species. The form species is based on one particular state and is therefore monomorphic by definition. Consequently pleomorphic imperfect fungi, i.e. organically connected form species, will necessarily be assigned to distinct form genera at the same time.

In the nomenclature of this system, the epithet given to the form species refers only to the state of the fungus represented in its type specimen. Just as in names of anatomical categories, the epithets given to morphologically distinct states cannot be synonymized even when these states are shown to be of the same fungus. The names of the form genera typified by these form species based on distinct states of the same fungus cannot be synonymized either. Only those form-generic names whose respective type species are typified by the same kind of spore or vegetative structure can be synonymized. This implies that the diagnostic characters of these form taxa are only emendable within the state and cannot be amplified to include those of another state of the same fungus. As a result, the transfer of an epithet based on one state to a form-generic name ultimately based on another state of the fungus is logically prohibited, and there can be several correct binomials for the same organism.

This system requires that all genera be monomorphic. Logically then, pleomorphic genera and species should be rendered monomorphic by the emendation of their diagnostic characters to circumscribe one state only. Sometimes pleomorphic taxa might be viewed as monomorphic when the different states of the fungus are so constantly and simultaneously associated that they must be regarded as a single, complex state.

The advantages of the anatomical system are: (i) the unequivocal morphological signification of the form genus, guaranteeing some stability of the classification and the nomenclature of the imperfects; (ii) a correct and unambiguous binomial for each state of these fungi; and (iii) the lack of necessity to emend generic diagnoses or to create new genera with each discovery of pleomorphy in type species, such pleomorphy entailing no changes in nomenclature.

The disadvantages of the anatomical system are: (i) the multiplication of binomials for the same pleomorphic fungus; (ii) the inclusion in form genera of unrelated fungi by virtue of their having a congeneric state, which may lead to unnatural groups; and (iii) the suppression of pleomorphic genera (if one holds these to be more natural than genera based on a single state).

Such an anatomical system is, I believe, not what is intended by botanists whose aim is to classify and name plants, not mere organs or states. However, the decision to allow the use of names for imperfects in parallel with the name for the whole perfect species (to satisfy applied scientists) has entailed the segregation of the nomenclature of the imperfects from that of the perfect fungi.

Further, the intention of those who wrote and rewrote the Codes has not been made clear. Do they want to classify and name imperfect states or imperfect fungi, making the segregated nomenclature a botanical one?

2. The Botanical System

In that segregated botanical system of the Fungi Imperfecti, the genus is conceived to be a botanical taxon, grouping real species: an "imperfect genus" for "imperfect species." Here the generic name is typified by a type imperfect species in its conceptual entirety. This will lead automatically to monomorphic and pleomorphic genera, but that distinction here is only a matter of circumscription of the generic characters. These are no longer limited to characters of one state, but may comprise those of as many states as are needed to establish the congenericity of the included species. The synonymizing of genera raises the possibility of synonymy between monomorphic genera where a previously monomorphic type species is discovered to be pleomorphic (the genera originally having been based on the now connected states). Also, if two species prove to be based on linked states of the same fungus, their names are simply synonymized. The only correct epithet, which is the oldest legitimate one, will be combined (i) with the oldest legitimate generic name whose type species is congeneric, or (ii) with the name of the most appropriate genus, or (iii) with that of a new genus for the pleomorphic species. As in other botanical systems, the genus may be emended by amplification to include the characters of the type species in each of its states. Here, pleomorphic genera have a normal status.

The advantages of this system are: (i) the provision of a more natural system for pleomorphic imperfect species, assuming that the association of more than one state in the same organism can show real affinity among species grouped in such a genus; and (ii) the provision for the Fungi Imperfecti of the one correct specific epithet each must bear in a botanical nomenclature.

The disadvantages seem numerous: (i) since the genera are characterized by an association of states (or by one state), overlapping, or even an actual reticulum of diagnostic features, will occur, making many generic concepts equivocal; (ii) each new discovery of a connection between states will require emendation of species and genera, and reconsideration of their synonymy; (iii) a monomorphic fungus whose sole state is held in common with some pleomorphic species, and whose name thereby falls into synonymy, will have to be renamed on the basis of its monomorphic character; (iv) the creation of new generic names becomes necessary either to accommodate pleomorphic species or to rename those genera whose names have gone into synonymy and have become unavailable for the residual monomorphic species. 

These so-called disadvantages are not real ones, but merely signs of an actively evolving botanical nomenclature, and are inherent in the nomenclature of any pleomorphic organisms, in the Phycomycetes and the other perfect fungi, as well as in the Fungi Imperfecti.

Applied scientists may object that this system does not provide a denomination of the states: the fungus is named, but the state remains unnamed, at least in the case of a pleomorphic fungus.

As stated in the introduction, Tulasne (1851a, b) recognized this problem and considered the use of a refined spore terminology as a solution. Certainly, in this system the only way to designate a morphological feature or state is by a term describing the type of spore or vegetative structure. The use of names of monomorphic genera is not appropriate, since all monomorphic genera are potentially pleomorphic.

3. The Botanico-anatomical System

This system provides a compromise between the two opposing positions just stated, and will, it is hoped, satisfy both botanists and applied scientists.

The genus, although apparently botanical, since its members are real imperfect species, is a form genus by virtue of the typification and application of its name. The generic name is not typified by a species as a whole, or by a species restricted to its asexual phase (imperfect species), but by an imperfect species in its sole state or in one selected state (lectostate). Like the form genera in the anatomical system, the name of such a genus is typified by a form species, which in the anatomical system was named but here remains unnamed, because it is included in the circumscription of the named type species. The system will require the exact designation of the typical state ("status anatomicus typicus") of the named type species.

The epithet of the species is typified, as in the botanical system, by a type specimen which shows the fungus in one or more states. The species bears only one correct epithet which, by emendation whenever necessary, will cover it in all its asexual or vegetative manifestations. Synonymy of specific epithets may thus occur beyond the states represented in their respective type specimens.

The application of the generic name is fixed by the typical state of the type species. To ensure the unequivocal meaning of the generic name, the diagnostic features of the genus cannot be emended beyond the "status anatomicus typicus." Generic names based on different typical states of the same type species will thus not be synonymized. Synonymy of generic names in this system is only allowed as a result of the identity or congenericity of the typical state of their type species. An imperfect species will be assigned to a form genus on the basis of the congenericity or identity of one of its states with the typical state of the genus.

The binomial of the imperfect species is thus composed of a botanical epithet, which covers the whole species and must be the oldest legitimate one, and of a form-generic name based on one of the states of the species. The form-generic name for a species will be, within the state, the oldest legitimate name. Between possible correct form-generic names for different states of the species, the choice will depend upon the judgment of the taxonomist.

The form genera are allowed to contain all kinds of imperfect species, monomorphic and pleomorphic, but they remain of monomorphic value by the typification of their names. The system does not prohibit the creation of genera for the accommodation of pleomorphic fungi, but it requires the designation by the original (or a later) author of the typical state (holostatus or lectostatus) of the type species. Thus *Chalaropsis* Peyronel, based on chlamydospores and phialidic conidia, requires designation of the lectostate of its type species (the chlamydosporic state) before a monomorphic concept of the generic name can be established (Hennebert 1968).

The advantages of the composite botanico-anatomical system appear evident: (i) it will allow more stable grouping of species with an increasing degree of affinity, by admitting typical states and genera for pleomorphic fungi; (ii) it will avoid the confusing synonymy of generic names whose types represent different states; (iii) it will ensure that there is only one correct name for the imperfect species; (iv) it will allow us to keep monomorphic species together with pleomorphic ones, or eventually to group together (on the basis of their common denominator) pleomorphic species whose pleomorphy is not identical but multidirectional; (v) it provides us with an accurate anatomical nomenclature to denote the separate states of the species and to designate the typical state of the generic type species.

Up to this point, the state as a morphological feature of the species is still unnamed, and the taxonomist as well as the applied scientist may wish to name it. The need thus arises for an adequate morphological terminology, or perhaps nomenclature. Already aware of that need, Tulasne (1851a, b) revised the spore terminology, but preferred to say "the *Cytospora* state of *Valsa ambiens*" rather than "the pycniospores of *Valsa ambiens*," for reasons of precision. After Tulasne, and Mason (1937), such expressions as "the *Cytospora* state of *Valsa ambiens*" or "the *Echinobotryum* state of *Cephalotrichum stemonitis*" came into common use (Hughes 1958). They have the advantages of being more precise than morphological terms, and of designating unequivocally any state of the imperfect species by avoiding confusion with specific binomials and by using generic names in their strict morphological meaning. It is precisely this unequivocal use that is provided by the "botanico-anatomical" system of nomenclature of the Fungi Imperfecti.

These expressions may be considered too long, or not easily enough translated into other languages, or even difficult to list in monographs of species having similar states. Hence a standardized Latin denotation, like a trinomial, is proposed here to designate asexual states of pleomorphic fungi. This denotation is composed of two parts, the first being the correct binomial of the species, perfect or imperfect, with or without an author citation. The second part consists of the term "status anatomicus," abbreviated to "st." and printed in roman typeface, followed by the generic name (in italics) of the correct genus of Fungi Imperfecti for that state. For example, in Figure 14.1 A, the two states of the species *Doratomyces stemonitis* now normally distinguished could be cited as *Doratomyces stemonitis* (Pers. ex.

Fr.) Morton & Smith st. *Echinobotryum* and *Doratomyces stemonitis* st. *Doratomyces,* for the chlamydosporic and annellidic states respectively. In the same way, the two states of *Chalaropsis punctulata* Hennebert may be denoted as *Chalaropsis punctulata* st. *Chalara* and *Chalaropsis punctulata* st. *Chalaropsis,* phialidic and chlamydosporic states respectively. 

These Latin denotations of states of fungi should be considered, not as botanical names of infraspecific rank, but only as anatomical connotations of the species, and should not be allowed to appear in synonymies.

## DISCUSSION

An analysis of the text of the Code and of taxonomic literature has demonstrated the existence of widely differing concepts of generic and specific taxa, and of the names of Fungi Imperfecti. An attempt to understand the basic differences between these concepts has led to an explanation of the principles underlying the various nomenclatural systems extant for the Fungi Imperfecti. That each system has its advantages and disadvantages has been briefly shown above. That all three systems are in current use, whether explicitly or by inference, will be shown below by a few examples selected from the literature.

An example of what I call the "anatomical" system was pointed out by Mason (1937). Wakefield (1918) described a parasitic fungus with acervuli under the name of *Cercosporella antirrhini* Wakef.; in 1929, Buddin and Wakefield transferred the name to *Pseudodiscosia antirrhini* (Wakef.) Budd. and Wakef. In the meantime, the same authors had observed a pycnidial form in the same host and, in spite of a pure culture demonstration of its connection with *P. antirrhini*, described it as a separate new species, "*Heteropatella antirrhini* Budd. and Wakef. sp. nov." Here it is clear that each name covers one and only one state of the fungus; these names, therefore, are the names of form species within form genera. This is one of the rare examples of the adoption of separate specific names for states of a single imperfect fungus when the connection between the states is already known.

The "botanical" treatment of imperfect taxa and of their names at both specific and generic levels is much more common among mycologists. *Trichosporonoides* Haskins and Spencer, with its type species *T. oedocephalis* Haskins and Spencer (1967), is a good example of the botanical system in operation, for the fungus develops arthric conidia together with blastic conidia and chlamydospores, the blastic conidia arising either as consecutive buds in a *Candida*-like arrangement or as simultaneous terminal buds of the *Oedocephalum* type. The whole pleomorphy is encompassed by the names of the species and the genus.

Another example of a pleomorphic imperfect genus based on a pleomorphic species is *Diheterospora* Kamyschko (Kamyschko 1962), a new genus erected for two dimorphic species, *D. heterospora* Kamyschko, the type species, and *D. catenulata* Kamyschko. In 1966, Barron and Onions, reconsidering this with respect to *Stemphyliopsis* A.L. Smith and one of its species, arrived at the following treatment:

*Diheterospora chlamydosporis* (Goddard) Barron and Onions, comb. nov.
≡ *Verticillium chlamydosporium* Goddard, Bot. Gaz. 56: 275-276, 1913.
= *Stemphyliopsis ovorum* Petch, Trans. Br. Mycol. Soc. 23: 147, 1939.
= *Diheterospora heterospora* Kamyschko, Bot. Mater. 15: 138, 1962.

This nomenclatural procedure - (i) placing in synonymy names given at one time to the phialidic state, and at another time to the chlamydosporic state, and (ii) choosing (by virtue of strict priority) the earliest epithet no matter what state is covered by the name - is certainly derived from a "botanical" approach to the Fungi Imperfecti, and is exactly that which Mason (1937, 1941) recommended for general use.

In the above example, it is only because *Stemphyliopsis ovorum* Petch is not the type species of *Stemphyliopsis* that *Stemphyliopsis* is not accepted as a synonym of *Diheterospora.* This final step, of actually accepting the synonymy of such generic names, was taken by Morton and Smith (1963) when they synonymized the type species of genera originally based on distinct states, and accepted the consequences of this with a synonymy of generic names as follows:

*Doratomyces* Corda, in Sturm's Deutschlands Flora, iii (Pilze), Bd. 3, Heft 7, p. 65, 1829.
= *Stysanus* Corda, in Icones Fungorum 1, p. 21, 1837.
= *Echinobotryum* Corda, in Sturm's Deutschlands Flora, iii (Pilze), Bd. 3, Heft 12, p. 51, 1831.

*Doratomyces stemonitis* (Pers. ex Fr.) comb. nov.
= *Doratomyces neesii* Corda, in Sturm's Deutschlands Flora, iii (Pilze), Bd. 2, Heft 7, p. 65, 1829.
= *Cephalotrichum stemonitis* (Pers.) Link ex Fr., in Syst. Myc. 3, p. 280.
≡ *Periconia stemonitis* Persoon, in Synopsis Meth. Fung., p. 687, 1801.
= *Echinobotryum atrum* Corda, in Sturm's Deutschlands Flora, iii (Pilze), Bd. 3, Heft 13, p. 51, 1831.

Schol-Schwarz (1968) took a similar botanical decision when she emended an originally monomorphic genus *Rhinocladiella* Nannf. (Melin and Nannfeldt 1934), and placed into synonymy with it several later generic names, based on the same pleomorphic type species, or species she considered congeneric with it, namely: *Fonsecaea* Negroni (1936), *Hormodendroides* Moore and Almeida (1936), *Carrionia* Briceno-Iragorry (1939), *Phialoconidiophora* Moore and Almeida (1936), *Botrytoides* Moore and Almeida (1936); all were proposed for one or another association of up to three of the four known states of these fungi.

This procedure, although strictly botanical, has the disadvantage mentioned earlier that useful names of monomorphic genera may disappear into synonymy, or by emendation; but this is true of any botanical system.

The "botanico-anatomical" system restricts the synonymy of generic names to individual states. This is the kind of option which seems to have been taken intuitively by Hughes (1958) when, although he synonymized *Echinobotryum atrum* Corda with *Cephalotrichum stemonitis* (Pers.) Link, both considered the nomenclatural types of their respective genera, he did not put the generic names into syno-

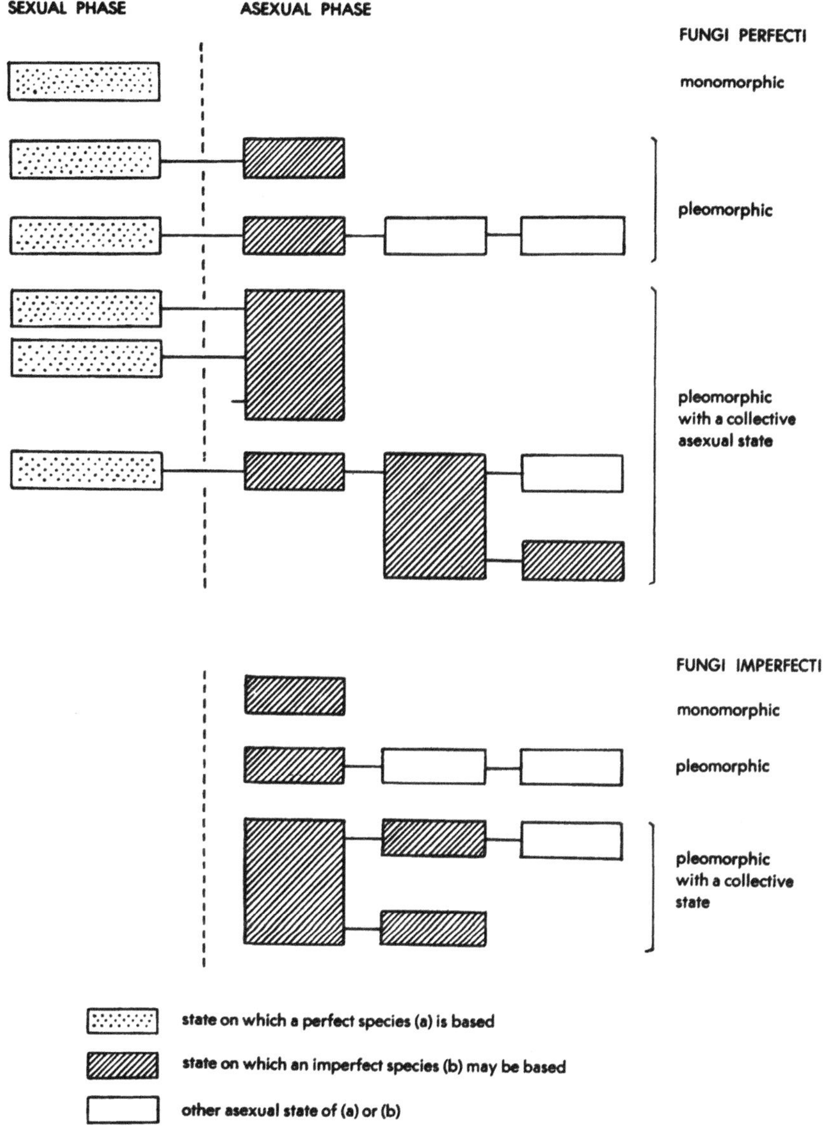


Figure 14.3 Schematic figure of organic connections between states and phases in perfect and imperfect species of fungi.

nymy. In this compromise system, no matter how pleomorphic an imperfect genus may be, its name must be typified by one characteristic state of a species; and therefore, for nomenclatural purposes, it must be considered monomorphic. This is the treatment given experimentally by Hennebert (1967) to *Chalaropsis* Peyr. in characterizing it by the chlamydosporic state of the type species, *C. thielavioides* Peyr., which is more readily recognized by its chlamydospores than its *Chalara* phialidic conidia.

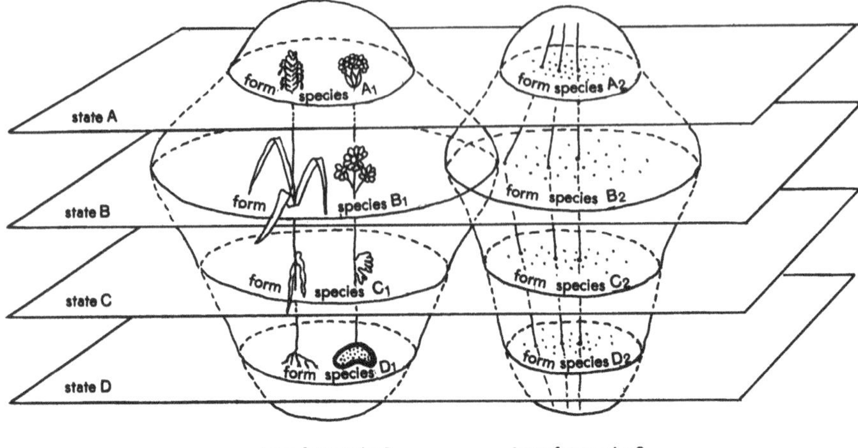


Figure 14.4 Diagrammatic figure of the imperfect species and form species in pleomorphic imperfect fungi. The pleomorphic imperfect species, represented as a volume (1 and 2), is a bundle of individuals (the vertical lines) like wheat plants in a field. The form species, characterized by states A, B, C, and D (roots, old leaves, green leaves, and immature ear of wheat, or sclerotia, appressoria, conidia, and spermatia of *Botrytis cinerea*), are represented by the sections of the volume made by the horizontal planes A, B, C, and D. They group individuals (the dots) as parts of the vertical lines at each level. Notice the overlapping of form species $B_1$ and $B_2$ of the imperfect species 1 and 2.

The use of an expression such as "the *Chalara* state of *Chalaropsis punctulata*" signifies that the species name is meant in the botanical sense to cover the whole pleomorphic phase of the fungus. In other words, the species is not a form species, and the name is not a state name. This approach to the species is common to the botanical and the botanico-anatomical systems.

Obstacles to the adoption of the latter system lie in the need for a clear definition of what a state is (in the Fungi Imperfecti), and for a special kind of typification of the generic names. Which of the three systems is worth exclusive adoption remains to be decided. But that this matter is worth discussion and thought seems obvious in the interests of establishing a consistent basis for the nomenclature and classification of Fungi Imperfecti.

## CONCLUSION

Pleomorphism in fungi has led, through the discovery of the multiplicity of associated asexual states, to two systems of nomenclature for species. One, the botanical, is horizontal, the individuals assembled side by side as in a field of wheat, the area of which delimits the species. The other, the anatomical, is vertical, cutting these same individuals into morphologically distinct, superimposed slices (Figure 14.4). These two classifications, apparently contradictory by virtue of the irreducible difference of their taxa (species and states), actually neatly intermesh to form a three-dimensional volume, the species concept. Practical problems arise from this situation in the nomenclature of the Fungi Imperfecti. The nomenclatural systems elaborated here for the species and for the genus, as well as for the denotation of morphological states, should permit greater uniformity in the nomenclature of the imperfect, pleomorphic fungi. 

## ACKNOWLEDGMENTS

I am grateful to Dr R. P. Korf, Cornell University, who with rare generosity proceeded with me for days along the difficult pathways of fungus nomenclature. His clear knowledge of the Code and its implications, and his unfailing amiability in the task of translating the text, are more reasons for warmest acknowledgment.

Deep gratitude is also expressed to Dr Luella K. Weresub, Plant Research Institute, Ottawa, for her well-elaborated and provoking comments on the first draft of the paper as presented at the Kananaskis Conference, and for stimulating the revision of the paper into its present form.

## REFERENCES

Ainsworth, G.C. 1961. Ainsworth and Bisby's Dictionary of the Fungi. 5th ed. Commonwealth Mycological Institute, Kew

Atkinson, G.F. 1909. Motions for additional articles to the rules for Nomenclature of the Fungi, proposed for action at the 3rd International Botanical Congress at Brussels, 14-22 May 1910. Ithaca

Barron, G.L. 1966. A new species of *Scopulariopsis* from soil. Ant. van Leeuwenhoek 32: 293-8

Barron, G.L., and Onions, A.H.S. 1966. *Verticillium chlamydosporium* and its relationship to *Diheterospora, Stemphyliopsis*, and *Paecilomyces*. Can. J. Botany 44: 861-9

Beauverie, J. 1899. Sur le polymorphisme de l'appareil conidien du *Sclerotinia fuckeliana* (de Bary) Fuckel, le *Botrytis cinerea* et la maladie de la toile. C.R. Acad. Sci., Paris 128: 846-9, 1251-3

Brefeld, O. 1893. Untersuchungen aus dem Gesamtgebiete der Mykologie XII, Hemibasidii. Münster. pp. 99-236

Briceno-Iragorry, L. 1939. Sobre chromoblastomicosis. Rev. Clin. Louis Razetti 1: 108-28 (Abstr. in Rev. Appl. Mycol. 19 [1940]: 277)

Briquet, J. 1912. International Rules of Botanical Nomenclature adopted by the International Botanical Congresses of Vienna (1905) and Brussels (1910). Jena

Buddin, W., and Wakefield, E.M. 1929. The fungus causing leaf spot of the carnation. Trans Brit. Mycol. Soc. 14: 215-21

De Bary, A. 1854. Ueber die Entwicklung und den Zusammenhang von *Aspergillus glaucus* und *Eurotium*. Bot. Zeit. 12: 425-34, 441-51

- 1887. Comparative morphology and biology of the Fungi, Mycetozoa, and Bacteria

Delpino, F. 1887. Equazione chimica e fisiologica del processo della fermentazione alcoolica. Nuovo Giorn. Bot. Ital. 19: 260

Dodge, C.W. 1935. Medical mycology. Mosby, St. Louis

Domsch, K.H., and Gams, W. 1970. Pilze aus Agrärboden. Stuttgart

El-Ani, A.S. 1968. The cytogenetics of the conidium in *Microsporum gypseum* and of pleomorphism and the dual phenomenon in fungi. Mycologia 60: 999-1015

Gams, W. 1968. Two new species of *Wardomyces*. Trans. Brit. Mycol. Soc. 51: 798-802
Gäumann, E.A. 1952. The Fungi. Transl. by F.L. Wynd. Hafner, New York
Gilkinet, A. 1875. Mémoire sur le polymorphisme des champignons. Acad. Roy. Belg., Bruxelles
Grigoraki, I. 1925. Recherches sur les dermatophytes. Ann. Sci. Nat. Botany 7: 165-444
Haskins, R.H., and Spencer, J.F.T. 1967. *Trichosporonoides oedocephalis* n. gen., n. sp. I. Morphology, development, and taxonomic position. Can. J. Botany 45: 515-20
Hennebert, G.L. 1967. *Chalaropsis punctulata*, a new hyphomycete. Ant. van Leeuwenhoek 33: 333-40
- 1968. *Echinobotryum, Wardomyces* and *Mammaria*. Trans Brit. Mycol. Soc. 51: 749-62
Hughes, S.J. 1958. Revisiones hyphomycetum aliquot cum appendice de nominibus rejiciendis. Can. J. Botany 36: 727-836
- 1959. Microfungi. IV. *Trichocladium canadense* n. sp. Can. J. Botany 37: 857-9
Kamyschko, O.P. 1962. De Monilialibus terrestribus novis notula. Bot. Mater. (Notul. Syst. Sect. Crypt. Inst. Bot. Acad. Sci. U.S.S.R.) 15: 137-41
Kobayashi, Y., and Tubaki, K. 1965. Studies on cultural characters and asexual reproduction of heterobasidiomycetes. I. Trans. Mycol. Soc. Japan 6: 29-34
Langeron, M., and Milochevitch. 1937. Sur l'irréversibilité du pléomorphisme et la faillité de tous les procédés de régénération. Ann. Parasitol. 15: 177-81
Langeron, M., and Talice, R.V. 1930. Nouveau type de lésion pilaire expérimentale produit par la culture purement pléomorphique de *Sabouraudites·felinens.* Ann Parasitol. 8: 419-21
Lanjouw, J. (ed.) 1952. International Code of Botanical Nomenclature adopted by the VIIth International Botanical Congress, Stockholm, 1950. Utrecht
- 1956. International Code of Botanical Nomenclature adopted by the VIIIth International Botanical Congress, Paris, 1954. Utrecht
- 1961. International Code of Botanical Nomenclature adopted by the IXth International Botanical Congress, Montreal, 1959. Utrecht
- 1966. International Code of Botanical Nomenclature adopted by the Xth International Botanical Congress, Edinburgh, August 1964. Utrecht
Mason, E.W. 1937. Annotated account of fungi received at the Imperial Mycological Institute. List II (Fascicle 3 - General Part). C.M.I. Mycol. Pap. 4: 69-99
- 1941. Annotated account of fungi received at the Imperial Mycological Institute. List II (Fascicle 3 - Special Part). C.M.I. Mycol. Pap. 5
Melin, E., and Nannfeldt, J.A. 1934. Researches into the blueing of ground woodpulp. Svenska Skogsvårdsforen. Tidskr. 32: 397-616
Moore, M., and de Almeida, F. 1936. New organisms of chromoblastomycosis. Ann. Miss. Bot. Gdn. 23: 543-52
Morton, F.J., and Smith, G. 1963. The genera *Scopulariopsis* Bainier, *Microascus* Zukal, and *Doratomyces* Corda. C.M.I. Mycol. Pap. 86
Negroni, P. 1936. Estudio micologico del primer caso argentino de cromomicosis *Fonsecaea* (n.g.) *pedrosoi* (Brumpt, 1921). Rev. Inst. Bacteriol. Dep. Nac. Higiene B. Aires 7: 419-26
Raper, K.B. 1957. Nomenclature in *Aspergillus* and *Penicillium*. Mycologia 49: 644-62
Sabouraud, R. 1900. Dermatophytes. La pratique dermatologie. Paris
Saccardo, P.A. 1877. Commentarium mycologicum fungos in primis italicos illustrans. Michelia I. Padua
- 1882. Sylloge fungorum. Vol. 1. Padua
Savile, D.B.O. 1969. The meaning of "pleomorphism." Mycologia 61: 1161-2

Schol-Schwarz, M.B. 1968. *Rhinocladiella*, its synonym *Fonsecaea* and its relation to *Phialophora*. Ant. van Leeuwenhoek 34: 119-52

Snell, W.H. 1936. Three thousand mycological terms. Publ. 2, Rhode Island Bot. Club, Providence

Snell, W.H., and Dick, W.A. 1957. A glossary of mycology. Harvard Univ. Press, Cambridge, Mass.

Traaen, A.E. 1914. Untersuchungen über Bodenpilze aus Norwegen. Nyt. Magaz. Naturvidensk. 52: 19-120

Tubaki, K. 1958. Studies on Japanese hyphomycetes. V. Leaf and stem group with a discussion of the classification of hyphomycetes and their perfect stages. J. Hattori Bot. Lab. 20: 142-244

Tulasne, L.R. 1851a. Note sur l'appareil reproducteur dans les lichens et les champignons (1ère partie). C.R. Acad. Sci., Paris 32: 427-30

- 1851b. Note sur l'appareil reproducteur dans les lichens et les champignons (2ème partie). C.R. Acad. Sci., Paris 32: 470-5

Tulasne, L.R., and Tulasne, C. 1861-1865. Selecta fungorum carpologia. Paris

Wakefield, E.M. 1918. New and rare British fungi. Kew Bull. 1918: 229-33

# 15
# Discussions on Terminology

"When I use a word," Humpty Dumpty said, in a rather scornful tone, "it means just what I choose it to mean - neither more nor less."

(*Through the Looking Glass*)

"Then you should say what you mean," the March Hare went on. "I do," Alice hastily replied; "At least - at least, I mean what I say - that's the same thing, you know." "Not the same thing a bit!" said the Hatter.

(*Alice's Adventures in Wonderland*)

*Here begins the most controversial segment of the book. New terms were introduced, old terms retired, as we worked gradually towards a unified scheme of terminology during the last day and a half of the conference. In the Preface it was suggested that, with the kind of conference set-up planned for Kananaskis, the discussions could be "just as valuable ... as the papers." The present chapter should bear this statement out. It is made up largely of discussion, and would seem to epitomize the give-and-take of the conference. Prejudice and polemics will be found. Ideas, too. And, finally, some recommendations, adopted either unanimously or by majority vote.*

DR ELLIS

One of the main aims of this conference is to reach agreement about the ways in which Fungi Imperfecti should be described. To ensure this, we have to be clear in our minds about the precise meaning of the terms we employ in our descriptions.

Descriptive writing lends itself to the use of long words; and in our efforts to be precise or to avoid repetition, we tend to use a lot of nouns such as sympodioconidium and botryoaleuriospore. These are excellent within limits, but their number is increasing at an alarming rate. It may, perhaps, be more economical in the long run, and just as precise, to use quite a small number of clearly defined adjectives in conjunction with nouns already in common usage. Some of these nouns, however, even ones we use most often (such as conidiophore), need to be defined exactly.

I believe that descriptions of Fungi Imperfecti should follow certain patterns and that they should be truly comparative. *Colony characters* may be followed by a description of *vegetative parts* - hyphae, setae, appressoria, stromata, sclerotia, etc.; next the type, arrangement, growth, shape, branching, swellings, colour, septa, and walls of *conidiophores*; then the type, arrangement, proliferation, scars, denticles, and separating cells of *conidiogenous cells*; lastly, the origin, arrangement, type, shape, hilum, colour, septa, walls, and germination of *conidia*. 

Many terms are already well defined in standard glossaries such as Snell and Dick's *A Glossary of Mycology*; some new ones appeared in a paper by Luttrell in *Mycologia*, 55 (1963): 643-74, and in Barron's *Genera of Hyphomycetes from Soil*, 1968. A few that I suggest we may discuss here, or which some of you may wish to see clarified, are: *conidiophore*, micronematous, macronematous; *conidiogenous cell*, phialidic, tretic, blastic, percurrent, sympodial; *conidium*, endogenous, exogenous, etc.

Dr Kendrick wrote on the blackboard "Think conidia." As chairman of this session on terminology, I'd like to second that recommendation. In our discussions I'd like us to avoid the word spore and substitute the word conidium whenever we are specifically mentioning a propagule produced by a Fungus Imperfectus. I'd also like to see us extend that preference to derivatives and compound forms: conidiogenous cell, phialoconidium, etc.

Now I'd like to read two definitions supplied by Dr Carmichael: (i) "The cell - any unit of a fungus thallus or spore which is morphologically separated from neighbouring units by a wall or septum." A footnote reads: "Contiguous cells do not necessarily contain individual protoplasts; cytoplasm and nuclei may pass from one cell to another." (ii) "The conidiogenous cell - any cell from which or within which conidia are directly produced."

Next is the conidiophore. We've had an excellent general account of this from Dr Pirozynski, but there was disagreement about a suitable definition, the main dissenter being Dr Carmichael. The original definition was "a hypha which bears conidia." We soon modified this to: "Conidiophore - a simple or branched hypha or hyphal system which forms conidia from either integrated or discrete conidiogenous cells." This enables us to use the term for both micronematous and macronematous conidiophores, whether the conidia are borne below, at, or above the surface of the substrate. Dr Carmichael's definition reads: "Conidiophore - a cell or cells differentiated to support the conidia away from the vegetative mycelium."

DR CARMICHAEL

The difference between these definitions is that the first includes both macronematous and micronematous forms, whereas the second says that micronematous forms do not have conidiophores as such.

DR S.J. HUGHES

Could the definition be expanded to include ascospores, the cells of which may sometimes bear conidiogenous cells? "A simple or branched hypha or hyphal system or spore ..."

DR CARMICHAEL

Yeasts have conidia but don't have conidiophores.

DR S.J. HUGHES

But they have a one-celled thallus which is itself a conidiogenous cell.

DR CARMICHAEL

Do you want to call this single conidiogenous cell a conidiophore?

DR S.J. HUGHES
Yes!
DR CARMICHAEL
A single conidiogenous cell *may* sometimes be a conidiophore. It may be differentiated so as to hold the conidia well up away from the mycelium. The proliferation of a conidiogenous cell may form a conidiophore.
EDITOR
*The meeting remains unconvinced. A straw vote rejects Dr Carmichael's strictures.*
DR S.J. HUGHES
There should be some indication in the definition that the conidiophore may be another conidium or an ascospore. It also might be possible to introduce the term sessile where a conidiogenous cell is borne directly on the mycelium.
DR CARROLL
Does this category include situations where the conidiogenous cell is part of the mycelium?
DR HUGHES
The term integrated covers that case.
DR KENDRICK
Could I ask where we could fit in something about a single conidiogenous cell acting as a conidiophore? That should be explicit rather than implicit.
DR CARMICHAEL
If we have a branched system of prostrate hyphae, and conidia are formed at the tips of the branches, where is the conidiophore? We have terms for the various elements - conidium, conidiogenous locus, conidiogenous cell, and hypha.
DR PIROZYNSKI
If prostrate hyphae bear isolated conidiogenous cells, these cells (for all intents and purposes) are conidiophores. If prostrate hyphae are made up of conidiogenous cells, the whole system constitutes a conidiophore. The distinction between an erect chain of conidiogenous cells and a prostrate one is most arbitrary.
DR CARMICHAEL
The whole thallus, then, is a conidiophore?
DR ELLIS
Only the parts bearing conidia.
DR CARMICHAEL
If the whole thallus, here and there, bears conidia, it would all be a conidiophore?
DR KENDRICK
My impression in the case Dr Ellis and Dr Carmichael are discussing is that the conidiophore is the conidiogenous cell. Otherwise you can't define it.
DR CRANE
In higher plants, a simple scale on a stem is recognized as a leaf - a modified leaf - a reduced leaf. Here we have a simple cell - a reduced conidiophore.
DR ELLIS
We need as simple a definition as possible, to be in line with the original meaning: a hypha which bears conidia. We have extended this because a conidiophore can be a single cell, a spore, or a hyphal system; but basically it is something which bears conidia.

EDITOR 

*At this point there was considerable in-fighting: several people spoke at once, others were busy rearranging several definitions on the blackboard.*

DR S.J. HUGHES

Traditionally, the term conidiophore has been applied to something that is different from the hypha that bears it, and I think there should be something in the definition to the effect that differentiated conidiophores may be absent. This would cover hyphae of the thallus which bear conidiogenous cells.

DR KENDRICK

It is still possible to interpret the whole of a hyphal system as a conidiophore - in other words, we can still disagree while using the same definition. It is not yet unequivocal.

DR SUBRAMANIAN

"A conidiophore is a simple or branched hypha, hyphal system, or spore which forms conidia from integrated or discrete conidiogenous cells. It may be reduced to a single sessile or integrated conidiogenous cell."

DR HENNEBERT

I would replace "simple or branched" by "more or less differentiated."

DR KENDRICK

This is probably the crux of Dr Carmichael's argument. Even if the conidiogenous structure was growing along in the same direction as the vegetative hyphae on the surface of the substrate, if it was differentiated from those hypae, it would be recognizable as a conidiophore. Here it may be reduced to a conidiogenous cell. We could say: "If there is no differentiated supporting structure, then the conidiophore is the conidiogenous cell." Is that reasonable? It is a strange blend of the two positions.

DR ELLIS

That is O.K. A conidiophore, then, is a supporting structure.

DR KENDRICK

"A conidiophore is a supporting structure, be it a spore, a cell, a hypha, or a hyphal system, which is differentiated from normal vegetative mycelium, either morphologically or by the presence of a conidiogenous locus or loci."

DR PIROZYNSKI

"A conidiophore is a system of conidiogenous cells, or a single conidiogenous cell, with or without differentiated supporting stuctures."

EDITOR

*The meeting, again with the exception of Dr Carmichael, provisionally accepts this definition.*

DR CARMICHAEL

My reason for not accepting it is that, if we are going to say that all our fungi have conidiophores, then the term is superfluous. All fungi have conidiogenous cells of one kind or another, and if we do not wish to restrict the term conidiophore to some specialized structure that holds the conidia away from the vegetative thallus, we don't need the term at all.

DR KENDRICK

All fungi have cells, so why do we need to define "cell"?

DR CARMICHAEL

We need to differentiate fungal cells from those of other organisms.

DR MÜLLER

There are fungi which don't have conidiophores - Phycomycetes, for example.

DR CARMICHAEL

When I say "fungus," I mean "hyphomycete"! [*Laughter*]

DR ELLIS

I'd like us now to proceed to the report of the Arthroconidia Committee. I want us to consider this first because the committee has made a very broad report dealing with a number of important points. I'll call upon the chairman, Dr Kendrick, to read the report.

## REPORT OF THE COMMITTEE ON ARTHROCONIDIA

Chairman: Dr Bryce Kendrick
Members: Dr Bill Carmichael, Dr Garry Cole, Dr Gregoire Hennebert, Mrs Flora Pollack, Dr Keisuke Tubaki

We consider that the term arthrospore (now arthroconidium) as used in the contemporary literature circumscribes a rather heterogeneous assemblage of developmental phenomena. Not only do arthroconidia, sensu stricto, differ ontogenetically from meristem arthroconidia, but the latter term encompasses two divergent groups. We decided that we could not treat these phenomena in isolation, but must examine and dissect the entire spectrum of conidiogenesis in an attempt to recognize its basic components, and to synthesize a rational and inclusive terminology.

We recognized six components:

1. The kind of development shown by the individual conidium. We noted a basic dichotomy between what Cole and Kendrick have described as "growth" and "conversion." The growth phenomenon we termed blastic and the conversion, thallic. Our first definitions were: (a) *blastic* - marked enlargement of a recognizable conidium initial (more than 2 times linear); (b) *thallic* - little or no enlargement of conidium initial (less than 2 times linear). [Conidium initial = a cell or part of a cell which will become a conidium by differentiation.] Further discussion revealed that our original conception had been too simple: while nicely defining the extremes, we had ignored many forms in which "growth" and "conversion" were both involved. For instance, some conidia that we would clearly regard as thallic - those of *Amblyosporium*, and the intercalary chlamydospores of a number of fungi - expand to several times their original size before maturity. Our definitions of blastic and thallic overlapped in a most important respect. We eventually formulated the following more precise definitions: (a) *blastic* - marked enlargement of a recognizable conidium initial *before* the initial is delimited by a septum, or septa; (b) *thallic* - if any enlargement of the recognizable conidium initial occurs, it occurs only *after* the initial has been delimited by a septum, or septa.

In the discussion which followed, these definitions were reduced to their simplest terms:
(a) *blastic* - conidium derived from part of a cell;
(b) *thallic* - conidium derived from an entire cell.

2. The kind of connection between the conidium and the conidiogenous cell or hypha:

| | Geotrichum candidum | Conidial Hysterium insidens | Conidial Monascus ruber |
|---|---|---|---|
| | Hughes's group VII | Group VA | Group VB |
| Conidium ontogeny | Thallic | Thallic | Blastic |
| Connection | Full width | Full width | Full width |
| Conidia | Successive irregular | Successive regular | Successive regular |
| Conidiogenous locus | Irregular | Diffuse | Retrogressive |
| Conidiogenous cell | Determinate | ± indeterminate | Determinate |
| | Shortening | ± stable | Shortening |
| Secession by | Fission | Fission | Fission |

(a) *minimal* - narrow, comparable in width with a septal pore (perhaps obviating the development of a double septum);
(b) *restricted* - wider than a septal pore, but noticeably narrower than the conidiogenous hypha;
(c) *full width* - the full width of the conidiogenous hypha.

3. The way in which a plurality of conidia develop:
(a) ± *synchronously;*
(b) ± *successively* (in time), *regularly* (in space);
(c) ± *successively* (in time), *irregularly* (in space).

4. The "conidiogenous locus" in 3 (b) and (c) may be:
(a) *stationary;*
(b) *retrogressive* (moving downward);
(c) *progressive* (moving upward);
(d) *irregular;*
(e) *diffuse* or *extended.*

5. The modifications of a conidiogenous cell or hypha to permit production of a succession of conidia:
(a) *stable* - no change in shape or size;
(b) *shortening* - cell being consumed;
(c) *growing* - (i) percurrently; (ii) sympodially, or by swelling; (iii) basauxically.

6. Methods of secession:
(a) *fission* of inner wall, always at a double septum;
(b) *fracture* of inner wall, not at a septum: (i) of an unmodified cell wall, (ii) of a wall with a thin area, (iii) of a wall weakened by lysis.

Having analysed "conidiogenesis" in the terms outlined above, we returned to a consideration of fungi producing the various kinds of "arthroconidia."

We recommend the rejection of the term arthrospore. If the recommended alternative arthroconidium (or, preferably, thalloconidium) is to be used, we would restrict it to conidia whose ontogeny is thallic, and whose connection is full width.

We recommend the rejection of the term meristem arthrospore for those conidia produced by *Monascus, Trichothecium*, and *Cladobotryum*, because these are essentially blastic conidia with full-width connection, which are produced by a linear, basipetal succession of conidiogenous loci which consume the conidiogenous cell.

The conidia of *Hysterium insidens* and *Phragmotrichum chailletii* clearly differ from those of *Monascus*, etc., in being thallic in origin and in involving a diffuse or extended meristem. We recommend the retention of the term meristem to refer to these conidia.

DR S.J. HUGHES

Have you numbered the concepts in the order of the importance you attribute to them?

DR KENDRICK

To some extent, but not altogether.

DR ELLIS

It is important to note all of these concepts, because they come into the discussion of the other sections. That is why I brought this report down first.

DR SUBRAMANIAN

Do these concepts in fact allow proper definition of such terms as blastospore and aleuriospore, as we have been using them?

DR ELLIS

It is a question of when the swelling takes place in relation to the laying down of the delimiting septum. Although conidia may look alike when mature, their ontogeny is the criterion here.

DR CARROLL

In view of the fact that septa are often pierced by septal pores, how valuable a criterion can the septum be?

DR KENDRICK

We know that in many fungi, for example, the *Basipetospora* state of *Monascus ruber*, the second conidium does not begin to form until the first conidium is delimited by a basal septum. Whether there is a septal pore or not, the actual laying down of the delimiting septum does seem to be a pivotal event.

DR ELLIS

May I proceed to some other recommendations of the Arthroconidia Committee: "We recommend the rejection of the term arthrospore. If the recommended alternative arthroconidium (or, preferably, thalloconidium) is to be used, we would restrict it to conidia whose ontogeny is thallic, and whose connection is full width."

Does the meeting wish to reject "arthrospore"? - Yes, unanimously.

Does the meeting wish to adopt one of the two substitutes proposed by the committee? - Yes, unanimously.

The committee has, I understand, a preference for "thalloconidium." Would members of the committee like to give their reasons?

DR CARMICHAEL

The reason for preferring "thalloconidium" was that some of these conidia swell considerably, whereas what we have become accustomed to calling arthrospores generally do not. Some conidia which fall into this group may swell after the septa are laid down and become large and multicellular. Single, terminal conidia may also be produced in this way. Neither of these match what we have been calling arthrospores. We preferred a more general term that covered all these cases.

DR KENDRICK

I agree with what Dr Carmichael has said. It all hinges on whether you like the idea of the "thallic-blastic" dichotomy. We tried for hours to think of an alternative, and couldn't. Everything else we tried broke down.

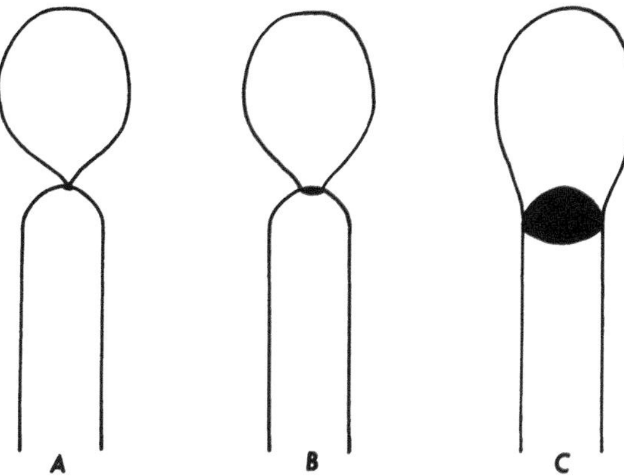


Figure 15.1

DR ELLIS
May we now vote on this issue? - 9 prefer "arthroconidium," 6 prefer "thalloconidium," and some members abstain.

The committee further recommends "rejection of the term meristem arthrospore for those conidia produced by *Monascus, Trichothecium,* and *Cladobotryum*, because these are essentially blastic conidia with full-width connection, which are produced by a linear, basipetal succession of conidiogenous loci which consume the conidiogenous cell."

Does the meeting wish to support this point of view? - Yes, unanimously.

We will have further discussion of the broader recommendations of the Arthroconidia Committee after we have heard the reports of the other committees. I would merely ask the meeting to decide now whether or not it approves of the "blastic-thallic" dichotomy, at least in principle. - Yes, all but two members approve.

Next comes the report of the Committee on Blastospores, Aleuriospores, and Chlamydospores (Chairman: Dr Bill Carmichael. Members: Dr Stan Hughes, Dr Gregoire Hennebert, Dr Gil Hughes, "and some visitors"). The keynote speaker, Dr Carmichael, early came into disagreement with the rest of the committee. [*Laughter*] The members wanted to proceed with a definition of the "blastoconidium," and to pay little attention to the aleuriospore and the chlamydospore. After five hours of discussion, I received from Dr Stan Hughes the following sentences - the quintessence. [*Laughter*] "A blastoconidium arises as a bud from a conidiogenous cell whose outer wall is continuous with that of the bud. Blastoconidia may arise solitarily, synchronously, or successively. Each blastoconidium may produce a chain of secondary conidia by acropetal budding."

EDITOR
*Dr Ellis and several other members considered that the "blastoconidium" as here defined is only one of several kinds of conidia which fall under the major ontogenetic heading "blastic" as proposed and defined by the Arthroconidia Commit-*

 *tee. Dr Kendrick pointed out that conidia with basically the same "blastic" ontogeny could be produced by the "blowing out" of a minute* point *on the conidiogenous cell, of a larger* area *of the apex, or of the* entire apex *of the cell (Figure 15.1). Dr Carmichael and others thought that "blastoconidia," rather than being defined in isolation, should now be placed in perspective, relative to other kinds of "blastic" conidia, by using the criteria outlined in the inclusive report of the Arthroconidia Committee.*

*The meeting decided to return to this discussion in the wider context of the final general session on terminology, but to give qualified approval to an interim definition of "blastoconidium" as follows: "Blastoconidium - a conidium which arises as a blown-out portion of the conidiogenous cell. Blastoconidia may arise solitarily, synchronously, or successively. Each blastoconidium may produce a simple or branched chain of secondary blastoconidia by acropetal budding."*

## REPORT OF THE COMMITTEE ON POROSPORES

Chairman: Dr Martin Ellis
Members: Dr George Carroll, Dr C.V. Subramanian

We note that the word pore and its derivatives are now used in widely differing contexts (macroscopic, microscopic) within mycology. In deference to well-established usage in descriptions of basidiocarp morphology in many Aphyllophorales, we recommend that the term not be used in describing the conidium ontogeny of Fungi Imperfecti like *Helminthosporium velutinum*. We recommend that the term *tretic* be accepted as an adjective suitable to define any conidiogenous cell which forms conidia solitarily or in acropetal chains through one or more minute channels in its outer wall. A conidiogenous cell with one channel may be described as *monotretic*, one with several channels as *polytretic*. A *tretoconidium* is any conidium formed by a tretic conidiogenous cell.

Examples of hyphomycete genera with tretic conidiogenous cells: *Curvularia, Dendryphiella, Dendryphiopsis, Drechslera, Exosporium, Spadicoides, Helminthosporium.*

DR ELLIS

"Treta" is a Greek word meaning "channel." The tiny opening through which tretoconidia emerge is a channel dissolved through the thick outer wall of the conidiogenous cell. Tretoconidia are clearly blastic. I am convinced that the use of "pore" and its derivatives to describe this phenomenon is unfortunate. To underscore the importance which I place on the need for a replacement, I may add that "tretic" is the first new term I have introduced in 25 years of mycological research.

DR PIROZYNSKI

I support the term tretic because it would be used with greater discrimination: many so-called porospores may not be tretic, as Dr Carroll's work on *Stemphylium* shows. In addition, the word pore is also applied to another microscopic phenomenon - the germ "pore" - which is found on many spores and has no ontogenetic significance.

EDITOR

*A discussion of the usage, etymology, and semantics involved here was terminated by a vote, which accepted the committee's recommendations by 11 to 7.*

## REPORT OF THE COMMITTEE ON PHIALIDES

Chairman: Dr C.V. Subramanian
Members: Dr Bryce Kendrick, Dr Garry Cole, Dr Gregoire Hennebert, Mme Jacqueline Nicot, Dr T.R. Nag Raj

A "phialide" is a conidiogenous cell which produces, from a fixed conidiogenous locus, a basipetal succession of conidia whose walls arise de novo. The wall of the phialide is broken when the first conidium is extruded or liberated. The phialide wall from the point of rupture to the fixed conidiogenous locus is the "collarette." The length of the phialide is not changed by the production of a succession of conidia, e.g. *Phialophora verrucosa* Medlar. The adjective phialidic derives from the term phialide.

Conidia produced from phialides (phialidic conidia) may be termed phialoconidia.

Phialides may proliferate vegetatively in a percurrent (e.g. *Catenularia*) or a sympodial (e.g. conidial *Lasiosphaeria hirsuta*) manner. A phialide with more than one open end formed successively (sympodially) or synchronously may be termed a polyphialide. A conidiophore which bears phialides may be termed a phialophore.

Although the nouns given above are acceptable, we recommend that an adjectival terminology be used wherever possible, since a description like "phialide proliferating sympodially" is more intelligible to the non-specialist and conveys more ontogenetic information than "polyphialide."

We consider the following terms unnecessary, and recommend their rejection: orthophialide (Gams), plagiophialide (Gams), schizophialide (Gams), pleurophialide (Gams), endoconidium (Brierley), ascoconidium (Seaver), endoconidiophore (Davidson), ascoconidiophore (Seaver).

We also recommend that "sterigma" should not be used for "phialide" (see Goos 1956, Proc. Iowa Acad. Sci. 63: 311-20), and that although "phialis" is the correct Latin singular, it should only be used in Latin diagnoses. The singular term originally proposed by Vuillemin (1910, C.R. Acad. Sci., Paris 150: 882-4), and ratified by more than half a century of usage, is "phialide."

DR S.J. HUGHES

The collarette is one of the characteristic things about most phialides, and should be mentioned in the definition, but I'd like to amend the sentence: "The phialide wall from the point of rupture to the fixed conidiogenous locus is the 'collarette.'" In some *Catenularia* and *Codinaea* species, some of the wall seems to disappear after rupture, and the collarette that remains is much smaller than it was originally. I suggest, "The phialide wall distal to the fixed conidiogenous locus is the collarette." Another important feature is that the first phialoconidium arises de novo, and its wall is free from that of the future collarette - but perhaps it would be safer not to be too specific about the origin of the conidium wall, and say instead, "the first conidium arises within the apical extension of the phialide and is liberated by the rupture of the outer phialide wall."

DR KENDRICK

We must remember that the first phialoconidium sometimes carries away with it part of the outer wall of the phialide apex, giving a false impression that phialoconidia are typically clad in that wall.

DR S.J. HUGHES

I'm trying to make the point that our definitions shouldn't be so cryptic that we fail to characterize our terms properly. We aren't sending a telegram! [*Laughter*]

DR KENDRICK

We can give two definitions, one condensed and one expanded.

DR PIROZYNSKI

Phialide - a conidiogenous cell which produces from a fixed conidiogenous locus a basipetal succession of conidia.

DR CARMICHAEL

Can any of those who want an expanded version suggest any kind of conidiogenous cell, other than a phialide, which fits our short definition? I doubt it.

DR S.J. HUGHES

I don't like "fixed"; do we need to specify the conidiogenous locus so rigidly?

DR KENDRICK

In those phialides we have investigated so far, it is fixed from the beginning, or soon becomes so. In the phialide of *Chalara*, the cross-walls (double septa) are laid down at almost exactly the same place for all conidia after the first one or few. We have confirmed this in our time-lapse film. I surmise that this is also the point at which the lateral walls of each conidium are laid down as the cytoplasm surges upward, but we haven't been able to verify that visually.

DR PIROZYNSKI

In *Wallemia ichthyophaga*, which in Barron's book does not have a place in the system, a phialide-like conidiogenous cell produces a filament which breaks up into arthroconidia. I would like to see the basauxic section broadened to include this and other odd taxa (see discussion on basauxic conidiophores). One can, of course, argue that *Wallemia* has a phialide in which the delimitation of conidia is delayed. If this is the case, an arthroconidial phialide in which the protoplast is fragmented into conidia (showing little or no increase in size outside the conidiogenous cell) and a blastoconidial phialide (in which conidia are blown out to exceed considerably the diameter of the conidiogenous cell) may be examples of evolutionary convergence rather than divergence.

DR KENDRICK

Dr Pirozynski has argued persuasively for the segregation of what he calls "arthroconidial phialides." However, in the conidial states of some *Ceratocystis* species - *Chalara, Chalaropsis, Thielaviopsis* (which Mr Nag Raj is grouping under *Chalara*) - often, before the apex of the phialide ruptures, a definite basipetal "endoarthroconidial" segmentation occurs, which stops at a certain point when the upwelling cytoplasm seems to strike a balance with the downward delimitation and maturation of conidia. So I'm not sure we can separate Dr Pirozynski's two kinds of phialide. The *Chalara* phialide appears to display both mechanisms.

DR SUBRAMANIAN

When arthroconidia form, I think the nuclei have already divided and been distributed along the hypha, since the hypha dissociates in an irregular way; but in the phialide the nucleus goes on dividing at one point.

DR KENDRICK

Yes, it's quite possible that we could finish up with a karyological definition of many of these mechanisms. [*But see Chapter 18.*]

DR ELLIS

Terms considered misleading or superfluous, and therefore proposed for rejection by the phialide committee: biphialide, sensu Swart (1959).

DR S.J. HUGHES

Are there two conidiogenous loci in this phialide which produce a double row of conidia? If so, wouldn't the term polyphialide cover it?

DR KENDRICK

Polyphialide is a strange term, because we are at present using it for only one specific kind of phialide, with more than one conidiogenous locus - the kind in which only one conidiogenous locus is functional at any given time, and in which vegetative proliferation is sympodial. Other phialides with more than one opening haven't been called polyphialides, and probably shouldn't be.

DR CARMICHAEL

I'm not really in favour of the term biphialide, but because we don't know enough about this phenomenon yet, we should reserve judgment. [*This was generally accepted.*]

DR ELLIS

The committee considers the term monophialide an unsuitable complement to polyphialide, because we already speak of monophialidic species in certain genera as those species in which the conidiophores bear, or consist of, only one phialide.

EDITOR

*The conference supported the rejection of a number of other terms considered misleading or superfluous by the committee. Unanimous approval was given to both the short definition of phialide, and to the long definition as amended by Dr S.J. Hughes to read as follows: "Phialide - a conidiogenous cell which produces at least its first conidium initial within the apical extension of the cell; this is liberated by the rupture of the upper cell wall. Thereafter, from a fixed conidiogenous locus, a basipetal succession of conidia is produced. Any phialide wall distal to the conidiogenous locus is the collarette. The length of the phialide does not change during the production of a succession of conidia. Example,* Phialophora verrucosa *Medlar."*

## REPORT OF THE COMMITTEE ON ANNELLOPHORES

Chairman: Dr Stan Hughes
Members: Dr Martin Ellis, Dr Gil Hughes, Dr Garry Cole

An annellophore (or annellate conidiogenous cell) is typically a conidiogenous cell producing a single terminal blastic conidium, and by percurrent vegetative proliferations (annellations), a succession of similar conidia, which may be called annelloconidia or annellidic conidia. Examples: *Spilocaea pomi, Doratomyces stemonitis, Sporidesmium atrum, Triposporium elegans, Scopulariopsis brevicaulis.*

EDITOR

*There was some discussion of this, and several rearrangements were suggested.*

DR ELLIS

Dr Carmichael has already suggested that we replace the term annellophore by annellide (to match phialide), for the specialized kind of conidiogenous cell we are now considering. It has also been suggested that if the term annellophore is to be retained, it should be applied to supporting structures which bear annellides (compare phialophore). The objections were: (i) the wide acceptance of the term annellophore in the literature, (ii) the similarity of "annellide" to a zoological term for a kind of worm.

EDITOR

*The meeting put off final consideration of this until the general closing session.*

DR SUBRAMANIAN

Annelloconidium is unsuitable and superfluous. We are not reducing the number of terms, though many of us had hoped to do so. Could we have a vote on that?

DR COLE

Tomorrow we'll have a similar vote on sympodioconidium. I hope people won't want to keep that term. It doesn't really describe any definite entity, but rather describes one thing in terms of another: the conidium in terms of the mode of proliferation of the conidiogenous cell. Annelloconidium is a similar irrelevant term.

DR KENDRICK

As the perpetrator of the term sympodioconidium, I would like to second Dr Cole's motion. Basic terms describing conidia should concern their individual ontogeny, *not* the mode of proliferation of the conidiogenous cells. Both characteristics would, of course, be teamed in a description of an entire fungus. I would reject both sympodioconidium and annelloconidium.

DR ELLIS

Perhaps we could include it in the description, but say that we do not recommend it. May we now vote on annelloconidium? - By a vote of 9 to 8, the conference recommends that the term annelloconidium should not be used.

EDITOR

*Returning to the basic definition proposed by the committee, and the alternatives put forward during discussion, the meeting unanimously agreed to accept a definition derived from two versions on the blackboard which differed in construction rather than in content. The Conference did not perform the necessary synthesis subsequently, so it has fallen to the editorial prerogative. The editor alone, therefore, must accept blame for any inadequacies in the following definition: "Annellophore - a conidiogenous cell which produces a single blastic conidium from the apex of each of a succession of percurrent vegetative proliferations (annellations)."*

*The session adjourned at 10:30 p.m.*

*Dr Ellis opened Tuesday morning's session by stressing the developmental, as opposed to the descriptive, morphological, point of view, and went on:*

DR ELLIS

I'd like to start this session with the report of the committee on sympodulae. This is one of the reports which will be most encouraging to the mycological public, because it actually recommends a *reduction* in the number of terms.

## REPORT OF THE COMMITTEE ON SYMPODIALLY PROLIFERATING CONIDIOGENOUS CELLS

Chairman: Dr Garry Cole
Members: Dr Gregoire Hennebert, Dr Bryce Kendrick, Dr C.V. Subramanian, Dr Keisuke Tubaki

The adjective sympodial may be applied to any conidiophore or conidiogenous cell in which the growth of each of a succession of subterminal, lateral, vegetative apices ends in the production of a conidiogenous locus. The net effect is an increase in length or swelling of the conidiogenous cell. We must emphasize the description of the conidiophore or conidiogenous cell, because several very different kinds of conidia can be produced on sympodially proliferating conidiogenous cells: *Sympodiella acicola* produces arthroconidia; *Beauveria globulifera*, blastoconidia; *Curvularia inaequalis*, tretoconidia.

Thus we consider that the term sympodioconidia is misleading, and we recommend that it be rejected. We also consider the term sympodula to be more or less superfluous.

DR KENDRICK

Whenever possible, the basic descriptive terms applied to conidia should describe their actual individual ontogeny, not the characteristics of the conidiogenous cell from which they arise. As a rider, I would suggest that, since the terms polyphialide and sympodula are open to frequent misinterpretation and are also superfluous, they should be rejected.

DR CARMICHAEL

I'd agree with our definition of "sympodial," which conforms to the way the word has always been used. This adjective can apply either to a conidiogenous cell, or to a whole conidiophore. I agree that we need general descriptive terms like this. However, in the fungi we find three very distinctive kinds of conidiogenous cell. It would be useful to have terms for all three; I have suggested phialide, annellide, and rachide.

DR S.J. HUGHES

And the ampulla? It is also very distinctive.

DR CARMICHAEL

Perhaps so, but each of the other three kinds produces a succession of conidia. This is why I wanted annellide rather than annellophore, because the latter can apply to a whole conidiophore as well as to a conidiogenous cell. In *Stysanus, Scopulari-*

*opsis, Leptographium,* etc., there is a very definite conidiogenous cell which differs significantly from the annellophore of *Annellophora* - just as the rachides of *Tritirachium* differ from the sympodially developed conidiophore in *Helminthosporium.*

DR S.J. HUGHES

Most of us agree that the phialide and the annellophore (annellide) are distinctive. But the term rachide is something different, being applied to a conidiogenous cell which may form conidia by several methods.

DR PIROZYNSKI

We should be consistent. If we are to use phialide, annellide, rachide, and ampulla, we should create nouns for conidiogenous cells which produce tretoconidia, aleuriospores, catenate blastoconidia, or are transformed into arthroconidia and meristem arthroconidia. Is this proliferation of terms necessary? Can't we, more logically, describe the various types as annellated conidiogenous cell, sympodial conidiogenous cell, etc. - i.e. using an adjectival system? If people insist on having nouns for everything, we should, to be consistent, coin more terms. If a phialophore is a conidiophore which bears or is composed of phialides, then an annellophore would be a conidiophore bearing or composed of annellides. What are we going to call conidiophores which bear or are composed of other kinds of conidiogenous cell?

DR COLE

Dr Nelson has shown genetically that the ontogenetic stability of the phialide and the annellide is not shared by the sympodula, so I don't think we should use a special restrictive term for the sympodially proliferating conidiogenous cell.

DR SUBRAMANIAN

I suppose we have four types: blastic, thallic, phialidic, and tretic. I don't understand all cases of phialidic ontogeny yet, and we may still have to change our ideas. But if we can agree on these four types, we could find terms for the conidiogenous cells. We now have only one widely accepted name: the phialide. When this was first used by Vuillemin, it denoted only the shape of the conidiogenous cell, and had no deep ontogenetic significance. We have restricted its application to suit ourselves. It is most improper to redefine terms like this, and we could easily dispense with the term phialide completely. We could use adjectives instead. Under each of the four headings I have given above, we'll probably have several different kinds of conidiogenous cell, depending on how it behaves or proliferates. For example, in the blastic type you can have percurrent proliferation and sympodial proliferation. These modes of proliferation can also be found in the tretic group. An annellide merely exhibits a form of percurrent proliferation. Sometimes the annellations can't be seen; so if we called these cells percurrent instead, little would be lost.

Another point concerning the annellide. In *Monosporella setosa* and others, we have a conidiophore with annellations separated by whole cells. Each cell showing one or more annellations is an annellide according to the original definition, so we have a conidiophore with several integrated annellides in it. It sounds strange, doesn't it? If the adjective annellate is used, one can say that the conidiophore has several integrated annellated conidiogenous cells. I feel that adjectives are best. As many as are needed: sympodial, percurrent - sympodial sometimes, percurrent sometimes. Let us remember that the fungi vary so much that, if we try to find a noun for every variation, we'll be in trouble. Adjectives are more useful than nouns.

DR ELLIS

I fully agree with everything Dr Subramanian has said.

DR CARMICHAEL

I hope Dr Subramanian doesn't intend those terms to be mutually exclusive, because both phialidic and tretic conidia are blastically produced.

DR ELLIS

You can use two, three, or four adjectives to modify a single noun, but you can't modify nouns with nouns in the same way. You can say, "blastic-phialidic," "blastic-tretic." You could only do the same kind of thing with nouns if you used very long compound nouns of the kind found in German.

DR CARMICHAEL

The term annellide is a shorthand way of saying "annellated conidiogenous cell," and I, for one, do not wish to write out three words where one will do.

DR COLE

If you pin a name like rachide on a conidiogenous cell that begins by proliferating sympodially and then proliferates percurrently, you have a problem.

DR CARMICHAEL

Give me an example.

DR PIROZYNSKI

*Eriomycopsis meliolicola*, which, together with related forms, Mr Deighton and I are going to put into a new genus, *Chionomyces*. The conidiogenous cell proliferates both sympodially and percurrently.

DR HENNEBERT

I think the various kinds of conidiogenous cell are organs which need names as well as adjectives. We should not abandon the terms phialide, sympodula, ampulla, annellide. Just as we say stroma or sclerotium, we need anatomical terms to designate these cells.

DR KENDRICK

We have one - conidiogenous cell. This is intelligible to the non-specialist, whereas many of our more arcane terms are not. They are the jargon with which we are accused of creating a smoke-screen, behind which to hide.

MRS POLLACK

I endorse the scheme we are constructing, but we need a hierarchy of terms. We have basic ontogenetic terms (blastic and thallic) and morphological terms (phialidic and tretic). They don't belong at the same level.

DR KENDRICK

I think what we need now is some kind of decision as to whether we would like to apply adjectives or nouns. That is the basic point at issue.

DR ELLIS

May we have a vote on this issue? Those who would prefer an adjectival system? - 14. Those who would prefer nouns? - 3. With the exceptions of Drs Hughes and

Hennebert and Mme Nicot, the meeting prefers the use of adjectives. This does not exclude the use of some nouns, but indicates a desire to curtail the proliferation of new nouns. In recent years this has been going on in our own field; and now, with the application of electron microscopy, the possibilities are endless.

MME NICOT

I did not support either proposition. I am not in favour of a strictly adjectival system, nor of a proliferation of useless nouns!

DR ELLIS

May we record, then, that Mme Nicot is in favour of a combined system using adjectives and some useful nouns.

DR S.J. HUGHES

With recent advances in electron microscopy, we have seen many necessary new terms proposed, but certainly not an "enormous proliferation"; new terms are to be expected with such a new tool as the electron microscope. The Fungi Imperfecti are a vast group, nevertheless we do not have a great number of terms available, or at least in common usage, for referring to and for directing attention to distinctions which the fungi themselves make.

DR ELLIS

Perhaps we are becoming a little more polarized over this than is necessary. I think most of us feel that a simple adjectival system is basically a good idea, but that certain nouns are useful and expressive, and should be retained. Could we consider the usefulness or otherwise of the adjectives first, then consider the nouns? It seems to me that we should certainly delete or not recommend such terms as botryoaleuriospore and sympodioconidium, and the terms proposed by Dr Gams. Could we now consider the adjectives?

DR GOOS

It seems to me that Vuillemin's distinction of "thallospores" and "conidia vera" is a very valid one. There are two kinds of conidium development, thallic and blastic, and we could very nicely use this old dichotomy, including the arthroconidia in the thallic group, and all the others (six groups) in the blastic.

DR ELLIS

Could we have some views on the adjectives thallic and blastic?

DR CRANE

I think our emphasis on the septum is wrong. If we simply use "thallic" in the sense of transformation or conversion, "blastic" as blown-out wall, we'd still have two fundamental types. We should disregard the septum because there may or may not be one there; or it may have a pore.

DR CARMICHAEL

Our definitions state whether the conidium initial is made recognizable as such by the formation of a cross-wall (thallic), or whether it is made recognizable as a conidium initial by morphological alteration of the shape of the conidiogenous cell (blastic). In *Stemphylium*, as Dr Carroll showed us earlier, considerable enlargement occurs before a cross-wall is laid down, and the conidium initial is recognized as such as soon as it begins to blow out, *not* when the cross-wall is formed. It is clearly blastic. In *Geotrichum* or conidial *Hysterium insidens,* the conidium as such is delimited by the formation of a cross-wall, and after that it may grow (as in *Hysterium*), or it may not. It is clearly thallic. In the blastic form there is always growth after the conidium initial is morphologically delimited and before it is cut off by a septum. 

EDITOR

*The terms thallic and blastic are accepted unanimously.*

DR G.C. HUGHES

I like the new scheme in many respects. The word that bothers me is "blastic." We seem to have a small-b "blastic" and a large-B "Blastic." A conidium can be Blastic and tretic; but can it be Blastic and blastic?

DR ELLIS

Is there any adjective that will more clearly define this? I used "blastic" previously in a restricted sense, but this concept has been enlarged and extended for me during this conference. I've seen and appreciated other people's points of view. It may be that we need an additional adjective or two.

DR SUBRAMANIAN

A line separates blastic and thallic. This is generally accepted by the meeting. But there are some objections to treating phialidic and tretic at the same level of significance. I think we are now applying blastic to cases where all the layers of the conidiogenous cell wall blow out. Thallic implies conversion rather than blowing out. Tretic implies that an inner wall layer blows out through the channel or pore. It is basically a blastic type, but only the inner wall blows out. The wall of the conidiogenous cell does not contribute to the average phialidic conidium; instead, a completely new wall is being laid down specifically for the conidium.

Where the total wall blows out, we could say that the process is "exoblastic"; where only an inner wall blows out, as in tretic and phialidic cases, the process could be called "endoblastic."

DR ELLIS

There is some agreement that we need to use the terms thallic and blastic, with phialidic and tretic as subdivisions of blastic.

DR CARMICHAEL

Dr Subramanian's point is well taken. It clarifies the difference between Blastic ontogeny and blastoconidium (as Dr S. J. Hughes has used the term). By blastoconidium we mean any exoblastic conidium - one where both inner and outer walls are involved. And perhaps "exoblastic" isn't the best term.

A VOICE, musingly

"Holoblastic"?

THE GROUP (one by one, and then in enthusiastic chorus)

"Holoblastic"! Yes!

DR ELLIS

"Holoblastic" would be very good indeed!

DR SUBRAMANIAN

Excellent term!

DR ELLIS

May I know who made that suggestion?

SEVERAL VOICES

Dr Müller.

DR ELLIS

Thank you, Dr Müller, a very good term indeed.

DR KENDRICK

"Endoblastic" isn't in that class.

DR SUBRAMANIAN

Would someone please suggest a substitute?

DR HENNEBERT

"Enteroblastic"?

DR ELLIS

Another good term. [*The group endorses it.*]

DR ELLIS

Perhaps now we can consider the interpretation of the term blastic, and of the two subdivisions of this: holoblastic (suggested by Dr Müller) and enteroblastic (suggested by Dr Hennebert). To me these seem excellent suggestions. They are most descriptive of what actually happens, and they eliminate the misinterpretations of the word blastic. What Dr Hughes, in 1953, called the blastoconidium is what we can now term holoblastic - the entire thickness of the conidiogenous cell wall is involved in the "blowing out" of the conidia. The term enteroblastic means that the "blowing out" takes place from inside and does not incorporate the outer wall layer(s) of the conidiogenous cell. This term covers both the blowing out from within of a phialide (phialidic), and the blowing out through a narrow channel or tunnel in the wall (tretic). This is a very clear and useful dichotomy. I think it also necessitates the retention of the term phialide, because without the term phialide we cannot have the adjective phialidic.

DR HENNEBERT

We must be quite clear that annellidic conidia are holoblastic, not enteroblastic. Each time the conidiogenous cell proliferates, it produces new outer wall as well as new inner wall, and both of those layers are involved in the formation of the new conidium.

DR S.J. HUGHES

In the long phialide of *Sporoschisma mirabile* and other similar fungi, the early conidia are "carved out" of the cytoplasm. This process starts when the phialide has already stopped growing, and is what I have always considered a sort of cleavage.

DR KENDRICK

I think our film sequence of *Chalara* shows that in cases like this a double septum is laid down between adjacent conidia. It doesn't look like simple cleavage. But I do agree that it begins after the phialide has stopped growing, and this suggests that the first few conidia here are a kind of "endoarthroconidium." Subsequent phialoconidia may look like arthroconidia, but this is merely due to the tube-like form of the collarette. They are actually formed at a fixed conidiogenous locus.

DR PIROZYNSKI

I wonder if "basauxic" really belongs in the holoblastic group? If we take the conidiophore mother cell as the conidiogenous cell, the filament that comes out could be considered, as Dr Müller pointed out, equivalent to an atypical first conidium which goes on to produce different types of secondary conidia. In that case it would be enteroblastic, because it was the inner wall that elongated. 

DR CARMICHAEL

The difficulties presented by the basauxic phenomenon arise because the so-called conidiophore mother cell sometimes produces a conidium first, then an additional conidiogenous cell or area which, in its turn, produces conidia. The filament is obviously enteroblastic, but it will form its own outer wall, and the subsequent conidia produced on it may be regarded as holoblastic.

DR ELLIS

I will now read out the short definition of a basauxic conidiophore, arrived at by Dr Tubaki and his committee, Mrs Pollack and Dr Pirozynski: "A basauxic conidiophore consists of a mother cell and an extending filament arising from it, which is conidiogenous."

DR HENNEBERT

I'd like to emphasize a point raised earlier: the similarity between the basauxic conidiophore and the phialide. I think they are both basauxic. I'd like to use the term basauxic for the combination of two kinds of growth.

DR KENDRICK

The term basauxic does not fit into our system of terms referring to the ontogeny of individual conidia. I think "basauxic" is a term comparable to "sympodial" in its level of significance.

DR CARMICHAEL

We must remember that our scheme of terms is *not* a classification of fungi. It is not even simply a classification of conidia. It is two separate and distinct things: (i) a classification of conidia on the basis of their individual ontogeny; (ii) a classification of various methods employed by the fungi to produce a succession or plurality of conidia. These two do not necessarily fall into any rigid relationship. The use of a term from (i) does not necessarily specify any particular term in (ii).

DR ELLIS

Perhaps we can now get the feelings of the Conference on our scheme: acceptance of thallic and blastic, the usefulness of holoblastic and enteroblastic, and the subdivision of enteroblastic into phialidic and tretic. All of these terms refer to individual conidium ontogeny. They do *not* refer to the shape or mode of proliferation of the conidiophore or conidiogenous cell, nor do they refer to the way in which a succession of conidia is produced. - Accepted unanimously.

DR SUBRAMANIAN

Since it is difficult to distinguish between blastospore in the sense in which it has been used, and gangliospore, because there are all intergradations between these two, I would now call them both holoblastic. I would also place conidia produced from sympodulae and annellides in this group. Then we are left with the tretic and phialidic conidia which both fall under the heading enteroblastic. Differences within each of these groups can be described in terms of the behaviour of the conidiogenous cell: basauxic, sympodial, percurrent, etc. The arthroconidia fall under the heading thallic. We can thus reduce the number of divisions and terms.

DR ELLIS

Since I shall not be with you this afternoon, but on my way back to England, I would like to say with what pleasure I have attended this conference. What has appealed to me most has been the give-and-take, the ready appreciation by each member of other members' views. We have, as a result, been able to discuss fully and frankly every point that has been raised. I have been put onto the right path on many subjects, and I think we have all learned a good deal. We have appreciated the new approaches presented to us, particularly by Drs Kendrick and Cole, and Drs Carroll and G. C. Hughes. We are entering a completely new stage in the study of the Hyphomycetes with these new and fascinating time-lapse sequences and electron micrographs. I have really enjoyed this conference. It is the best thing of its kind that I have ever attended. [*Prolonged applause*]

EDITOR

*Pause for lunch, after which Dr Kendrick resumed the chair.*

DR KENDRICK

We made considerable progress this morning in resolving some of our differences of opinion. Heslop-Harrison wrote that a natural classification is one based on all attributes of an organism, or one expressing relatedness in the genetic or evolutionary sense. When we look at Fungi Imperfecti, we can't examine all attributes because we are only looking at parts of organisms. Nor can we consider relatedness in the genetic or evolutionary sense. Thus, if Heslop-Harrison is to be believed, we cannot expect to achieve a natural classification of Fungi Imperfecti in the foreseeable future. We are trying here to establish an artificial system - I won't even call it a classification yet - which will incorporate as many as possible of the fungi we know, with a minimum of overlap or confusion. Our new adjectival scheme is based ontogenetically, taking its cue from Dr S.J. Hughes's original postulates. With each other's help, we have perceived basic developmental similarities between conidia, and have devised some serendipitous terms to describe the ontogeny of individual conidia. I hope we will continue to separate these from terms describing those features of conidiogenous cells that permit them to produce a succession of conidia. May we begin, this afternoon, by examining the case of what I call the "solitary terminal thallic" conidium, in which the septum is laid down first and then the conidium differentiates and grows. Is this really thallic? I think it is. A whole cell is involved; the basal septum is there before swelling begins. The idea of thallic development does not exclude growth subsequent to the laying down of the septa. In many cases, e.g. *Amblyosporium* and conidial *Hysterium insidens*, there is considerable growth after the septa are laid down.

DR HENNEBERT

Perhaps this solitary terminal thallic conidium and the solitary holoblastic "chlamydospores" of some members of Dr Hughes's group III are really the same. Perhaps we should put these together as a transition between holoblastic and thallic.

DR KENDRICK

In the group III "chlamydospores," there is a differentiation of a recognizable conidium initial before the septum is laid down. By our definition, these conidia are blastic.

DR CRANE

We have stressed the concept of the septum quite well and quite fundamentally in the two definitions we've arrived at. Both definitions are based on the idea of a septum that either is or isn't being laid down, and not on the idea of a vegetative structure that is swelling.

DR KENDRICK

Another way of looking at it is to say that in blastic development, *part* of a cell is becoming a conidium, and in thallic development a *whole* cell is becoming a conidium.

DR G.C. HUGHES

The criterion that we used to distinguish these was the septum, the same thing upon which we based the very first definition we agreed upon, that of the cell.

DR HENNEBERT

I propose, then, that we adopt this principle of the septum, but on an experimental basis.

DR KENDRICK

It *is* experimental. Dr Hughes in 1953 said that *his* groupings were experimental. Only time will tell if our idea works. In 1971, at the first International Mycological Congress, we can meet again and reassess these characters.

DR SUBRAMANIAN

Perhaps a thallic conidium could be defined as "a conidium formed by the transformation of a pre-existing cell or cells."

DR CARMICHAEL

You have said, in different words, exactly the same thing that we said in our definition of thallic. Your definition may be a little less precise, but it is simpler.

DR S.J. HUGHES

I'd like to ask Dr Crane if there is any variation in the pattern of development exhibited by his aquatic fungi?

DR CRANE

No! There's always a septum laid down first to delimit the apical cell, which then develops into a conidium, either tetraradiate or sigmoid. We aquatic mycologists (Dr Tubaki and I) would like to call these "aleuroconidia."

DR CARMICHAEL

That is incompatible with Vuillemin's original definition of aleuriospore.

DR GOOS

Aleuriospore has been widely used in the literature. It was, and is, a confused term, and could only be rescued by a restrictive definition. It was originally coined for spores of some fungi which are pathogenic to humans, and has been used extensively in the literature of medical mycology for a very different kind of conidium.

DR CARMICHAEL

The term aleuriospore does not refer to whether a spore is thallic or blastic. It refers to the manner in which the spore secedes. Vuillemin described, in what he called *Aleurisma flavissimum* (which we now know as *Chrysosporium merdarium*), secession by lysis followed by fracture of the supporting or separating cell, or of the whole spore-bearing part of the mycelium. If you wish to use the term aleuriospore, it must refer to this method of dehiscence. If you don't want to consider dehiscence, but rather ontogeny, then you must use thallic or blastic, according to how the spore develops. Aleuriospore is not a developmental term, either by origin or by its subsequent usage in medical mycology.

DR CRANE

I will admit that the prefix "aleurio" was taken from medical mycology and has been wrongly used. My only argument is that there are perhaps 25 genera whose generic description begins with the word "Aleuriosporae." Regardless of what we decide here, we'll have to contend with that. There are only two aquatic mycologists here, and I don't know whether Ingold, Nilsson, and the others will be willing to drop the term aleuriospore. I thought perhaps the modification "aleuroconidium" would make some accommodation possible.

DR CARMICHAEL

There are an equal number of genera where aleuriospore is used in its original meaning. Should we change the original ones, or the new ones that were wrongly named?

DR KENDRICK

We've now heard both sides of this argument very well expounded, and should be ready to vote - to make recommendations.

(i) Do we wish to retain the term aleuriospore in the sense in which it has been used in aquatic mycology - 5 for; 10 against.

(ii) Should the term be retained in its original sense, as referring to the manner in which a conidium secedes, and as used in medical mycology? - 2 for; 13 against.

(iii) Do we wish to retain the term aleuriospore or aleuroconidium? - 3 for; 15 against.

Perhaps these aquatic genera, which have hitherto been called "aleuriosporic," could be fitted into our new terminology under some heading like "terminal thallic conidia."

Let us now consider "chlamydospore" and "chlamydoconidium."

DR CARMICHAEL

I propose that chlamydoconidium be defined as a thallic or blastic conidium which is released by fracture or dissolution of a separating cell or cells. Examples, *Coremiella, Chrysosporium, Helicoma isiola.*

DR CRANE

I don't like the definition because it mixes blastic and thallic, two things we had very clearly and decisively separated earlier.

DR GOOS

I believe this term should be reserved for thallic conidia whose function is one of survival. It should not be a term based on ontogeny or dehiscence, but on function.

DR KENDRICK

In other words, you don't think this term should have any part in our scheme?

MR BHATT

What will we call the chlamydospores of *Thielaviopsis basicola*? I can cite at least 25 references to these that have appeared during the past year.

DR KENDRICK

I consider them arthroconidia.

DR CARMICHAEL
I think they are chlamydospores in this sense, and later break up into conidia. They secede as a unit, and then break up.

DR KENDRICK
This is another case of a term which has been applied in several different senses. Are we to give it an ecological definition, a functional definition, or an ontogenetic definition, or are we to consign it to oblivion?

DR MÜLLER
We should remember that it is sometimes difficult to differentiate between a chlamydospore and a sclerotium.

DR KENDRICK
I would prefer to call them "thick-walled thallic" conidia.

DR G.C. HUGHES
That is a lot safer than trying to say something precise about something we don't know enough about.

DR S.J. HUGHES
There seems to me to be a complete series, from the intercalary chlamydospore, to the one which expands beyond the confines of the cell, to a shortly stalked one. Some may break off by fracture, some by lysis, but they form a group. I'd include *Humicola, Allescheriella, Trichocladium, Bactridium, Bactrodesmium,* some species of *Sporidesmium, Sepedonium, Fusarium, Hypomyces*, and some others. I'd like to see chlamydospore retained for these. It could be defined as a terminal, shortly stalked, sessile or intercalary, blastic, thick-walled conidium.

DR CARMICHAEL
Most of the spores you have cited are thallic in the sense in which we defined that term.

DR HENNEBERT
One thing connecting all these spores is that they are developed from portions of undifferentiated vegetative hyphae.

MR BHATT
In the same fungus we may have intercalary chlamydospores which are thallic, and terminal chlamydospores which are blastic - e.g. *Humicola* and *Fusarium*.

DR KENDRICK
I don't know how you can be sure that the terminal chlamydospore is blastic if you haven't followed its development.

DR GOOS
In *Fusarium* I would say it is thallic.

DR HENNEBERT
Usually it is separate before the swelling begins. In *Mycogone*, too, and *Sepedonium*, and *Chlamydomyces* - all are septate at the beginning. This is a good character.

DR CARMICHAEL
Do we want to include solitary, thick-walled, blastic conidia in the definition, or do we want to restrict it to those with thallic development? It is a matter of choice; the

 term chlamydospore is not a part of our main scheme, so it doesn't really matter what we decide.

DR GOOS

I would like to move that we retain chlamydospore in its original sense, as defined by Barron: a thick-walled resting spore, frequently intercalary, which is formed by modification of pre-existing cells.

DR CARMICHAEL

Barron apparently restricted the term to cases of thallic development, and perhaps this would be a good idea.

DR PIROZYNSKI

Barron's definition covers all sorts of cells, which in some cases may not even be spores and have no taxonomic significance. They may be a way of disposing of insoluble waste products; they may store food. Most fungi produce such cells in abnormal environments - of which laboratory culture is one.

DR KENDRICK

A vote on the motion that we wish to retain the prefix "chlamydo-" with the suffix "conidium." Those in favour? (8 in favour; 8 against.) That means we can't do a darn thing! Another vote: we wish to retain the term chlamydospore. (11 in favour.)

How about a definition of chlamydospore as follows: a thick-walled, thallic, terminal, shortly stalked, sessile or intercalary spore?

DR CARMICHAEL

Most of the things that were previously called aleuriospores and chlamydospores are now included in this definition.

DR G.C. HUGHES

This is the one term we have dealt with that is used in all areas of mycology, and in other groups too.

DR KENDRICK

Could we consider section VB as erected by Cole and Kendrick in 1968? At present this comprises three fungi: the *Basipetospora* state of *Monascus ruber*, *Trichothecium roseum*, and *Cladobotryum variospermum*. The conidiogenous cell is gradually consumed basipetally during the production of a series of blastic conidia. We have mentioned that the conidiogenous locus is retrogressive.

DR HENNEBERT

I propose the term retroconidia.

DR CARMICHAEL

These are typical holoblastic conidia; are we now going to give them a special name, not because they are a different kind of conidium, but because they are produced on a different kind of conidiogenous cell?

DR KENDRICK

How about "retrograde blastic conidia," then?

DR CARMICHAEL

Fine. But *not* retroconidia.

DR SUBRAMANIAN

Retrogressive is better than retrograde.

DR KENDRICK

Would the Conference like to adopt the term retrogressive to describe the method by which a succession of the holoblastic conidia of the former section VB are formed? Unanimously adopted.

May we discuss the term annellidic?

DR SUBRAMANIAN

I suggest that percurrent is a good descriptive alternative.

DR HENNEBERT

Percurrent is too general; it applies to any kind of vegetative proliferation through the opening at the tip of the conidiogenous cell, like a phialide; or through the end of a broken-off hypha.

DR G.C. HUGHES

Should we not now consider whether annellophore is really the term we want?

DR KENDRICK

Does this mean we want to suggest an alternative noun, or that we want to use adjectives? First, may we discuss annellophore?

DR CARMICHAEL

Do we wish to distinguish between a percurrent conidiophore which does have annellations, and a specialized annellated conidiogenous cell as it occurs in *Stysanus* or *Scopulariopsis*? Do we wish to use annellophore for the conidiophore, and annellide for the conidiogenous cell?

DR KENDRICK

Or do we wish to use merely "annellated conidiophore" and "annellated conidiogenous cell"? First, I'd like to see the term annellation approved. Are we all in favour of its adoption? Yes.

Now the term annellate: does everyone approve of it? Yes.

The next question is, do we want to call this kind of conidiogenous cell an annellophore, or an annellide, or merely an annellated conidiogenous cell? Several people (Drs Carmichael, Pirozynski, Hennebert, Goos) have suggested that we should have counterparts for phialophore and phialide. My first question is: Does the Conference wish to accept the term annellated conidiogenous cell as an adequate description for the type now under consideration (percurrently proliferating - Hughes's group IIIB)? (5 for; 10 against.)

DR S.J. HUGHES

We are not rejecting the idea of adjectives here: we have already accepted annellate and conidiogenous cell, and can combine them whenever it is appropriate. We merely think a noun should also be available for occasions when *it* is appropriate.

DR KENDRICK

May we now vote to accept either annellide or annellophore as being equivalent to annellated conidiogenous cell? Those in favour of annellide? - 12. Those in favour of annellophore? - 0. Several people abstained because they prefer an entirely adjectival terminology, but almost half the members of the Conference still consider nouns like annelloconidium to be useful, not so much in the diagnoses, but in more general discussions and comparative accounts of species. They can save a lot of repetition.

DR PIROZYNSKI
I would like at this point to present a recommendation which Dr Tubaki, Mrs Pollack, and I have drawn up: "The Committee on Basauxic Conidiophores recommends that this section be considered and classified independently of the main scheme proposed, in view of the fact that it is the entire conidiophore complex that is distinctive, rather than the type of conidium or conidiogenous cell."

EDITOR
*The meeting gave its approval to this suggestion, neatly side-stepping perhaps the most problematical, but fortunately also the smallest, group under consideration at Kananaskis. Despite the considerable discussion this group had received, the many things still unknown about the basauxic phenomenon clearly made non-commitment the most prudent course.*

DR KENDRICK
May we now consider a thallic group, Dr Hughes's section VII. One of the names proposed for this kind of conidium ontogeny is "arthric"; another one I might suggest is "disjunctive." I think we are fairly well agreed on what this section is all about, at least in general terms, and we have definitions read into the proceedings.

DR HENNEBERT
Arthric is an appropriate adjective, and the conidia may be called arthroconidia instead of arthrospores.

DR S.J. HUGHES
I'll go along with that.

DR KENDRICK
Does the Conference recommend the use of the terms arthric and arthroconidium? Almost unanimous; one abstention.

Now we have section VA, Dr Hughes's original section V: "meristem arthrospores," he called them. The meeting is open to suggestions concerning this name.

DR COOKE
At the time this group was discussed, it was suggested that the word arthrospore be dropped and that they be called "meristem conidia."

DR KENDRICK
That seems reasonable, because growth is going on over the whole series of maturing conidia, and there seems to be persistent cytoplasmic connection.

DR CARMICHAEL
This is simply a thallic, basauxic conidiophore, and the conidia are arthric thalloconidia.

DR HENNEBERT
They are certainly thallic. The conidium is initiated by the septation of a portion of the hypha. It differentiates with swelling and may become septate, and is liberated by the separation of the components of a double septum.

DR CARROLL
We've been considering conidium initiation to be a process different from conidium maturation. I wonder if, after the first septum is formed in the meristem, the rest of the observed growth could not be considered a form of maturation. In that case we'd have to consider the meristem as localized at the base.

DR KENDRICK

It's very difficult to localize it. Dr Hughes's original suggestion was that it was spread out, and I haven't yet met anyone who could point and say: "*There's* where the meristem is."

DR S.J. HUGHES

Dr Müller told me that there is a pore in the septum of the conidium in the chain of *Erysiphe*.

DR KENDRICK

There must be in all these cases, because the absolute amount of growth is very large in some cases (not so much in *Erysiphe*), and food must be continuously translocated up the chain to achieve this.

DR CARROLL

Then it really is a diffuse meristem.

DR KENDRICK

Yes, it's different from most other fungi whose meristematic zone is restricted to one conidium at a time, even in those which form chains of conidia.

DR S.J. HUGHES

It differs from arthric in retaining its meristematic connections. In a way, it's like a phialide without the outer wall.

DR KENDRICK

Can we, for the time being, put down "thallic meristem conidia," and then perhaps someone will come up with an appropriate word later? Do we all agree on that? - Yes, more or less unanimously.

All we have to do now is to deal with what the committee called blastoconidia, which would now be regarded as holoblastic. How are we to separate the various ways in which a succession of conidia is produced? "Retrogressively successive" is one. Annellidic is also distinctive because of the repeated percurrent proliferations. Holoblastic conidia are also formed on sympodial conidiogenous cells. They may be formed synchronously on swollen conidiogenous cells often known as ampullae, and they form in unbranched or branched acropetal chains.

*It was now late. We had come to the end of our third session of the day, and the last of the Conference. Our energy, if not enthusiasm, was running out. We had tried to touch on all the subjects germane to our theme, and we had almost succeeded. But as we left the library for the last time we realized that an acceptable synthesis of our deliberations was still lacking. Determined to finish the job as far as was humanly possible, a group of nocturnal mycologists met in the lodge and doggedly proceeded to hammer out a scheme.*

*Proposals and counter-proposals followed in rapid succession. Pencils flashed as the air once again grew thick with smoke and controversy. The zealous proponents of a revolutionary system wrestled verbally with the scepticism of the more conservative element. It would not be true to say that all doubts were laid to rest that night, but at an advanced hour the group agreed upon a scheme. This scheme is provisional and experimental, and will doubtless evolve (or become extinct) as it faces the selection pressure of future mycological thought, observation, and experimentation. It is presented in the next chapter, along with the other major decisions and recommendations of the Conference.*

# 16
# Conclusions and Recommendations

BRYCE KENDRICK

It has fallen to the lot of the editor to compile this chapter from the recorded and unrecorded discussions at Kananaskis, and from subsequent correspondence or personal meetings with members of the Conference. It would be confusing, if not impracticable, to present all shades of opinion (these have already had a good airing in the discussion) so, in most cases, only unanimous or majority decisions are given here.

By a vote of 14 to 3 the Conference recommended the use of primarily, though not exclusively, adjectival terminology (much of it new) to describe the ontogenetic features of conidia and conidiogenous cells. That decision is reflected in the priorities of this chapter. Nevertheless, there are a number of indispensable nouns. These the Conference retained, even proposing one or two new ones of its own, and they are listed and defined below. Each list is in alphabetic order for easy reference.

I have made no attempt to compile a comprehensive glossary of all terms used in describing Fungi Imperfecti; that would be to exceed my mandate, and would obscure the line of thought developed at Kananaskis. I have included in the lists of approved terms only those which are directly applicable to the main issues dealt with by the Conference.

These terms, as I have already indicated in Chapter 15, were assembled by members of the Conference into a tentative hierarchical scheme. A few members were opposed to the idea of erecting *any* hierarchical system, but the majority gave their qualified or tacit approval to much of the scheme presented below. For any misinterpretations, and for the extrapolations represented by categories 3, 4, and 5 of the scheme, I must accept responsibility.

This chapter concludes with a listing of terms which the Conference rejected outright or proscribed with varying degrees of severity.

Department of Biology, University of Waterloo, Waterloo, Ontario.

ACROPETAL: condition of a chain of conidia in which the youngest is at the tip, the oldest at the base. In this kind of chain (found in *Cladosporium harknessii, Bispora antennata, Septonema secedens,* some *Alternaria* spp., etc.) the *conidiogenous loci* are *serial*, and cytoplasmic continuity must be maintained between members of the chain to permit the translocation of food material to the developing apical conidium (unless, of course, the apex is in contact with a food substrate, and can absorb its own nourishment).

ALTERNATE: a condition in which *thallic-arthric* conidia are differentiated from approximately half the cells of a conidiogenous hypha, while the other cells degenerate, their cytoplasm and nuclei largely migrating into the developing conidia. This alternation between fertile and sterile cells may be regular, as in *Amblyosporium*, or less regular, as in *Sporendonema*.

ANNELLATE(D): condition of a conidiogenous cell which has undergone a number of very short *percurrent* vegetative proliferations, each of which terminated in the production of a single *holoblastic* conidium. Each conidium in turn has been dislodged by the next apex and has left behind the vegetative portion of the proliferation which, from its usual appearance as a narrow band of wall material encircling the conidiogenous cell, is called an *annellation.*

ANNELLIDIC: 1) a specialized form of *holoblastic* conidium ontogeny in which the growth of successive, very short, *percurrent,* vegetative proliferations of a conidiogenous cell is terminated by the development of a *conidiogenous locus* which gives rise to a single conidium. The conidiogenous cell usually becomes longer as it produces a *basipetal* sequence of conidia. What remains of each vegetative proliferation after the resulting conidium has seceded may be termed an *annellation,* and the conidiogenous cell itself may be described as *annellated*, or termed an *annellide.* This method of producing a plurality of conidia from a single conidiogenous cell may be exemplified by the *Spilocaea* state of *Venturia inaequalis, Doratomyces stemonitis, Scopulariopsis brevicaulis, Sporidesmium atrum, Triposporium elegans.*

2) less appropriately, describes the conidia produced by such a conidiogenous cell.

ARTHRIC: a form of *thallic* conidium ontogeny characterized by conversion and disarticulation of a pre-existing, *determinate* hyphal element (i.e. one whose extension growth has ceased). Secession of conidia may be by *fission* of double septa laid down across the full width of the hypha, or by *fracture* or *lysis* of the walls of adjacent, degenerated cells.

[*Since a widely accepted definition of growth is "irreversible increase in volume" (usually, though not always, accompanied by increase in dry matter, especially wall material), it is impossible to dissociate the concept of growth, albeit growth of a secondary nature, from what we now call* arthric *ontogeny. Although the conidia of* Geotrichum *approach the "ideal" of conversion without growth, the* alternate arthroconidia *of, say,* Amblyosporium *are also accepted without question as fitting our* thallic-arthric *concept, despite the fact that they may swell considerably during differentiation after they have been clearly delimited by septa. (This may be at least partly because of immigration of cytoplasm and nuclei from adjacent cells.) Thus we cannot entirely separate the concepts of "growth" and "conversion" in our categorizations of conidium ontogeny.*] 

BASAUXIC: [*Hughes (Can. J. Botany 31* [*1953*]*: 577-659) coined this term to describe conidiophores which "elongate by a basal growing point. The conidiophore itself may arise from a barrel-shaped or flask-shaped conidiophore mother-cell" (e.g.* Arthrinium, Endocalyx, Dictyoarthrinium, Cordella, Spegazzinia, Isthmospora*). Our discussions of this group were severely handicapped by a lack of observational data, and although some interesting comparisons were drawn, and some suggestions made for extending the basauxic concept, the mood of the Conference remained cautious. Eventually it was recommended that the basauxic phenomenon should, for the present, be considered and classified independently of the main scheme proposed, in view of the fact that it is the entire conidiophore complex that is distinctive, rather than the type of conidium or conidiogenous cell.*]

BASIPETAL: condition of a chain of conidia in which the youngest is at the base, the oldest at the tip. This kind of chain may be produced by several different types of conidium ontogeny: *enteroblastic-phialidic*, as in *Phialophora lagerbergii*, where the conidiogenous locus is *fixed*; *holoblastic-annellidic*, as in the *Spilocaea* state of *Venturia inaequalis*, where the *serial* conidiogenous loci are *progressive*; *holoblastic-retrogressive*, as in the *Basipetospora* state of *Monascus ruber*, where the *serial* conidiogenous loci are *retrogressive; thallic-arthric* as in some species of *Oidiodendron*, where the *concurrent* conidiogenous loci are *retrogressive*. Conidia of basipetal chains are not cytoplasmically connected, except for those with thallic-"meristem" ontogeny; e.g. the *Oidium* state of *Erysiphe polygoni.*

BLASTIC: one of the two basic modes of conidium development; there is marked enlargement of a recognizable conidium initial *before* the initial is delimited by a septum. The conidium differentiates from *part of* the cell.

CONCURRENT: the simultaneous differentiation of a series of conidia which were initiated successively (as in the *thallic-meristem* conidia of *Phragmotrichum chailletii*).

CONIDIOGENOUS: literally "giving rise to conidia"; replaces "sporogenous" in references to the formation of non-motile, asexual fungal spores of all kinds except sporangiospores.

DETERMINATE: applied to conidiophores or conidiogenous cells which cease extension growth at or before the onset of conidiogenesis, and which do not resume vegetative growth during or between the formation of successive conidia. (e.g. most phialides, conidiogenous hyphae of *Oidiodendron* and *Geotrichum*, conidiogenous hyphae of the *Basipetospora* state of *Monascus ruber*). Some "determinate" conidiogenous cells may proliferate vegetatively, percurrently or sympodially, between periods of reproductive activity (e.g. the phialides of *Catenularia* and *Codinaea*), but this is interpreted as an aberration (albeit often a consistent one) rather than a necessary part of conidiogenesis.

ENTEROBLASTIC: a mode of *blastic* conidium ontogeny in which the outer layer(s) of the wall of the conidiogenous cell is (are) not involved in the formation of the conidium wall. This rather negative style of definition encompasses two major kinds of blastic ontogeny, the *tretic* and the *phialidic* (q.v.).

ENTEROTHALLIC: the mode of *thallic* spore ontogeny in which the outer wall of the sporogenous cell is not involved in the formation of the spore wall (e.g. sporangiospores). It is uncertain whether any thallic conidia of Fungi Imperfecti fit into this category.

FIXED: a condition in which the conidiogenous locus remains in one place, though it may give rise to a succession of conidia (as in phialides).

HOLOBLASTIC: a mode of *blastic* conidium ontogeny in which all wall layers of the conidiogenous cell are involved in the formation of the conidium wall.

HOLOTHALLIC: a mode of *thallic* conidium ontogeny in which all wall layers of the conidiogenous cell are involved in the formation of the conidium wall. (This appears to be the norm in thallic Fungi Imperfecti.)

INDETERMINATE: describes conidiophores or conidiogenous cells which continue extension growth during or between the formation of successive conidia (e.g. *annellides, sympodial* conidiogenous cells, the conidiophore of the *Oidium* state of *Erysiphe polygoni*, basauxic conidiophores).

MERISTEM: a mode of *thallic* conidium ontogeny in which there is continuous *retrogressive* conversion of an *indeterminate* conidiophore whose apical region is continuously extending. There is a linear series of cytoplasmically connected conidia maturing *concurrently*, the distal conidium being the most mature, the proximal conidium hardly yet differentiated from the cells of the conidiophore (e.g. the *Oidium* state of *Erysiphe polygoni, Phragmotrichum chailletii*). [*A more satisfactory term is urgently needed here, but we have so far been unable to find a single word which explains the complex phenomenon of the diffuse, extended meristem, the sequential initiation of conidia, and their concurrent, staggered maturation and secession.*]

PERCURRENT: a mode of vegetative proliferation of conidiophores or conidiogenous cells in which each successive apex arises *through* the previous apex. Annellated conidiogenous cells (*annellides*) characteristically proliferate percurrently, but other kinds of conidiogenous cell may do so on occasion (e.g. the phialides of *Catenularia* and *Chloridium*, and the sympodial conidiogenous cells of *Spiropes melanoplaca*).

PHIALIDIC: 1) a mode of *enteroblastic* conidium ontogeny in which the conidia are clad in an entirely new wall, which is not derived from any existing layers of the wall of the conidiogenous cell. A *basipetal* succession of conidia is formed from a *fixed* conidiogenous locus. 

2) may also describe the conidia produced in this way.

PROGRESSIVE: describes a conidiogenous locus which moves forward or advances, either in a *sympodially* or *percurrently* proliferating conidiogenous cell, or into successive conidia of an *acropetal* chain.

RANDOM: lacking any obvious pattern of occurrence in space or time.

RETROGRESSIVE: 1) conidiogenous cells or hyphae which become converted into conidia, basipetally, randomly, or synchronously.

2) conidiogenous loci which appear in a *serial* or *concurrent* basipetal sequence, either along a stable conidiogenous cell or because the serial or concurrent formation of conidia involves conversion of the conidiogenous cell.

SERIAL: 1) conidia produced one after another, each being differentiated and delimited before the next is initiated (as in *Phialophora, Annellophora, Basipetospora*).

2) conidiogenous loci which arise one after another, each completing its function before the next comes into action.

SINGLE: 1) solitary conidia, produced directly from a conidiogenous cell, which do not themselves proliferate or form part of a basipetal chain. Many single conidia may be produced from one conidiogenous cell, but each must be formed by its own, exclusive, conidiogenous locus; e.g. *Nodulisporium hinnuleum, Beauveria globulifera, Oidium conspersum, Helminthosporium velutinum.*

2) a solitary conidiogenous locus.

STABLE: determinate conidiogenous cells which are not retrogressively converted into conidia.

SYMPODIAL: a method of vegetative proliferation of conidiophores or conidiogenous cells in which the growth of each of a succession of subterminal lateral vegetative apices ends with the production of a conidiogenous locus. The net effect is an increase in length or a swelling of the conidiophore or conidiogenous cell. Sympodial conidiogenous cells may support different kinds of conidium ontogeny: *holothallic-arthric* in *Sympodiella acicola, holoblastic* in *Beauveria globulifera, enteroblastic-tretic* in *Curvularia inaequalis.*

SYNCHRONOUS: a condition in which many conidia are initiated simultaneously and differentiated simultaneously, as in the *holoblastic* conidia of the *Chromelosporium* state of *Peziza ostracoderma,* the primary *enteroblastic* conidia of *Gonatobotryum apiculatum*, the *thallic-arthric* conidia of *Amblyosporium spongiosum.*

THALLIC: one of the two basic modes of conidium ontogeny; if there is any enlargement of the recognizable conidium initial, it occurs only *after* the initial has been delimited by a septum or septa. The conidium differentiates from a *whole* cell.

TRETIC: 1) a mode of *enteroblastic* conidium ontogeny in which the conidia are clad in an extension of the inner wall of the conidiogenous cell. Conidia are solitary or in acropetal chains, and emerge through a narrow channel which is secondarily induced in the usually thick, darkly pigmented outer wall of the conidiogenous cell.

2) may also describe the conidia produced in this way.

RECOMMENDED NOUNS

AMPULLA: a swollen or vesicular conidiogenous cell with *multiple* conidiogenous loci from which many *blastic* conidia arise more or less *synchronously.*

ANNELLATION: the ring-like or cylindrical portion of an annellide produced between successive conidia. [*See Chapters 9 and 19.*]

ANNELLIDE: a conidiogenous cell which produces a single *blastic* conidium from the apex of each of a succession of *percurrent* vegetative proliferations (annellations) involving the half septum remaining after secession of the previous conidium.

ARTHROCONIDIUM: a *thallic-arthric* conidium.

CELL: any unit of a fungus thallus or spore which is morphologically delimited from neighbouring units by a wall or septum. Contiguous cells do not necessarily contain individual protoplasts; cytoplasm and nuclei may pass from one cell to another.

CHLAMYDOSPORE: a thick-walled, thallic, terminal or intercalary spore. [*This term has been very widely used in a variety of senses; we have therefore given it a very broad definition and regard it as of marginal value to our scheme.*]

CONIDIOGENOUS CELL: any cell from which, or within which, conidia are directly produced. Conidiogenous cells may be *integrated* (morphologically indistinguishable from vegetative cells) or may exhibit all degrees of morphological differentiation.

CONIDIOGENOUS LOCUS: a point, area, or zone of a conidiogenous cell at which a conidium arises.

CONIDIUM: a specialized, non-motile, asexual propagule, usually caducous, not developing by cytoplasmic cleavage or free-cell formation (compare sporangiospores, bacterial spores). Conidia are produced by some higher phycomycetes, by many ascomycetes and basidiomycetes, and by many fungi whose sexual state (if any exists) is unknown.

CONIDIUM INITIAL: a cell, or part of a cell, which will become a conidium by differentiation.

DOUBLE SEPTUM: a cross-wall which delimits a caducous conidium; it consists of two layers of wall material with a very thin "abscission layer" between them; this lamella appears to break down (by autolysis?) and allow the two halves of the septum to separate. 

FISSION: the kind of conidium secession which involves separation of the components of a *double septum*. Fission is the most common method by which Fungi Imperfecti release their conidia.

FRACTURE: the kind of conidium secession which involves the rupture of the wall of an adjacent vegetative or degenerated cell at some point removed from the septum. This kind of secession depends on the application of some external mechanical stress; as in *Sporendonema purpurascens, Endophragmia biconstituta.*

LYSIS: the kind of conidium secession which involves the dissolution of the wall of an adjacent degenerated cell. The dissolution may sometimes be autolytic, and sometimes brought about by the enzymes of other microorganisms in a natural habitat like soil (e.g. *Coremiella ulmariae, Trichophyton*).

MERISTEM: any place on a hypha, conidiogenous cell, or conidium at which growth (i.e., irreversible increase in volume) occurs. Usually the activities of a meristem result in cell extension, or in the formation of a new cell or cells (as at a conidiogenous locus).

PHIALIDE: a conidiogenous cell which produces, from a *fixed* conidiogenous locus, a basipetal succession of *enteroblastic* conidia whose walls arise de novo; cf. phialidic (e.g. *Penicillium, Phialophora, Chalara, Sporoschisma*). A more extended definition of phialide is as follows: a conidiogenous cell in which at least the first conidium initial is produced within an apical extension of the cell, but is liberated sooner or later by the rupture or dissolution of the upper wall of the parent cell. Thereafter, from a fixed conidiogenous locus, a basipetal succession of enteroblastic conidia is produced, each clad in a newly-laid-down wall to which the wall of the conidiogenous cell does not contribute. Any phialide wall distal to the conidiogenous locus is the *collarette.* The length of the phialide does not change during the production of a succession of conidia, though some phialides undergo intermittent vegetative proliferation, either *percurrent* (as in *Catenularia*) or *sympodial* (as in *Codinaea*) between conidiogenous episodes.

POLYPHIALIDE: a phialide which intermittently undergoes sympodial vegetative proliferation, and may become secondarily septate as it extends. Each conidiogenous locus with its collarette is in turn terminal, and each produces a plurality of conidia before being displaced to one side by the growth of the next subterminal vegetative apex and ceasing to function. It may be suggested that there is a single conidiogenous locus which alternates between the *fixed* and the *progressive* modes, since the only active site of conidiogenesis is that at the apex of the structure. This is a term which, though established in the literature, could with advantage be replaced by a short, descriptive phrase like *sympodial phialide.*

Tentative ontogenetically-based system for Fungi Imperfecti

| 1<br>Origin of conidia | 2<br>Origin of conidium wall | 3<br>Conidiogenous cells or hyphae | 4<br>Conidia | 5<br>Conidiogenous loci |
|---|---|---|---|---|
| b. Blastic | a. Holoblastic<br>b. Enteroblastic-tretic<br>c. Enteroblastic-phialidic | a. Indeterminate annellidic (percurrent)<br>b. Indeterminate sympodial<br>c. Indeterminate basauxic<br>d. Determinate stable<br>e. Determinate retrogressive | a. Single<br>b. Acropetal chains<br>c. Basipetal chains<br>d. "Alternate" chains<br>e. "Random" chains | a. Multiple synchronous<br>b. Multiple concurrent progressive<br>c. Multiple concurrent retrogressive<br>d. Multiple serial progressive<br>e. Multiple serial retrogressive<br>f. Multiple random<br>g. Single fixed<br>h. Diffuse or extended |
| b. Thallic | d. Holothallic | f. Indeterminate retrogressive ("meristem")<br>g. Determinate retrogressive (arthric)<br>h. Determinate stable | | |
| | e. Enterothallic (sporangic?) | | | |

In the accompanying table an attempt has been made to analyse the blastic and thallic groups separately in columns 2 and 3. Components of columns 4 and 5, however, may be applied wherever appropriate to members of both blastic and thallic groups. Perhaps the best way to explain the scheme is to apply it to a number of examples.

The *Chromelosporium* state of *Peziza ostracoderma* is 1a, 2a, 3d, 4a, 5a: i.e. blastic, holoblastic, with determinate, stable conidiogenous cells which produce conidia singly from multiple synchronous conidiogenous loci.

*Amblyosporium spongiosum* is 1b, 2d, 3g, 4d, 5a: i.e. thallic, holothallic, arthric, with determinate, retrogressive conidiogenous hyphae which produce conidia in synchronous "alternate" chains at synchronous conidiogenous loci.

*Nodulisporium hinnuleum* is 1a, 2a, 3d, 4a, 5e: i.e. blastic, holoblastic, with determinate, stable conidiogenous cells which produce conidia singly from retrogressive conidiogenous loci.

*Penicillium corylophilum* is 1a, 2c, 3d, 4c, 5g: i.e. blastic, enteroblastic-phialidic, with determinate, stable conidiogenous cells producing conidia in a basipetal chain from a fixed conidiogenous locus.

The *Basipetospora* state of *Monascus ruber* is 1a, 2a, 3e, 4c, 5d: i.e. blastic, holoblastic, with determinate, retrogressive conidiogenous cells producing basipetal chains of conidia from serial, retrogressive conidiogenous loci.

*Geotrichum candidum* is 1b, 2d, 3g, 4e, 5f: i.e. thallic, holothallic, arthric, producing random chains of conidia at random loci.

*Beauveria globulifera* is 1a, 2a, 3b, 4a, 5d: i.e. blastic, holoblastic, with indeterminate, sympodial conidiogenous cells producing conidia singly from serial, progressive conidiogenous loci.

*Scopulariopsis brevicaulis* is 1a, 2a, 3a, 4c, 5d: i.e. blastic, holoblastic, with indeterminate, annellidic conidiogenous cells producing basipetal chains of conidia from serial, progressive conidiogenous loci.

*Diplococcium spicatum* is 1a, 2b, 3d, 4b, 5f: i.e. blastic, enteroblastic-tretic, with determinate, stable conidiogenous cells giving rise to acropetal chains of conidia from random loci.

## TERMS REJECTED, REPLACED, OR NOT RECOMMENDED

ALEURIOSPORE: confused term, rejected.

ANNELLOCONIDIUM = holoblastic conidium produced on an annellated conidiogenous cell (annellide).

ANNELLOPHORE = annellated conidiogenous cell (annellide). [*Although there was a suggestion at the conference that annellophore should be used to designate a supporting structure bearing one or more annellides, I think this would be both inappropriate and confusing, since the word has been widely used in the broader sense of a "conidiophore with annellations" for 17 years.*]

ANNELLOSPORE = holoblastic conidium produced on an annellated conidiogenous cell (annellide).

ARTHROSPORE = thallic-arthric conidium (arthroconidium).
ASCOCONIDIOPHORE = phialide.
ASCOCONIDIUM = enteroblastic-phialidic conidium (phialoconidium).
BIPHIALIDE: judgement reserved; phenomenon insufficiently known to us.
BLASTOCONIDIUM = holoblastic conidium produced solitarily, synchronously, or in acropetal chains.
ENDOCONIDIOPHORE = phialide.
ENDOCONIDIUM = enteroblastic-phialidic conidium (phialoconidium).
GANGLIOSPORE = holoblastic conidium.
MERISTEM ARTHROSPORE (pro parte) = thallic-meristem conidium.
MERISTEM ARTHROSPORE (pro parte) = holoblastic retrogressive conidium.
MONOPHIALIDE = phialide.
ORTHOPHIALIDE = phialide.
PHIALOPHORE = supporting structure bearing one or more phialides; conidiophore, pro parte.
PHIALOSPORE = enteroblastic-phialidic conidium (phialoconidium).
PLAGIOPHIALIDE = phialide.
PLEUROPHIALIDE = phialide.
POROCONIDIUM = enteroblastic-tretic conidium (tretoconidium).
POROSPORE = enteroblastic-tretic conidium (tretoconidium).
PROPHIALIDE = cell subtending conidiogenous cell in *Zygosporium* (which is *not* a phialide).
SCHIZOPHIALIDE = phialide.
SPORE (pro parte) = conidium (in Fungi Imperfecti).
SPOROGENOUS CELL = conidiogenous cell.
STERIGMA: restricted to Basidiomycetes.
SUBENDOGENOUS: superfluous term referring to the position of the conidiogenous locus in a phialide in relation to the distal end of the collarette. A relative term concerning collarette length.
SYMPODIOCONIDIUM = holoblastic conidium produced on a sympodial conidiogenous cell.
SYMPODULA = sympodial conidiogenous cell.

# 17
# Conidium Ontogeny in Pycnidial and Acervular Fungi

B.C. SUTTON

*This and the next three chapters concern subjects which, although not actually dealt with at the Conference, are germane to its theme. They were written with our Kananaskis discussions in mind, and much of the terminology approved by the Conference is incorporated in them. They may in some sense be considered the first fruits of Kananaskis, and are therefore presented as an integral part of this book.*

The Sphaeropsidales and Melanconiales have not been extensively incorporated into the several accounts of deuteromycete classification based on developmental criteria. Costantin (1888) included *Ellisiella* with *Zygosporium, Beltrania,* and *Circinotrichum* in a group circumscribed by two types of hyphae: long sterile (setae) and short fertile (conidiophores). *Pestalotia* was equated with *Mastigosporium* but placed in the Melanconiales. Vuillemin (1910a, b, 1911) omitted the Sphaeropsidales and Melanconiales from his study, confining his attention to Hyphomycetes. Moreau (1953) adopted Vuillemin's system of classification, somewhat amended by Langeron and Vanbreuseghem (1952), and included *Pestalotia* in the Thallispori, Aleuromycetales, and *Phialophorophoma* in the Conidiospori, Phialidales. The majority of the pycnidial and acervular forms were placed in families of the Conidiospori, Sporophorales. The dissociation of the Melanconiales was advocated on the basis of conidium type, but no details were given.

Hughes (1953) placed *Alysisporium, Septotrullula,* and *Phragmotrichum* in section V of his scheme, *Melanconium* in section III, and the arthroconidium state of *Hendersonula toruloidea* in section VII. He suggested that the annellides typical of section III were probably common in the Coelomycetes (Sphaeropsidales and Melanconiales) and that, when the range of developmental types in the Melanconiales became known, the latter could be incorporated with the Hyphomycetes for

Commonwealth Mycological Institute, Kew, England.

 the purposes of classification. Subsequent work in the Sphaeropsidales and Melanconiales by a variety of authors, largely stimulated by the concepts of Hughes, has shown his predictions to be correct. Conidia *are* formed from definite conidiogenous cells in ways comparable to those outlined for Moniliales.

Most of the literature on coelomycete taxonomy before 1953 is glaringly deficient in accurate data on conidium and conidiophore ontogeny. However, there is a limited amount of useful information in isolated accounts and illustrations, though the authors did not realize the significance of their observations. Corda, for example, figured the annellophores of *Melanconium juglandis* Cda. in 1839. One important contribution was the highly individual attempt by Höhnel (1923) to provide a comprehensive system for the Fungi Imperfecti. Three major types of sporulating structure were acknowledged: separate conidiophores, synnemata, and plectenchymatic-parenchymatic fructifications (incorporating pycnidia, acervuli, and sporodochia). The last (Histiomyceten) was divided into the Endogenosporae, in which conidia were formed by histolysis of the inner cells of the fructification, no conidiophores being involved, and the Exogenosporae, in which conidia were produced by budding or division. This work was largely ignored after Petrak (1925) severely criticized Höhnel's taxonomy and nomenclature. Only Klebahn (1933) and Goidanich and Ruggieri (1947) attempted to put Höhnel's suggestions into practice. However, placed in perspective, Höhnel's system not only anticipated developments taking place in classification of the Fungi Imperfecti today, but was also the first attempt at incorporating genera of the Melanconiales, Sphaeropsidales, and Moniliales into a system based, at least in part, on developmental features of conidia and conidiophores.

To formulate a workable ontogenetically based classification for the genera of Sphaeropsidales and Melanconiales and incorporate it into the schemes already outlined for the Moniliales is particularly difficult because of the sparsity of information on most genera. Accurate data on conidium development are known for representatives of only 200 genera of a possible 1,300, and frequently this information is not about the type species. Many genera, such as *Ascochyta, Phyllosticta, Septoria, Phoma, Diplodia, Gloeosporium,* and *Septogloeum*, contain several hundred species, and a cursory glance at conidium development and fructification structure in a few species reveals considerable heterogeneity within genera as now circumscribed, and it is therefore usually impossible to define these genera in modern terms. Moreover, the ontogenetic data have not been fully correlated with fructification structure, a feature of considerable importance for the Sphaeropsidales (though less so for the Melanconiales and Moniliales). To apply developmental concepts in the Sphaeropsidales, and to circumscribe genera without due regard for the features of the pycnidia, can only lead to more confusion than exists already.

For these reasons I consider it premature to formulate any comprehensive system. An analysis of existing data, however, will demonstrate the range of diversity so far found in the Sphaeropsidales and Melanconiales and will also indicate problems which may have to be solved before a comprehensive system for the Fungi Imperfecti can be formulated.

In the Sphaeropsidales and Melanconiales, fungi with conidia and conidiophores typical of Hughes's sections IA, II, III, IV, V and VII have been found. So far there are no representatives of sections VI, VIII, and IX (Tubaki 1958). The majority of genera show phialidic (section IV) or annellidic (section III) ontogeny, very few belonging to the remaining sections. 

## CATENATE CONIDIA

One of the fundamental differences between the various types of conidium chain and the conidiogenous cells from which they arise is the point at which meristematic activity, if any, occurs. In section IA new conidia develop from the preceding one in the chain, the youngest conidia therefore being at the apex. Conidiophore growth is terminated by the onset of conidium formation unless branching occurs lower down the conidiophore. In some fungi belonging to section V it is similarly clear that conidia mature in the reverse manner, with the oldest conidium at the apex of the chain and the younger ones down towards the poorly differentiated conidiogenous cell. The meristem is at the apex of the conidiophore, not of the conidial chain, so the conidiophore is of indeterminate length. In section VII the conidiophore is of a determinate length with no meristematic zone, but conidia are often differentiated basipetally by simple fragmentation of the conidiophore, and in some fungi there can be a similar gradation in morphology between mature conidia at the apex and immature conidia towards the base. Unless the point can be determined at which the meristem, if any, is operative, it may not be easy to decide whether the fungus belongs to section V or section VII. This is particularly so in the Sphaeropsidales and Melanconiales, where the conidiophores and conidia are enclosed in either fungal or host tissue and developmental details are often obscured. If there is a continuous gradient of morphological change from younger to older conidia, as in *Phragmotrichum* Kunze and Schmidt, the fungus can be readily assigned to section V. However, the majority of Sphaeropsidales and Melanconiales with catenate conidia are more inscrutable, the conidia being relatively small, unicellular, sometimes truncate, and with very little morphological variation between mature and immature conidia. The problem here is to determine whether a conidiophore of determinate length is simply fragmenting (with or without subsequent changes in conidium morphology as in *Oidiodendron* Robak), or whether additional conidia are being abstricted from a continuing meristem at the apex of the conidiogenous cell. There are also to be borne in mind the subtle differences between species in section VA, which produce meristem thalloconidia, and those in what Cole and Kendrick (1968) called section VB (the *Basipetospora* state of *Monascus ruber, Trichothecium roseum, Cladobotryum variospermum*), which produce retrogressive blastoconidia.

In the pycnidial fungus *Desmopatella salicis* Höhn. long chains of identical, fusiform, guttulate conidia are formed from indistinct conidiophores which branch only at the base. There is no clear meristematic region at the apex of either the conidiogenous cell or the conidial chain. Similarly in *Siropatella stenospora* (Berk. and Curt.) Höhn. the pycnidia are poorly differentiated and convoluted, with such very variable septate, catenate conidia merging into ill-defined conidiophores that the basipetal or acropetal nature of the chains is hard to determine. *Acarosporium sympodiale* Bub. and Vleugel and *A. hederae* (Syd.) Petrak display a remarkable similarity to *Phragmotrichum* species and *Trullula olivascens* (Sacc.) Sacc. in having excipuliform fructifications and catenate conidia, but the chains of conidia branch dichotomously and conidia are of identical morphology throughout the chain. The presence of variously inserted conidium appendages prevents placement of *Acarosporium* species in section VII, but the absence of a clear meristem precludes disposition in either section IA or V with any certainty.

There are species in the Sphaeropsidales and Melanconiales where the ontogeny of catenate conidia has been unequivocally established. In describing basipetal maturation in section V, Hughes noted the large number of coelomycete genera with conidia in chains, but he was unaware of any in which pycnidiospores develop in acropetal succession. In *Hormococcus conorum* (Sacc. and Roum.) Robak, however, chains of hyaline, globose conidia from determinate, elongated or doliiform conidiogenous cells are produced in pycnidia. Successive conidia are formed from the preceding conidium and maturation is clearly acropetal, placing the species in section IA. The conidia of *Siroligniella salicicola* Naumov are also formed in acropetal chains which are often branched. In section V Hughes (1953) placed *Alysisporium rivoclarinum* Peyr. (=*Phragmotrichum rivoclarinum* (Peyr.) Sutton and Pirozynski), *Phragmotrichum chailletii* Kunze, and *Septotrullula bacilligera* Höhn. var. *cambrica* Grove and Rhodes (=*Phragmotrichum rivoclarinum*), all of which Sutton and Pirozynski (1965) characterized by excipuliform fructifications and dark brown, basipetally maturing chains of septate conidia formed from differentiated conidiogenous cells. *P. pini* (W. B. Cooke) Sutton and Sandhu (1969a) has the same kind of conidium ontogeny. In the Melanconiales, *Septogloeum populiperdum* Moesz and Smarods belongs to this section, but the chains are relatively short, the conidiophores branched near the base, and the conidia variable in both shape and septation, features quite different from the type species of *Septogloeum* Sacc.

Only three representatives of section VII are known so far in the Sphaeropsidales and Melanconiales. *Sphaerographium induratum* Syd. (Sydow and Sydow 1913) seems to produce conidia by basipetal fragmentation of determinate conidiophores lacking any meristematic zone. Conidiophores were described and figured as branched and fragmenting into cylindrical truncate conidia, a feature not shared by the 10 species originally placed in the genus by Saccardo. Pirozynski and Morgan-Jones (1968), in describing *Trullula olivascens* (Sacc.) Sacc., reported that the branched conidiophores developed transverse septa at regular intervals in basipetal succession. Conidia were regularly oblong, truncate, and little different from the conidiophore cells. *Vouauxiella lichenicola* Petr. and Syd. resembles *T. olivascens* in all features of conidium development except that the conidia are formed in definite pycnidia rather than excipuliform fructifications.

One feature common to these arthroconidia, meristem thalloconidia, retrogressive blastoconidia, and acropetal solitary blastoconidia is that individual conidia are delimited from one another by double septa. Secession is effected by transverse splitting across the median line of the septum and coincident rupture of the outer periclinal wall, which must be continuous between all conidia in the chain.

In most descriptions of pycnidial and acervular fungi, if conidium development is mentioned at all, it is assumed that ontogeny is blastic. This is far from true, but since 1953 a number of different types of holoblastic development have been found in the Sphaeropsidales and Melanconiales. The simplest form is that of the conidiogenous cell which is broadly doliiform and hardly differentiated from the inner cells of the pycnidial or acervular wall (Figure 17.1 A). Frequently the problem is to decide if such cells are holoblastic with solitary conidia or enteroblastic with a succession of phialoconidia; but, in common with similar examples in the Moniliales, blastoconidia bear a flattened base encircled by a marginal frill. The latter may be marked or hardly discernible in mature conidia. The possibility that the conidiogenous cells are the precursors of annellides cannot be excluded, although this is unlikely because annellides are generally elongated. Many fungi which produce a few, large conidia in pycnidia, such as *Tiarospora* Sacc. and March., *Prosthemium stellariae* Riess, *Scolecosporiella* species (Sutton 1968a), *Aristastoma* species (Sutton 1946a), *Kellermania* species (Sutton 1968a), *Apiocarpella macrospora* (Speg.) Sydow, and *Camarographium abietis* (Wils. and Anders.) Grove, tend to do so in this manner, although species of *Ascochytula* and *Microdiplodia* with conidia not exceeding 15 $\mu$ in length were shown by Dickinson and Morgan-Jones (1966) to develop in the same way. In *Tetranacrium graminum* Hudson and Sutton (1964) the solitary staurospores are formed holoblastically in hysteriiform pycnidia.

Elongated conidiogenous cells which are quite distinct from the pseudoparenchyma forming the pycnidial or acervular wall are the second recognizable category (Figure 17.1 B). Blastoconidia are produced solitarily from their apices. Such development may be the expression of solitary holoblastic ontogeny or the precursor (first-formed conidium) of annellidic holoblastic ontogeny as Hughes (1953) has indicated in his discussion of section III. Again, conidia tend to be rather large and complicated as is seen in *Septogloeum rhopaloideum* Dearn. and Bisby, *Mycohypallage congesta* (Berk. and Br.) Sutton (1963a), *Labridella cornucervae* Brenckle (Shoemaker 1963), *Toxosporiopsis capitata* Sutton and Sellar (1966), *Steganosporium muricatum* Bonorden, *Asterosporium asterospermum* (Pers. ex Gray) Hughes, *Entomosporium* Lév. species, *Chaetospermum chaetosporum* (Pat.) Smith and Ramsb. (Fonseka 1960), *Readeriella mirabilis* Syd., and *Stevensonula ciliata* Petrak. Genera range between the Sphaeropsidales and Melanconiales. *Phyllostictina* Syd. is a well-known example and it seems likely that the species of *Ascochyta* Lib. and *Stagonospora* (Sacc.) Sacc. described by Cunnell (1956, 1957) also belong here. With large conidia it is fairly easy to determine the presence of a marginal frill at the conidium base. Since the apex of the conidiogenous cell is comparatively wide there should be little confusion with phialidic ontogeny. Such interpretative problems as do arise here are concerned with the possibility that growth of the conidiogenous cell is not terminated by formation of a conidium but that percurrent growth occurs to produce an annellide and annelloconidia. *Discosia* Lib. is a typical example. Most related genera produce conidia from annellides, but in *Discosia* no more than a single conidium appears to be formed from each conidiogenous cell.

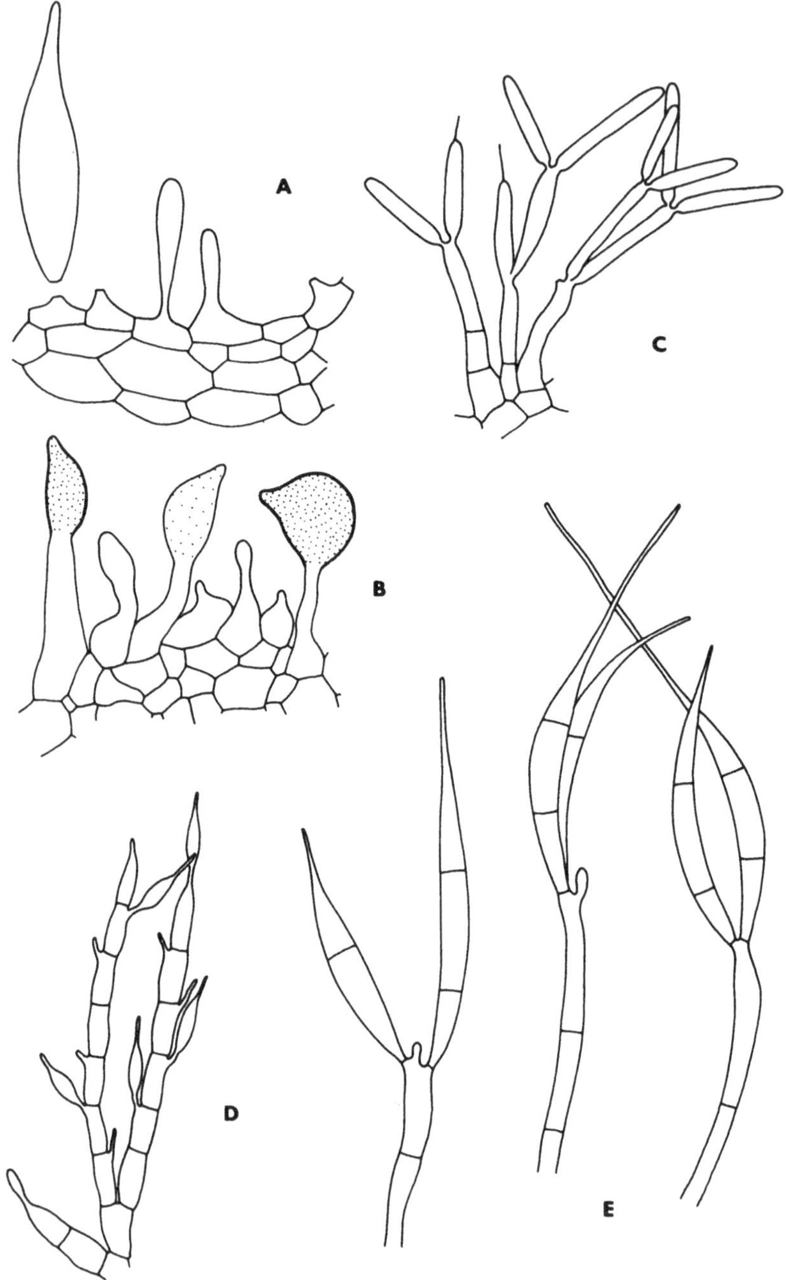


Figure 17.1 Holoblastic ontogeny (solitary, synchronous, sympodial): A, *Scolecosporiella kranzii* Sutton; B, *Readeriella mirabilis* Syd.; C, *Furcaspora pinicola* Bonar; D, *Eleutheromyces subulatus* (Fr.) Fckl. (after Seeler 1943); E, *Heteropatella alpina* (Ell. and Ev.) W.B. Cooke.

A type of conidiophore commonly associated with enteroblastic ontogeny, but rarely seen in connection with holoblastic development, is the elongated conidiophore which is formed of individual conidiogenous cells each producing a conidium from a point immediately below the septum separating contiguous cells (Figure 17.1 D). These are frequently referred to as pleurogenous, or acropleurogenous if a conidium is also formed at the apex of the terminal cell. A few examples are known in this group, but interpretation is tentative since the data have not been confirmed. Development may well be phialidic. Seeler (1943) illustrated *Eleutheromyces* Fuckel and *Eleutheromycella* Höhn., both with appendaged blastoconidia formed from a single, short lateral branch of each conidiogenous cell of the conidiophore, and *Catenophora* Luttrell (1940) was described with a similar type of development.

A conidiogenous cell may directly produce more than a single holoblastic conidium. This is accomplished in at least three ways. (1) Synchronous holoblastic development; many conidia may be formed more or less simultaneously near the apex of the conidiogenous cell (Figure 17.1 C). In *Furcaspora pinicola* Bonar up to three appendaged, Y-shaped conidia are formed like this from elongated conidiogenous cells. Sometimes, instead of producing a conidium, a new functional conidiogenous cell is formed. Such synchronous formation is doubtfully separated from (2) sympodial holoblastic development, where the growth of the conidiogenous cell is terminated by formation of the first conidium, subsequent growth being achieved by lateral development from below the first conidium scar (Figure 17.1 E). The sympodial growth, if that is what it is, takes place in a restricted region near the apex of the conidiogenous cell, so the precise sequence in development is not often clear. *Alveophoma* Alcalde (Sutton 1964b), *Septoria didyma* Fuckel, and *Septocyta ramealis* (Desm. and Rob.) Petrak show this in the Sphaeropsidales, and *Oramasia hirsuta* Urries (Sutton and Pirozynski 1963) in the Melanconiales. Species of *Heteropatella* Fuckel frequently show two or more conidia attached to the conidiogenous cell apex, and are perhaps, with the conidial state of *Colpoma quercinum* (Fr.) Wallr. (Twyman 1947) and *Conostroma* Moesz, the best examples in this group. *Monochaetiella cymbopogonis* Punithalingam and Sarwar apud Punithalingam (1969), *M. themedae* Kandas. and Sundar., and several *Septoria* and *Septogloeum* species show conidiogenous cells which are nodular near the apex as if bearing the scars left by successive conidia. (3) Annellidic holoblastic development; the term annellide has superseded annellophore, coined by Hughes (1953) to describe the conidiogenous cells or conidiophores typical of his section III. The original concept of section III was purposely broad and subsequently Tubaki (1958) divided it into three subsections, of which IIIB contained fungi with characteristic annellides. Recent studies with time-lapse photomicrography of *Scopulariopsis brevicaulis* (Sacc.) Bain. by Cole and Kendrick (1969a) and electron microscopy of *Melanconium bicolor* Nees, *M. apiocarpum* Link, and *Cryptosporiopsis* sp. by Sutton and Sandhu (1969b) have resulted in a very precise definition of annellidic development in terms of gross conidium-conidiophore ontogeny, conidium secession, and the relationships of the walls involved. [*See also Chapters 9 and 19.*]

As Hughes (1953) has suggested, annellides are common throughout pycnidial and acervular fungi, though not as prevalent as what are at present termed phialides. The typical annellide with its closely annellate apex is hard to confuse with other types of conidiogenous cell. However, there is one area where problems in interpretation can arise. Sutton and Sandhu (1969b) showed that percurrent growth of the annellides in a *Cryptosporiopsis* species did not result in secession of successive conidia at increasingly higher levels but that the point of secession could be below, at, or above the level at which the first conidium seceded. Under optical microscopy the conidiogenous cells of this *Cryptosporiopsis* showed the constricted apical cytoplasmic channels seen with various phialides in the Coelomycetes. The extent to which this occasionally retrogressive annellide is present in this group is unknown. Since constricted cytoplasmic channels are very common it seems most probable that the phenomenon is widespread.

Most annellides are hyaline to very pale brown with a few apical annellations (Figure 17.2 A). The complex of acervular genera including *Monochaetia* (Sacc.) Sacc., *Pestalotia* de Not., *Pestalotiopsis* Stey., *Seimatosporium* Cda., *Seiridium* Nees ex Cda., *Toxosporium* Vuill., and *Truncatella* Stey. was shown by Sutton and Sellar (1966) to be characteristic of this group. *Scolecosporium* Lib., *Coryneum* Nees ex Fr. (Sutton 1963b), and some *Steganosporium* species are another natural group with wide, elongated, branched conidiophores and sparse annellations. Other genera in both the Sphaeropsidales and Melanconiales with comparable annellides are *Bleptosporium* Stey. em. Sutton (1963a), *Chondroplea* Kleb. (Sutton 1964b), *Monostichella* Höhn., *Dichomera* Cooke, *Cylindrosporium* auct., *Septogloeum* auct., *Phleospora* Wallr., *Ahmadia* Syd., *Disculina* Höhn., *Cryptosporium* Kze. ex Fr., *Sirococcus* Preuss, *Cryptocline* Petrak, and *Colletogloeum* Petrak (Sutton 1964a), to name but a few. *Siroscyphella succinea* (Fr.) Höhn. has hyaline annellides with up to four closely spaced annellations. Some species, instead of showing the usual extremely short percurrent proliferations, produce widely spaced annellations (Figure 17.2 B). In *Idiocercus* Sutton (1967a) the flared cupulate annellations, reminiscent in size, though not morphology, of those of *Stigmina platani* (Fuckel) Sacc., are widely spaced. Amongst the Sphaeropsidales and Melanconiales, fungi with pigmented conidia often bear annellides which become dark brown towards the apex and occasionally show ornamentation similar to that of the conidia they bear (Figure 17.2 C). In *Didymosporina* Höhn., *Obstipipilus* Sutton (1968a), and *Annellolacinia* Sutton (1964a) many closely spaced annellations are formed from glabrous conidiogenous cells of similar colour to the conidia, whereas in *Cryptomela* Sacc., *Leptomelanconium* Petrak, *Melanconium* Cda. (Sutton 1964c), *Gloeocoryneum* Weindlm., *Schonbornia basidio-annulata* Bubak (1906), *Stilbospora terminaliae* Hughes, and *Hendersonia eucalypti* Hansf. (Walker 1962), although annellations are not so numerous, the conidiogenous cells occasionally become deeply pigmented and rough towards the apices (Figure 17.2 D).

Deviations from strictly annellidic development are rare. In *Seimatosporium kriegerianum* (Bres.) Morgan-Jones and Sutton apud Sutton (1964a) and *Ajrekarella polychaetriae* Kamat and Kalani (Sutton 1967b), however, not only are annellations produced, but sympodial growth of the conidiogenous cell also occurs. In neither case is it as marked as in *Graphium calicioides* (Fr.) Cke. and Massee, where there is considerable lability between sympodial and percurrent proliferation. These cases provide interesting links between holoblastic annellidic, holoblastic solitary and holoblastic sympodial developments.

## ENTEROBLASTIC CONIDIA

Conidiogenous cells that have been termed phialides in the Sphaeropsidales and Melanconiales are perhaps the most intractable to interpret. The apices are rarely more than 2$\mu$ wide, with the channel and aperture often less than 1$\mu$ wide. Though Cole and Kendrick (1969b) have made time-lapse light microscope analyses of the larger phialides of some hyphomycetes, the electron microscope offers a potentially more precise approach to the Coelomycetes. The endogenous nature of the first conidium in *Phoma exigua* Desm. and *P. fumosa* Ell. and Ev. has been demonstrated with the electron microscope by Brewer and Boerema (1965) and Sutton and Sandhu (1969b), respectively, and the unpublished results of Sutton and Chao on *Hendersonia pinicola* Wehm. are similar. It has not, however, been established whether conidiogenous cell wall is incorporated into the wall of this first conidium, but there is limited evidence for *P. exigua* and *P. fumosa* that conidium secession is by separation of a double septum.

Similar secession in phialidic members of the Moniliales has been shown with the electron microscope for *Penicillium claviforme* Bain. by Zachariah and Fitz-James (1967), and for *Verticillium albo-atrum* Reinke and Berth. by Buckley, Wyllie, and de Vay (1969). The conidia of such "budding" organisms as *Aureobasidium pullulans* (de Bary) Cif., Ribaldi and Corte, *Saccharomyces cerevisiae*, and *S. ludwigii* were shown by Durrell (1968), Agar and Douglas (1955), and Streiblová and Beran (1965), respectively, to secede by septation also. It seems likely that budding (secession without septum formation) is of dubious occurrence in the Deuteromycotina, and that conidia from all types of holoblastic ontogeny as well as enteroblastic phialidic ontogeny secede by separation of a double septum. Although the extent of this phenomenon in phialides of the Fungi Imperfecti is not really known, the implications are far-reaching.

The points of secession of subsequent conidia in relation to the point at which the first one seceded often cannot be resolved without the aid of the electron microscope. Sutton and Sandhu (1969b) showed varying levels of secession in the annellides of *Cryptosporiopsis*, until then considered to be phialidic, and suggested that the periclinal striations and ridges in the apical walls of conidiogenous cells in *P. fumosa* probably corresponded to annellations. Brewer and Boerema (1965) and Boerema (1965) also described successive ridges formed at the conidiophore apex. This is an area where considerably more study is required. Some conidiogenous cells that have been termed phialides on the basis of optical microscopy may be no more than annellides in which there is little or no conidiogenous cell growth following conidium secession. It is in the light of these limited observations that the range of "phialide" types in pycnidial and acervular fungi is tentatively discussed.

Phialides show systems of arrangement parallel to those of holoblastic conidiogenous cells in that there is a definite progression from simple to complex. Al-

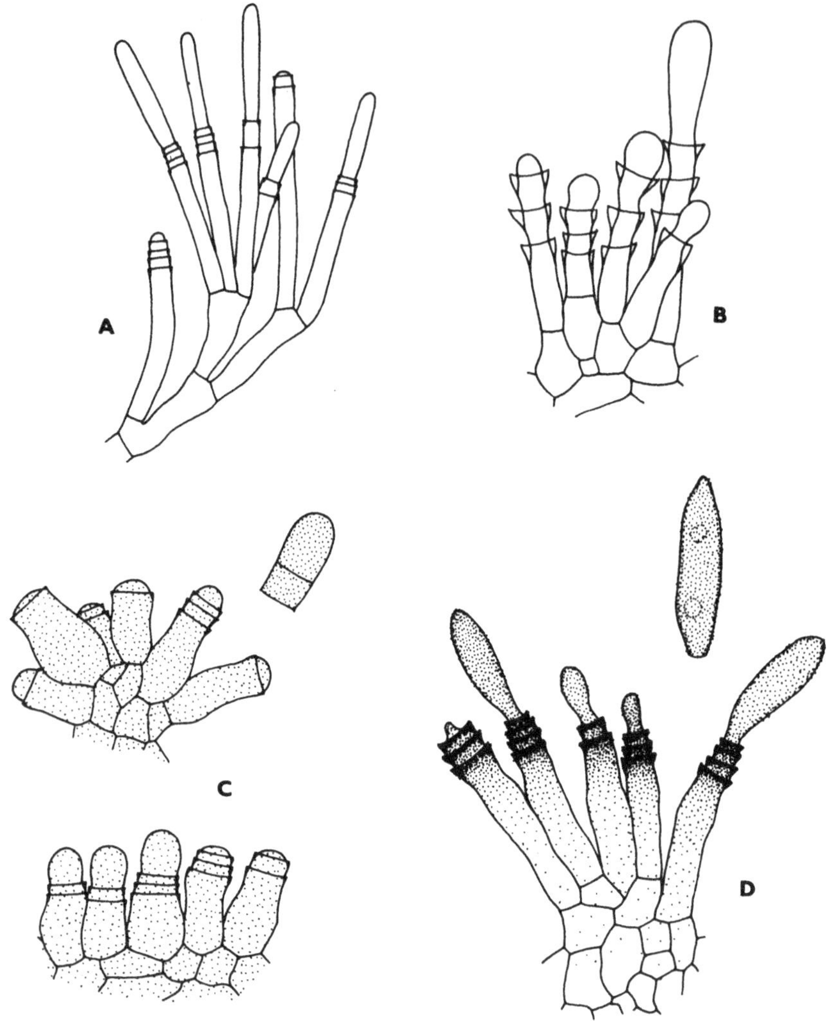


Figure 17.2 Holoblastic annellidic ontogeny: A, *Siroscyphella succinea* (Fr.) Höhn.; B, *Idiocercus pirozynskii* Sutton; C, *Didymosporina aceris* (Lib.) Höhn.; D, *Cryptomela typhae* (Pk.) Died.

though no satisfactory line can be drawn between the different types, because of their variability, it is possible to place them in recognizable groups. The commonest sort is the conidiogenous cell which is indistinguishable from the inner cells of the pycnidial wall but for a single apical aperture (Figure 17.3 B). This has been interpreted in *Phoma* and *Plenodomus* as producing "porospores" and "murospores" respectively by Boerema and Kersteren (1964) and phialoconidia by Sutton

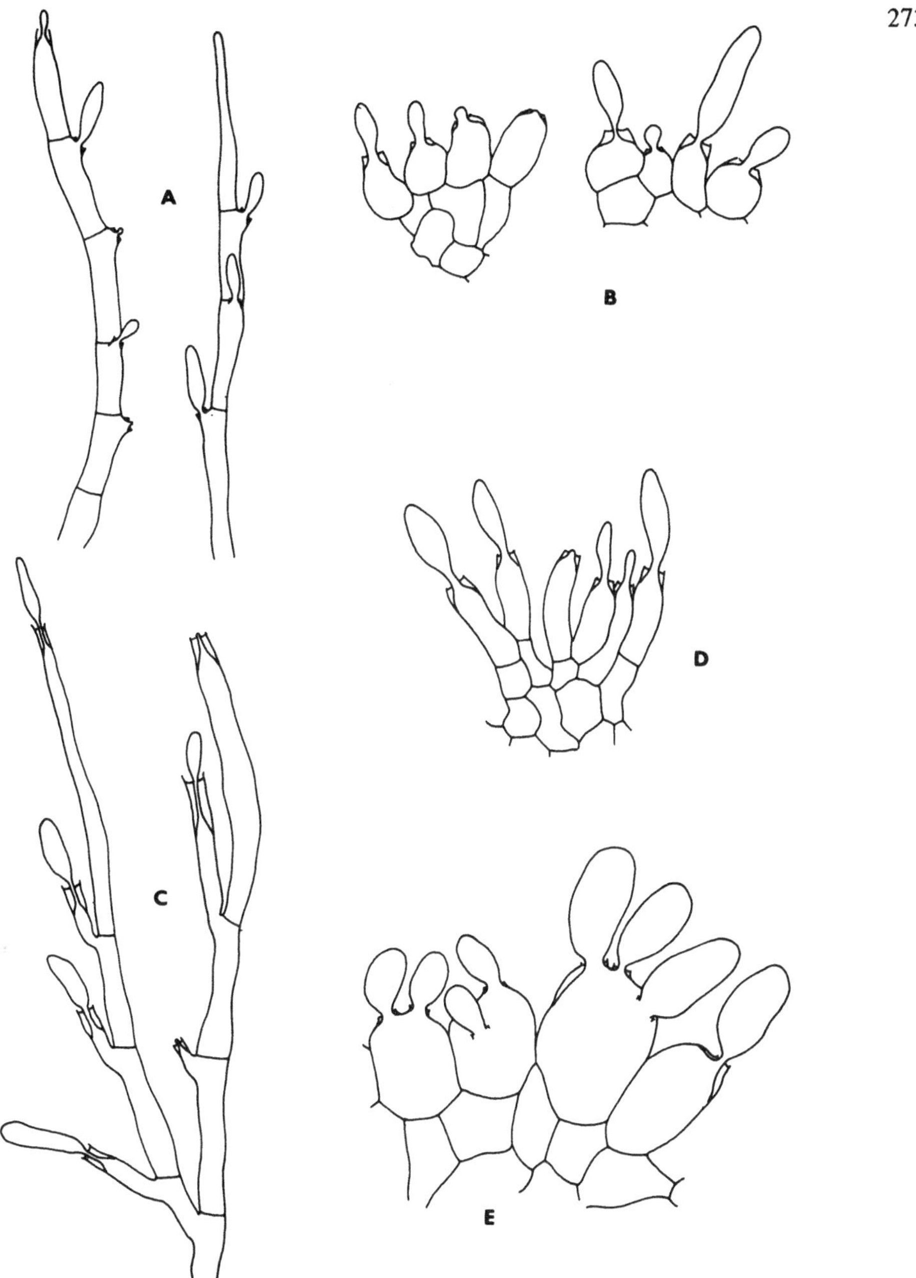


Figure 17.3 Enteroblastic ontogeny: A, *Pyrenochaeta fallax* Bres.; B, *Cytoplacosphaeria rimosa* (Oud.) Petrak; C, *Cytogloeum tiliae* Petrak; D, *Coniella musaiaensis* Sutton; E, *Sarcophoma miribelii* (Fr.) Höhn.

(1964b). Pycnidial fungi with minute hyaline conidia tend to belong to this group, including *Sclerophoma pithyophila* (Cda.) Höhn. (Sutton 1964b), *Plenodomus* Preuss, *Rhizosphaera* Mang. and Har., and most *Pyrenochaeta* species. In some species of *Coniothyrium*, the phaeosporous analogue of *Phoma* Westd., conidia are produced in the same way, as are the septate, coloured or ornamented conidia of many other genera belonging to the Sphaeropsidales and Melanconiales. These include *Camarosporium* species, *Monostichella robergei* (Desm.) Höhn., *Monochaetiella hyparrheniae* Cast., *Apiocarpella agropyri* Sprague and Greene and *A. minor* Sprague and Greene, *Ciliochorella mangiferae* Syd. (Sutton 1964a), *Septogloeum oxysporum* Sacc., Bomm. and Rouss., *Cytoplacosphaeria rimosa* (Oud.) Petrak, *Diplozythiella bambusina* Died., and *Wojnowicia* Sacc. and D. Sacc.

The simple elongated conidiogenous cell comprises the second group (Figure 17.3 D). These have a single apical aperture and are quite distinct from the tissue from which they arise. Chesters (1938) suggested that the elongated conidiophores of *Aposphaeria agminalis* Sacc., *A. fuscidula* Sacc., and *Cytoplea juglandis* (Schum.) Petrak were phialidic, and subsequently many pycnidial and acervular fungi have been found with similar conidiogenous cells. The natural group of genera with setose cupulate fructifications and appendaged conidia, including *Dinemasporium* Lév., *Pseudolachnea* Ranojevic, *Chaetopatella* Hino and Katumoto, *Stauronema* (Sacc.) Syd. and Butl., *Polynema* Lév., and *Amerosporium* Speg., belongs here although there is variation within the group. Other genera with appendaged conidia include *Ellisiella* Sacc., *Dilophospora* Desm., and *Neottiospora* Desm. In *Strasseria carpophila* Bres. and Sacc. (Sutton 1967a) and *Phellostroma tsugae* Kobayashi the body of the phialoconidium is formed first and the appendage last. Both hyaline and brown-spored species form conidia from similar conidiophores, a selection of which includes *Phaeocytostroma* Petrak (Sutton 1964a), *Pleurophomopsis* Petrak, *Sirococcus strobilinus* Preuss, *Titaeosporina* van Luijk, *Gloeosporidiella ribis* (Lib.) Petrak, *G. bartholomei* (Dearn.) Arx, *Cryptocline* species, *Cylindrogloeum trillii* (Ell. and Ev.) Arx, *Discogloeum veronicae* (Lib.) Petrak, *Diplodia* species, *Coniella* Höhn. (Sutton 1969), *Colletotrichum* Corda, and *Melanophora cercocarpi* (Ell. and Ev.) Arx.

Occasionally conidiophores become septate, each cell becoming a functional conidiogenous cell with a single phialidic aperture; or branching occurs near the base so that the conidiophores become fasciculate. Such divergences lead into the next group, which in its simplest state comprises elongated conidiophores composed of individual conidiogenous cells each producing conidia from one phialidic aperture located immediately below the delimiting septum (Figure 17.3 A). Sometimes considerable lateral or basal branching occurs, giving a more complex system (Figure 17.3 C). Phialoconidia are then formed from the apices of short and long branches. Linder (1944) proposed the name *Phialophorophoma* for a phomoid fungus with acropleurogenous conidiophores, and the phenomenon has since been found in several genera allied to *Phoma*, including *Chaetomella* Fuckel, *Pleurophomella* Höhn., *Discella acerina* (Westd.) Arx, *Selenophoma bromigena* (Sacc.) Sprague and Johnson, *S. moravica* Petrak, *Pleurosticta lichenicola* Petrak, *Phacostromella* Petrak, *Subramanella* Srivast., Zakia, and Govind., and *Chaetoconis* Clem. (Sutton 1968a).

In *Pyrenochaeta fallax* Bres. and *P. rubi-idaei* Cav., instead of the doliiform phialides found in the majority of *Pyrenochaeta* species, the conidiophores are elongated, septate, and with a lateral phialidic aperture below the septum delimiting each cell. The multibranched, more complex conidiophore is typical of many pycnidial and acervular genera including *Volutella* Tode ex Fr., *Phomopsis* Sacc. (Sutton 1967c, 1968b), *Rhabdospora* (Sacc.) Sacc., *Dendrophoma* auct., *Crocicreas atroviride* (Berk. and Br.) Höhn., *Cytogloeum tiliae* Petrak, *Patellina caesia* Elliott and Stanf., *Cytonema spinella* (Kalchbr.) Höhn., *Lemalis aurea* (Fr.) Sacc., *Pseudopatellina conigena* (Niessl) Höhn., *Cytospora* Ehrenb. ex Fr., and *Cytosporella* Sacc., to name but a few. 

Some species produce more than a single aperture simultaneously from near the apex of the conidiogenous cell (Figure 17.3 D). These are the polyphialides of Hughes (1953). *Suttoniella* Ahmad, which is almost a phialidic analogue of the holoblastic *Furcaspora*, produces several Y-shaped conidia from different phialide points at the apex of the conidiogenous cell. Apart from *Sphaerographium squarrosum* (Riess) Sacc., where conidia are hyaline and multiseptate, and *Selenophomopsis* species, which have large falcate conidia, other members of the group have relatively small conidia and are found in both Sphaeropsidales and Melanconiales. Conidiogenous cells are sparingly septate and branched. Fungi included here are *Kabatiella microsticta* Bubak, *Melanophora crataegi* (Dearn. and Barth.) Arx, *Diachorella onobrychidis* (DC. ex Fr.) Höhn., and *Sporonema nigrificans* Petrak.

All the phialides discussed hitherto may be grouped into Tubaki's section IVA. There is but a single example in the Melanconiales belonging to IVB. This is *Bloxamia truncata* Berk. and Br., described by Pirozynski and Morgan-Jones (1968) with endogenous conidia formed in basipetal succession within the wall of an elongated phialide.

## THALLIC CONIDIA

Some types of thallic ontogeny have been discussed under the section on catenate conidia because of the problems in differentiating between holothallic and holoblastic development in pycnidial and acervular fungi. In addition to holothallic ontogeny there are examples of what appears to be enterothallic development in some pycnidial fungi. Higgins (1929, 1936) described endogenous spermatia in *Mycosphaerella personata* Higgins and *M. tulipiferae* (Schw.) Higgins, and Dring (1961) described and illustrated the "schizospores" of *Asteromella brassicae* (Chev.) Boerema and Kesteren, the spermatial state of *Mycosphaerella brassicicola* (Duby) Oud. The protoplast within the conidiogenous cell divides into four or less, the resultant protoplasts being extruded through one or two "sterigmata" per cell.

## CONCLUSIONS

It is clear from the foregoing account that this approach to identification and classification of the Sphaeropsidales and Melanconiales is in its infancy and that in many areas critical work is urgently needed. The most important point to determine is the nature of what hitherto have been termed phialides. Their relationships with some types of annellide and the sort of phialide shown by *Penicillium* must be clarified before further advances can be made. Such progress can only come through critical electron microscopy. Other points high on the list of priorities are (1) the analysis of conidium development in many more species and genera, and (2) the correlation of different ontogenetic types with variations in fructification structure, particularly pycnidia.

## REFERENCES

Agar, H.D., and Douglas, H.C. 1955. Studies of budding and cell wall structure of yeast. Electron microscopy of thin sections. J. Bacteriol. 70: 427-34

Boerema, G.H. 1965. Spore development in the form-genus *Phoma*. Persoonia 3: 413-17

Boerema, G.H., and Kesteren, H.A. van. 1964. The nomenclature of two fungi parasitizing *Brassica*. Persoonia 3: 17-28

Brewer, J.G., and Boerema, G.H. 1965. Electron microscope observations on the development of pycnidiospores in *Phoma* and *Ascochyta* spp. Proc. K. ned. Akad. Wet. Amst. sect. C, 68: 86-97

Bubak, F. 1906. Zweiter Beitrag zur Pilzflora von Montenegro. Bull. Herb. Boiss., 2 sér. 6: 473-88

Buckley, P.M., Wyllie, T.D., and de Vay, J.E. 1969. Fine structure of conidia and conidium formation in *Verticillium albo-atrum* and *V. nigrescens*. Mycologia 61: 240-50

Chesters, C.G.C. 1938. Studies on British pyrenomycetes. II. A comparative study of *Melanomma pulvis-pyrius* (Pers.) Fuckel, *Melanomma fuscidulum* Sacc. and *Thyridaria rubronotata* (B. and Br.) Sacc. Trans. Brit. Mycol. Soc. 22: 116-50

Cole, G.T., and Kendrick, W.B. 1968. Conidium ontogeny in hyphomycetes. The imperfect state of *Monascus ruber* and its meristem arthrospores. Can. J. Botany 46: 987-92

- 1969a. Conidium ontogeny in hyphomycetes. The annellophores of *Scopulariopsis brevicaulis*. Can. J. Botany 47: 925-9

- 1969b. Conidium ontogeny in hyphomycetes. The phialides of *Phialophora, Penicillium*, and *Ceratocystis*. Can. J. Botany 47: 770-89

Corda, A.C.J. 1839. Icones fungorum hucusque cognitorum. 3. Prague

Costantin, J. 1888. Les Mucédinées simples. Matériaux pour l'histoire des Champignons. Vol. 2, pp. 1-210. Paris

Cunnell, G.J. 1956. Some pycnidial fungi on *Carex*. Trans. Brit. Mycol. Soc. 39: 21-47

- 1957. *Stagonospora* spp. on *Phragmites communis* Trin. Trans. Brit. Mycol. Soc. 40: 443-55

Dickinson, G.H., and Morgan-Jones, G. 1966. The mycoflora associated with *Halimione portulacoides*. IV. Observations on some species of Sphaeropsidales. Trans. Brit. Mycol. Soc. 49: 43-55

Dring, D.M. 1961. Studies on *Mycosphaerella brassicicola* (Duby) Oudem. Trans. Brit. Mycol. Soc. 44: 253-64

Durrell, L.W. 1968. Studies of *Aureobasidium pullulans* (de Bary) Arnaud. Mycopath. Mycol. Appl. 35: 113-20

Fonseka, R.N. de. 1960. The morphology of *Chaetospermum chaetosporum*. Trans. Brit. Mycol. Soc. 43: 631-6

Goidanich, G., and Ruggieri, G. 1947. Le Deuterophomaceae di Petri. Ann. Sper. agr., N.S. 1: 431-48

Higgins, B.B. 1929. Morphology and life history of some ascomycetes. II. Am. J. Botany 16: 287-96
- 1936. Morphology and life history of some ascomycetes. III. Am. J. Botany 23: 598-602
Höhnel, F. von. 1923. System der Fungi Imperfecti Fuckel. Mykol. Unters. und Berichte von R. Falck 3: 301-69
Hudson, H.J., and Sutton, B.C. 1964. *Trisulcosporium* and *Tetranacrium*, two new genera of Fungi Imperfecti. Trans. Brit. Mycol. Soc. 47: 197-203
Hughes, S.J. 1953. Conidiophores, conidia, and classification. Can. J. Botany 31: 577-659
Klebahn, H. 1933. Über Bau und Konidien-bildung bei einigen stromatischen Sphaeropsideen. Phytopath. Z. 6: 229-304
Langeron, M., and Vanbreuseghem, R. 1952. Précis de mycologie. Masson et Cie, Paris
Linder, D.H. 1944. I. Classification of the marine fungi. Farlowia 1: 401-33
Luttrell, E.S. 1940. An undescribed fungus on japanese cherry. Mycologia 32: 530-6
Moreau, F. 1953. Les Champignons. Vol. 2, pp. 941-2120. Paris
Petrak, F. 1925. Mykologische Notizen. Ueber v. Höhnel's neues System der Fungi Imperfecti. Ann. Mycol. Berl. 23: 1-11
Pirozynski, K.A., and Morgan-Jones, G. 1968. Notes on microfungi. III. Trans. Brit. Mycol. Soc. 51: 185-206
Punithalingam, E. 1969. New species of *Monochaetiella* and *Septoria*. Trans. Brit. Mycol. Soc. 53: 311-15
Seeler, E.V. 1943. Several fungicolous fungi. Farlowia 1: 119-33
Shoemaker, R.A. 1963. Generic correlations and concepts: *Griphosphaeria* and *Labridella*. Can. J. Botany 41: 1419-23
Streiblová, E., and Beran, K. 1965. On the question of vegetative reproduction in apiculate yeasts. Folia Microbiol. 10: 352-6
Streiblová, E., Beran, K., and Pokorny, V. 1964. Multiple scars, a new type of yeast scar in apiculate yeasts. J. Bacteriol. 88: 1104-11
Sutton, B.C. 1963a. Coelomycetes. II. *Neobarclaya, Mycohypallage, Bleptosporium*, and *Cryptostictis*. C.M.I. Mycol. Pap. 88
- 1963b. Two new species of *Coryneum* Nees ex Fries. Kew Bull. 17: 309-14
- 1964a. Coelomycetes. III. *Annellolacinia* gen. nov., *Aristastoma, Phaeocytostroma, Seimatosporium* etc. C.M.I. Mycol. Pap. 97
- 1964b. *Phoma* and related genera. Trans. Brit. Mycol. Soc. 47: 497-509
- 1964c. *Melanconium* Link ex Fries. Persoonia 3: 193-8
- 1967a. Two new genera of the Sphaeropsidales and their relationships with *Diachorella, Strasseria*, and *Plagiorhabdus*. Can. J. Botany 45: 1249-63
- 1967b. Redescription of *Ajrekarella* Kamat and Kalani. Mycopath. Mycol. Appl. 33: 76-80
- 1967c. *Libertina*, a synonym of *Phomopsis*. Trans. Brit. Mycol. Soc. 50: 355-8
- 1968a. *Kellermania* and its generic segregates. Can. J. Botany 46: 181-96
- 1968b. *Phomopsis schini* (Carranza) comb. nov. Trans. Brit. Mycol. Soc. 51: 616-18
- 1969. Type studies of *Coniella, Anthasthoopa*, and *Cyclodomella*. Can. J. Botany 47: 603-8
Sutton, B.C., and Pirozynski, K.A. 1963. Notes on British microfungi. I. Trans. Brit. Mycol. Soc. 46: 505-22
- 1965. Notes on microfungi. II. Trans. Brit. Mycol. Soc. 48: 349-66
Sutton, B.C., and Sandhu, D.K. 1969a. *Phragmotrichum pini* (W.B. Cooke) comb. nov. Trans. Brit. Mycol. Soc. 52: 67-71

- 1969b. Electron microscopy of conidium development and secession in *Cryptosporiopsis* sp., *Phoma fumosa, Melanconium bicolor,* and *M. apiocarpum*. Can. J. Botany 47: 745-9

Sutton, B.C., and Sellar, P.W. 1966. *Toxosporiopsis* n. gen., an unusual member of the Melanconiales. Can. J. Botany 44: 1505-13

Sydow, H. and Sydow, P. 1913. Novae fungorum species. IX. Ann. Mycol. Berl. 11: 54-65

Tubaki, K. 1958. Studies on the Japanese hyphomycetes. V. Leaf and stem group with a discussion of the classification of hyphomycetes and their perfect stages. J. Hattori Bot. Lab. 20: 142-244

Twyman, E.S. 1947. Notes on the die-back of oak caused by *Colpoma quercinum* (Fr.) Wallr. Trans. Brit. Mycol. Soc. 29: 234-41

Vuillemin, P. 1910a. Matériaux pour une classification rationelle des Fungi Imperfecti. C.R. Acad. Sci., Paris 150: 882-4

- 1910b. Les conidiosporés. Bull. Soc. Sci. Nancy, Sér. 3, 11: 129-72

- 1911. Les aleuriosporés. Bull. Soc. Sci. Nancy, Sér. 3, 12: 151-75

Walker, J. 1962. Notes on plant parasitic fungi. I. Proc. Linn. Soc. N.S.W. 87: 162-76

Zachariah, K., and Fitz-James, P.C. 1967. The structure of phialides in *Penicillium claviforme*. Can. J. Microbiol. 13: 249-56

# 18
# Karyology of Conidiogenesis in Some Hyphomycetes

BRYCE KENDRICK and M. G. CHANG

One of the points repeatedly raised during the discussions at Kananaskis was the possibility that there might be karyological differences between the various methods of conidium ontogeny, with the logical corollary that we might be able to define certain of these processes in karyological as well as, or instead of, morphological terms.

With this possibility in mind, we have begun to retrace the steps of Kendrick and Cole, who carefully followed the sequence of morphological events involved in the conidium ontogeny of a number of hyphomycetes. We are examining these same fungi using a modification of the Giemsa nuclear staining procedure of Robinow (1961). This chapter reports on three fungi: *Scopulariopsis brevicaulis* [UW 236] (see Cole and Kendrick 1969a), *Gonatobotryum apiculatum* [UW 200] (see Kendrick, Cole and Bhatt 1968), and *Geotrichum candidum* [UW 203] (see Cole and Kendrick 1969b).

## MATERIALS AND METHODS

Each of the three organisms was grown in coverslip culture till the appropriate developmental stages were present (the times differed for each species, and were originally discovered by a comprehensive experiment in which many slides were inoculated concurrently and then fixed sequentially at half-hour intervals). The coverslip was then ripped off the agar block, immediately immersed in modified Helly solution (Robinow 1961), and fixed for 20 minutes. Two or three rinses in 70 per cent alcohol, then in water, were followed by hydrolysis in 1*N* hydrochloric acid for 10 minutes at 60°C. After three rinses in water, the preparation was stained in a Columbia jar for 10-30 minutes in a solution made up from 10 ml of Gurr's

Department of Biology, University of Waterloo, Waterloo, Ontario, Canada.

 Giemsa buffer plus 18 drops of Gurr's Giemsa "R66" stock solution. The preparation was then rinsed and mounted in buffer for examination.

## OBSERVATIONS

Our first two fungi, *Scopulariopsis* and *Gonatobotryum*, both form "blastic" conidia: there is marked enlargement of a recognizable conidium initial *before* the initial is delimited by a septum. In the simplest terms each conidium is derived from part of a cell. We hoped that the comparison between these two blastic forms and the thallic *Geotrichum* would be informative.

Zachariah and Metitiri (Chapter 8) have shown that the normal phialide of *Penicillium claviforme* contains a single nucleus. This divides once during the formation of each conidium, one of the daughter nuclei entering the conidium, the other remaining in the venter of the phialide where it will divide again each time a new conidium is formed. Thus the status quo is maintained: one nucleus in the phialide, one in each conidium of a basipetal succession. This confirms the general picture outlined by Pontefract (1956 Univ. of Western Ontario M.Sc. Thesis, unpub.).

The annellide superficially resembles the phialide. It has a deceptive air of possessing a definite length, and its contours frequently match those of phialides: an elongated, often somewhat widened body, and a narrower neck. Apparently genuine ontogenetic differences between phialides and annellides exist and are gradually being explored (see Chapters 7, 9, 19). The fixed endogenous meristem of the characteristic phialide and its disinclination to increase in length with conidium production (any percurrent or sympodial proliferation which may occur is not directly involved in conidium formation) clearly delineate it from the average annellide, with its repeated, short percurrent proliferations (annellations: one per conidium) and its consequent multiplicity of wall lamellae (Figure 9.1), though annellations are often difficult to see with the light microscope and the wall lamellae are revealed only by the transmission electron microscope.

As far as we know, the only previous karyological study of an annellidic fungus was carried out by Backus and Keitt (1940), who investigated conidial *Venturia inaequalis*. They reported that individual cells of this organism, whether vegetative hyphal cells, conidiogenous cells, or conidia, each contained only one nucleus. The nucleus in the conidiogenous cell apparently divided once for each conidium produced. This picture is essentially the same as that found in the phialide. *Scopulariopsis brevicaulis*, however, diverges markedly from this pattern.

Nuclear division in *S. brevicaulis* is of relatively short duration, and certain stages, such as anaphase, proceed very rapidly indeed. Although we have no absolute figures for *S. brevicaulis*, Robinow and Caten (1969) found that the entire process of nuclear division in vegetative hyphae of diploid *Aspergillus nidulans* lasted only 18 minutes, and that their stages II (late metaphase) and III (anaphase) were extremely ephemeral.

The photomicrographs of *S. brevicaulis* are all of 24-hour-old cultures. The vegetative hyphae give rise to short, lateral branches which are the conidiogenous cell (annellide) initials (Figure 18.1 A). One or two nuclei migrate into each of these branches, and soon divide. In Figure 18.1 A, they are seen in the condition designated as stage II by Robinow and Caten (1969). Sometimes the young annellide is not delimited by a basal septum, and nuclei may continue to migrate into it quite freely. More usually, the annellide is cut off by a basal septum, nuclear migration ceases, and the narrowed apex soon blows out to form the first conidium initial. In B a nucleus (indicated by an arrow) is moving into the young conidium. In C, the nucleus in the conidium and those nuclei remaining in the conidiogenous cell are all dividing simultaneously (here seen in stage I, early metaphase, each nucleus a clump of irregularly disposed, arm-like chromatinic masses). D shows a young annellide with an almost completely differentiated first conidium; all four nuclei are in stage II of division. In E, the four dividing nuclei are all in telophase (stage IV), the eight daughter nuclei appearing at their most condensed, and the members of each pair still connected by a thin strand of chromatin (not to be confused with the spindle, which is not seen in Giemsa preparations.) In F, the first conidium has been delimited by a basal septum and contains two nuclei, probably the products of a division similar to that just described. However, if the conidium is not cut off by a septum, nuclei may continue to migrate into it. G shows a third nucleus (indicated by the arrow) entering the conidium, and H shows the likely product of this process after the basal septum has been laid down. Note that in F, G, and H several nuclei remain in the conidiogenous cell in every case. On occasion the migration may extend to the entrance of a fourth nucleus, as shown in Figure 18.2 A (arrow) before the septum apparently puts a stop to this movement. Migration is not the only way of arriving at a multinucleate conidium. Sometimes the number of nuclei is increased by division of those already inside the conidium (Figure 18.2 B). 

After the delimitation of the first conidium, a very short percurrent proliferation of the conidiogenous cell takes place and the second conidium initial begins to blow out. As with the first, a nucleus migrates into it (Figure 18.2 C) and then divides (D, stage I, early metaphase). Note that although the nuclei in the conidiogenous cell divide synchronously with that in the second conidium, the nuclei in the first conidium are no longer affected by divisions below them. This may be a sign that all cytoplasmic connection between the first and second conidium has been lost. In E, the nuclei in the conidiogenous cell and in the young conidium (arrows) are in anaphase (entering stage III as described by Robinow and Caten).

Subsequently the third conidium initial forms. In Figure 18.2 F, one nucleus has already entered it and a second is actively migrating towards it. In G, a conidium has been delimited and a nucleus is moving up to take its place in the next conidium initial, which is just forming.

From this sequence of photomicrographs, it is clear that no single well-defined invariant pattern of nuclear division, migration, and population of conidia exists in this fungus.

Examination of large numbers of conidia and conidiogenous cells of *S. brevicaulis* has shown that there is indeed considerable variation in the nuclear number of both annellides and conidia. The histograms (Figure 18.3) show the frequency distribution of the different numbers of nuclei in 1,000 conidiogenous cells. The

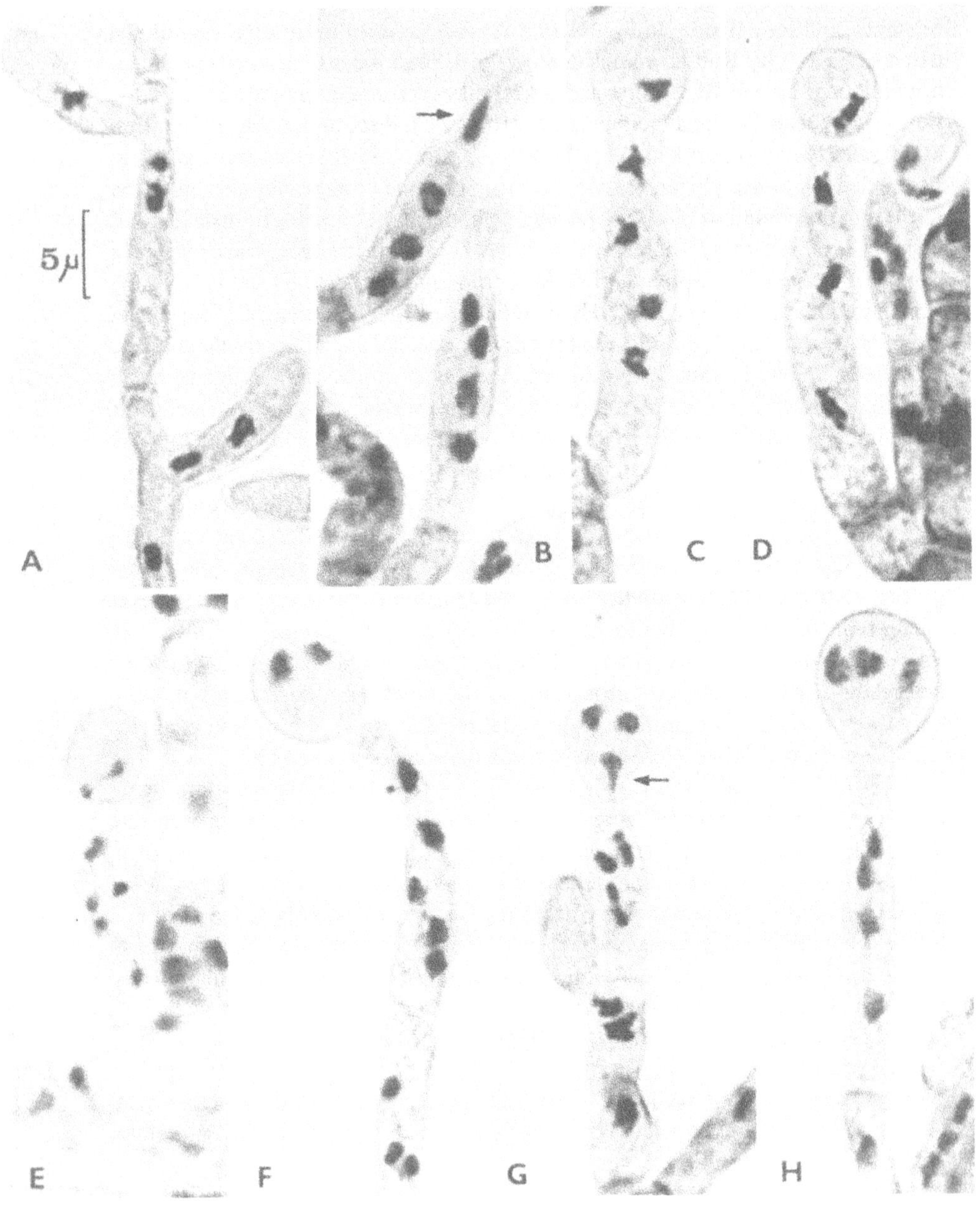


Figure 18.1 *Scopulariopsis brevicaulis*, karyology of conidiogenesis. Giemsa staining. For explanation, see text.

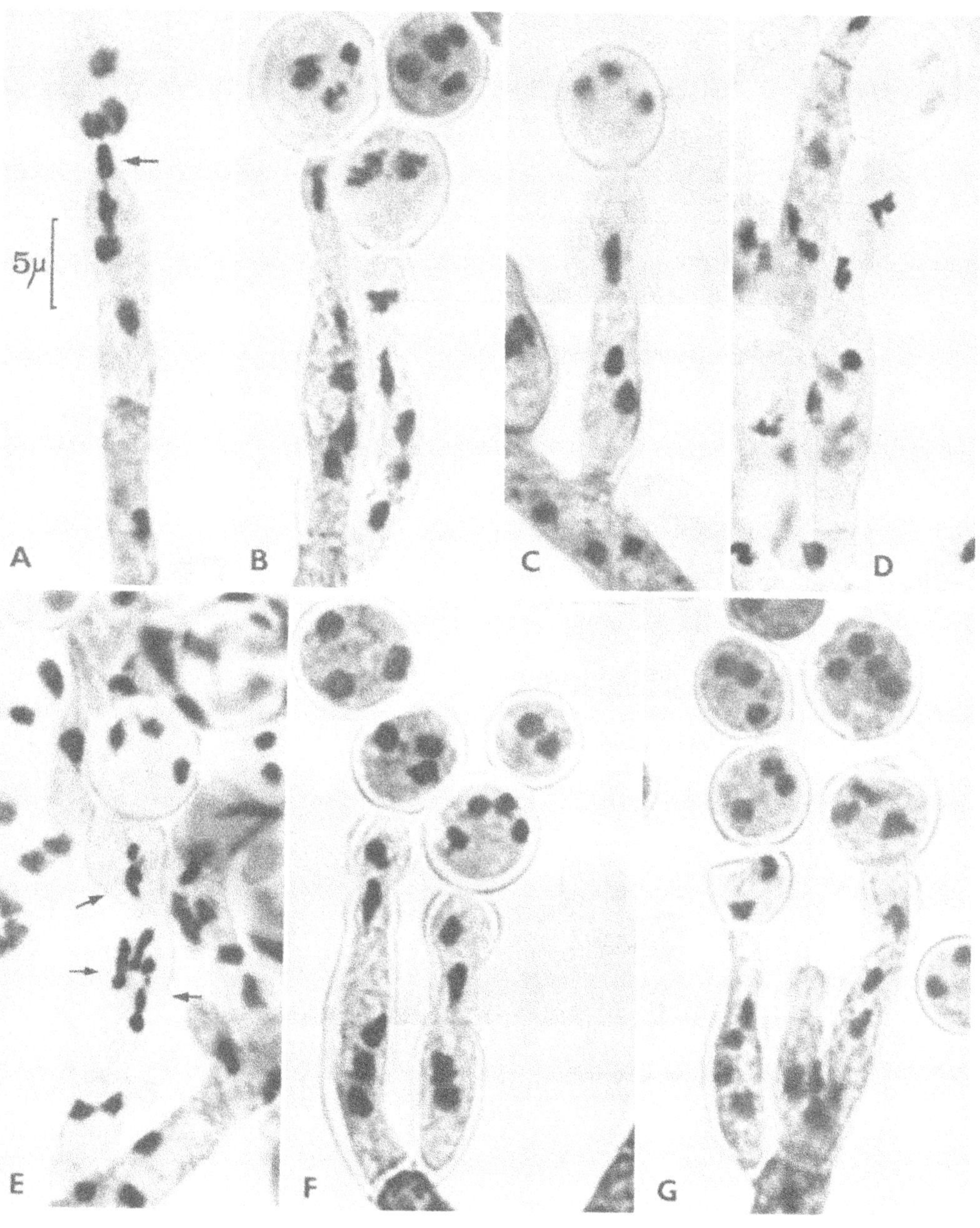


Figure 18.2 *Scopulariopsis brevicaulis,* karyology of conidiogenesis. Giemsa staining. For explanation, see text.

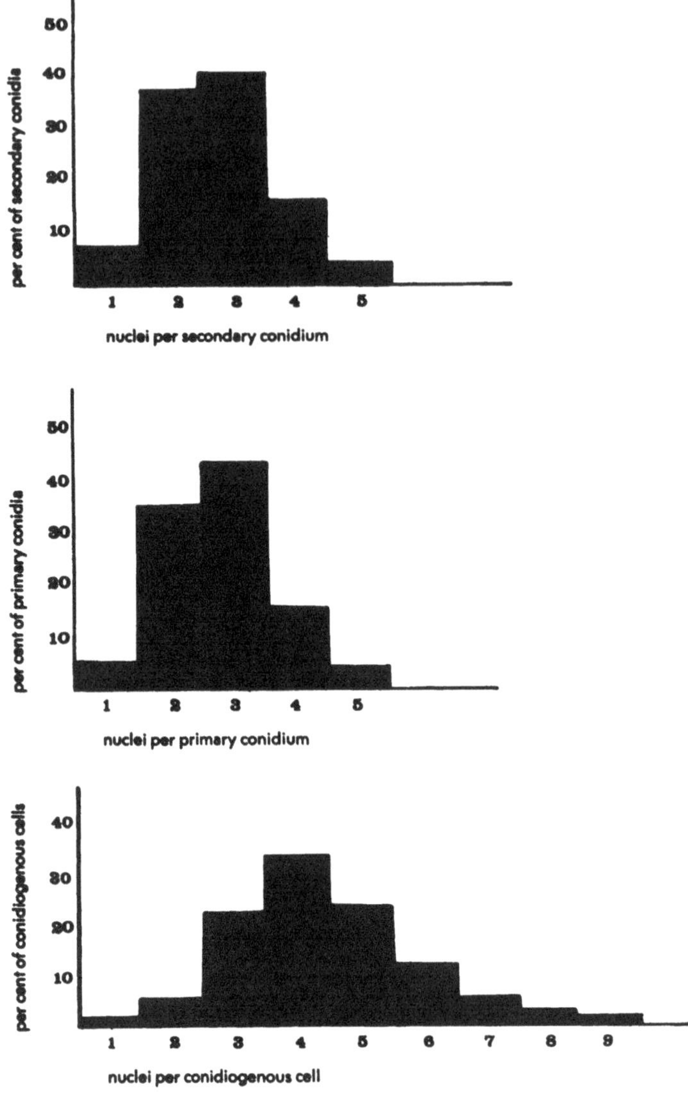


Figure 18.3 Histograms of nuclear number of conidiogenous cells and conidia in *Scopulariopsis brevicaulis.*

number of nuclei per annellide ranged from 1 to 9; 4 was the commonest number (33 per cent) but many cells possessed 3 (21.3 per cent) or 5 (22.9 per cent) nuclei, and a significant fraction contained 6 (10.8 per cent). 

Comparison of 1,000 first and 1,000 subsequent conidia shows that there is little difference in nuclear number between them. The range in both cases is from 1 to 5 nuclei, and in both cases the 3-nucleate condition is commonest (42.2 and 40.1 per cent) with the 2-nucleate condition running it a close second (35.6 and 36.4 per cent). The only other nuclear complement present in a significant proportion of the conidia was 4 (15.1 and 15.8 per cent).

From this study we may conclude that the annellide of *S. brevicaulis* is significantly less regular in its karyological mechanisms than that of *Venturia inaequalis*, and than the few phialides whose karyology has been reported in the literature. Not only does nuclear number in both annellide and conidium vary considerably, but the origin of the nuclei also varies. It appears at first sight as if a 3-nucleate conidium may owe one, two, or all three of its nuclei to migration from the conidiogenous cell. But there is one generalization we feel fairly confident in making. In most cases when one nucleus in a conidiogenous cell or conidium goes into division, all nuclei in that cell or conidium will divide. Thus if there are two nuclei in a conidium, it is unlikely to become 3-nucleate except by the immigration of a third nucleus. We can draw up a tentative table of origins (see Table 18.1).

It seems, therefore, impossible to categorize the annellide of *S. brevicaulis* rigorously from the karyological point of view. We shall be interested to see the results obtained with the annellides of other organisms.

Our second example is a very different fungus. *Gonatobotryum apiculatum* produces "blastic" conidia, not from a phialide or annellide, but from an ampulla (a conidiogenous vesicle). It produces conidia, not singly in basipetal succession, but first in a synchronous cluster, then in asynchronously developing acropetal chains. After the primary conidia have been blown out, a conidiogenous locus moves into each, and a secondary conidium is produced from its distal end. The locus then moves into the second conidium, a third conidium is produced from its apex, and so on.

Clearly, we can expect to discover few parallels with the phialide or the annellide. The logistics of the basipetal and acropetal processes are very different. Once a phialide or annellide has elaborated and delimited a conidium, that conidium loses all cytoplasmic connection with the conidiogenous cell. The mature conidium is ready to be dispersed, and the conidiogenous cell is free to proceed with the formation of the next conidium.

In *Gonatobotryum*, all building material for each new secondary conidium must either pass through all previously formed conidia of the chain, or be derived from the previous conidium. If a chain of several conidia is formed, it is clear that food material must be coming from the parent hypha. This means that, if the chain is broken off by some accident, the conidiogenous locus will be cut off from its food supply and will cease to function. It is apparent to us that the acropetal chain is a less sophisticated mechanism than the basipetal chain.

Nobles (1935) reported that in the ampulliform conidiogenous cell of the ba-

TABLE 18.1

| Nuclear complement of conidium | Origin of nuclei |
|---|---|
| 1 | Migration |
| 2 | 2 migrated, *or* one migrated and divided |
| 3 | 1 migrated and divided and then a third migrated, *or* 3 migrated |
| 4 | 1 migrated and divided, both daughter nuclei divided, *or* 2 migrated and divided, *or* 4 migrated |

sidiomycete *Peniophora allescheri* one nucleus divides repeatedly to give 16 or more nuclei and then the nuclei "take up their positions around the inner surface of the distal half" of the ampulla. "At this stage the slender tapering sterigmata push out and a swelling, the developing conidium, appears on the end of each. While the spore is still small a nucleus migrates into it ..."

This neat and logical procedure, which is apparently also found in *Corticium effuscatum* (Nobles 1942), was more or less what we expected to find in *G. apiculatum*, though we knew that variations existed. Loveland, reported in Hughes (1953), found that in *Botrytis allii* one nucleus migrates into each conidium produced on an ampulla, but then divides once to produce binucleate conidia. She also observed that, in *Botrytis cinerea* and *Botrytis streptothrix*, several nuclei migrate into each conidium initial. Despite this foreknowledge, *G. apiculatum* managed to surprise us.

The conidiophores of *G. apiculatum* arise as lateral branches or apical extensions of vegetative hyphae. Conidiophore initials are wider, thicker-walled, and more darkly pigmented than the parent hyphae. At this stage the long apical cell of the conidiophore already contains many nuclei, sometimes more than 20. Some of these migrate in before the basal septum of the cell is laid down, others migrate through the septal pore (Figure 18.4 A, arrow), and others arise as products of nuclear division within the apical cell. Figure 18.4 B shows a young, extending conidiophore initial in which the nuclei appear to be at metaphase. In C, the nuclei are at telophase, the daughter nuclei very condensed. D shows the much larger interphase nuclei that are eventually reconstituted. In E, the apex of the cell, now densely populated with nuclei, has ceased elongation and has become slightly swollen. This slight ballooning is the ampulla, and its appearance heralds conidiogenesis, which soon begins with the precisely synchronous appearance of 17-25 tiny conical protuberances regularly spaced over the surface of the ampulla (F). Although in E and F nuclei crowd the ampulla, and might be assumed to be individually responsible for coordinating the development of adjacent denticles, and, subsequently, conidia, G shows that this is not necessarily the case. In this figure the formation of primary conidia is at an advanced stage, but the cluster of nuclei remains far below the ampulla (the nearest nucleus is 20$\mu$ beneath it), yet conidium formation is proceeding normally.

Earlier observations (Kendrick, Cole, and Bhatt 1968) suggested that the apex of each denticle ruptures and that the primary conidium which blows out is clad in an extension of the inner wall of the ampulla. That would make these conidia en-

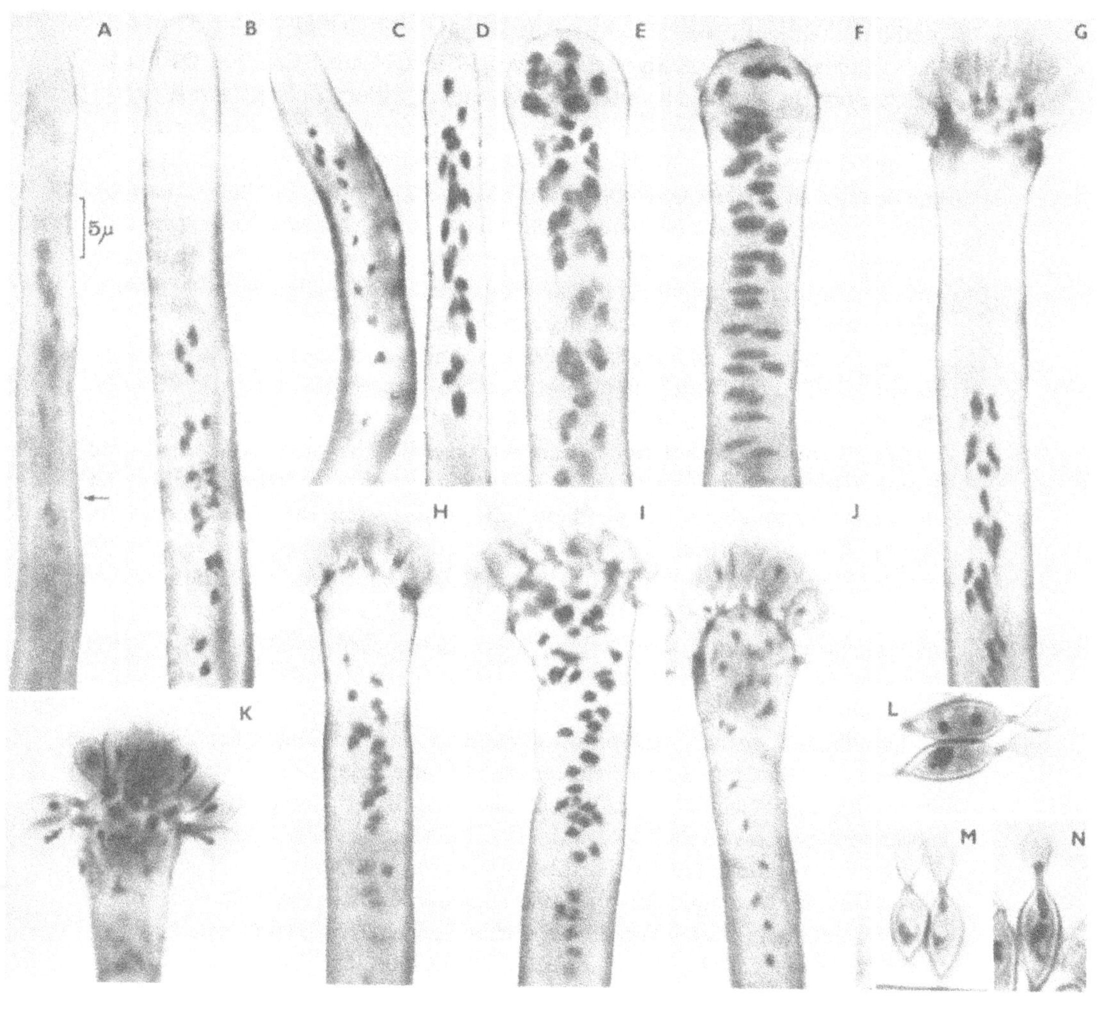


Figure 18.4 *Gonatobotryum apiculatum*, karyology of conidiogenesis. Giemsa staining. For explanation, see text.

teroblastic. Recent observations by Cole and Aldrich (see Figure 19.2 A, B) with the scanning electron microscope lend weight to this suggestion, but the matter must ultimately be settled by reference to the transmission electron microscope.

At this stage, if the nuclei of the apical cell of the conidiophore have not recently divided, they may do so. Figure 18.4 H and I show most nuclei at early metaphase, and J finds them in telophase and sees them migrating into the primary conidia. K clearly shows attenuated nuclei migrating into conidia. Nuclei generally move into the conidia only when the latter are almost full-size. Only one nucleus enters each primary conidium, and although nuclei do not populate all conidia simultaneously, all become nucleate in a relatively short time.

The secondary conidia are in many ways similar to the primary conidia. They form from denticles which develop at the distal end of the primary conidia, and they also are suspected of being enteroblastic. But they are not produced synchronously - the master control mechanism, whatever it is, has been fragmented - and their karyology seems to be less haphazard than that of the primary conidia. After the secondary conidium begins to blow out, the nucleus in the primary conidium divides. The upper primary conidium in Figure 18.4L is at telophase. In M, one of the daughter nuclei is just about to migrate into the secondary conidium, and N shows a nucleus which is in process of migrating into the young secondary conidium. We assume that, as an acropetal chain of conidia develops, each new apical conidium receives its nucleus as a result of a similar regular process of division and migration.

Conidia of *G. apiculatum* typically contain one nucleus. Conidia containing two nuclei are not uncommon, but are often about to give rise to a new apical conidium, or to initiate a branching of the conidial chain. Conidia with three or four nuclei are observed only occasionally.

The first two fungi discussed in this chapter produce their conidia blastically, de novo. Our third example, *Geotrichum candidum,* displays the thallic type of conidium ontogeny. Here, the main diagnostic characteristic is that if any enlargement of the recognizable conidium initial occurs, it occurs only after the initial has been delimited by a septum or septa. In the simplest terms, a thallic conidium is one derived from an entire cell, and its initial abutment with other conidia or the adjacent portion of the fertile hypha will always be the full width of that hypha.

The only previous cytological studies on thallic-arthric conidia of which we are aware were carried out by Brodie (1936), who described the formation of uninucleate conidia ("oidia") in *Collybia velutipes*. He reported that the single nucleus present in a hypha divided repeatedly, and that the hypha subsequently segmented basipetally. Again, as in the published work cited with reference to phialides, annellides, and ampullae, there is an over-all impression of neat and efficient organization: a state of affairs which seems to be much less in evidence in the fungi we have studied.

Cole and Kendrick (1969b) have already demonstrated unequivocally that in *Geotrichum* the process of hyphal segmentation and conversion into conidia does not follow a smooth basipetal sequence as had long been assumed, and we were prepared to discover less regularity in the karyological aspects of conidiogenesis than was reported by Brodie. We were not disappointed.

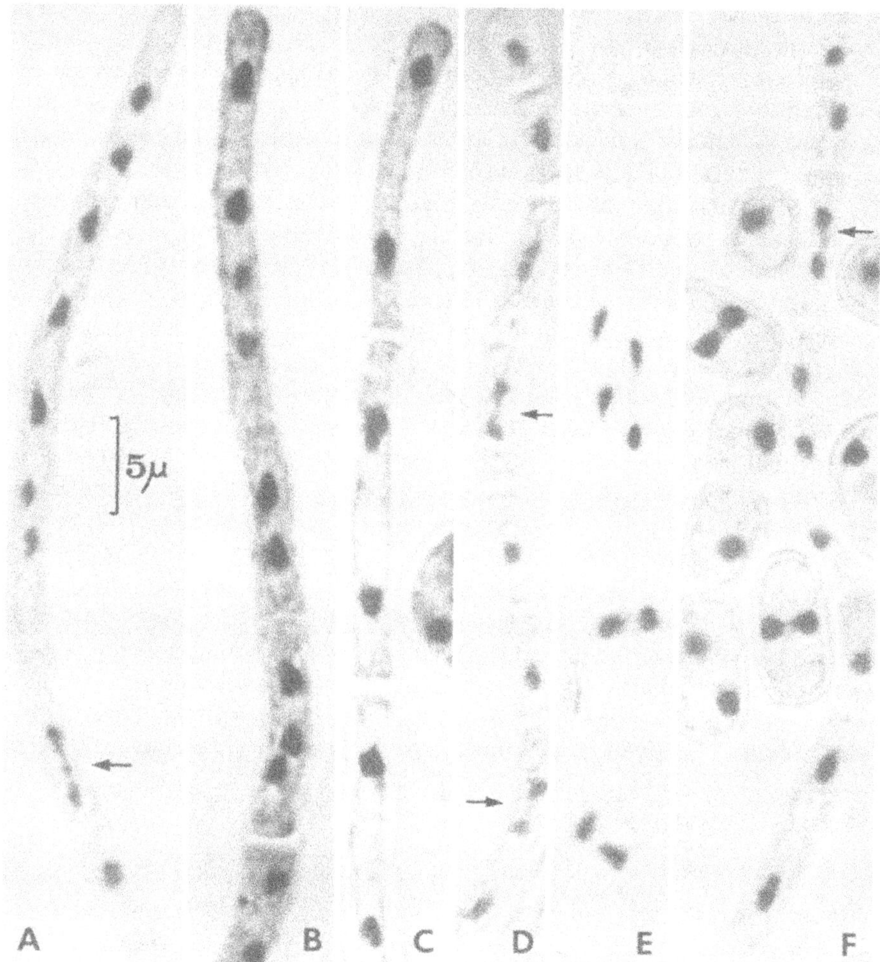


Figure 18.5 *Geotrichum candidum*, karyology of conidiogenesis. Giemsa staining. For explanation, see text.

The photomicrographs of *G. candidum* (Figure 18.5) are all of 12-hour-old cultures. The elongating hyphae are not septate for some distance behind their apex (Figure 18.5 A), and are populated with fairly regularly spaced nuclei which divide as required to maintain the appropriate spacing (arrowed nucleus in A is at late anaphase). In B, three septa, relatively widely spaced, are being laid down in an acropetal succession.

Each of these primary subdivisions of the hypha typically contains two nuclei (Figure 18.5 C). Subsequently, over a short period, each of the primary units becomes further subdivided by the deposition of a more or less median septum (Figure 18.5 D). Each unit now usually contains only a single nucleus. In many cases the units are to undergo further subdivision, and the nucleus in each of these segments now divides (arrows in D and F). If the daughter nuclei migrate to opposite ends of the segment (D, E, F) the deposition of yet another median septum may be anticipated. If, however, they remain in relatively close proximity to one another, often at the equator of wider cells (E, F), it may be assumed that the ultimate unit size has been attained. This very soon becomes evident as the units, which may now be called thallic-arthric conidia, secede from one another, the thin outer wall which connected them rupturing easily as their end walls (formerly halves of double septa) become convex (F). The nuclear complement of *G. candidum* conidia is almost always either one or two, with uninucleate conidia predominating.

As we might have suspected, the karyological specialization found in the phialide and, to a lesser extent, the annellide is not present in the fertile hypha of *G. candidum*. In *Geotrichum* there does seem, however, to be some relationship between the size of the conidium and its nuclear complement. Figure 18.5F shows clearly that larger conidia tend to be 2-nucleate and smaller conidia, 1-nucleate. Each nucleus appears to have a definite sphere of action, and if it should be granted more than its allotted volume of cytoplasm, it apparently divides to restore the "volume per nucleus" to the appropriate level.

Clearly, with such a limited sampling as we have made, it is imprudent if not impossible to draw any general conclusion regarding the karyology of conidiogenesis in hyphomycetes, but we hope that, from work continuing in our laboratory and similar studies elsewhere, patterns may ultimately emerge which will allow us to formulate some useful generalizations. Any such source of insight into the labyrinths of hyphomycete ontogeny and systematics will be more than welcome.

## ACKNOWLEDGMENT

The authors thank the National Research Council of Canada for financial support.

## REFERENCES

Backus, E.J., and Keitt, G.W. 1940. Some nuclear phenomena in *Venturia inaequalis*. Búll. Torrey Bot. Club 67: 768-9

Brodie, H.J. 1936. The occurrence and function of oidia in the Hymenomycetes. Am. J. Botany 23: 309-27

Cole, G.T., and Kendrick, W.B. 1969a. Conidium ontogeny in hyphomycetes. The annellophores of *Scopulariopsis brevicaulis*. Can. J. Botany 47: 925-9

- 1969b. Conidium ontogeny in hyphomycetes. The arthrospores of *Oidiodendron* and *Geotrichum*, and the endoarthrospores of *Sporendonema*. Can. J. Botany 47: 1773-80

Hughes, S.J. 1953. Conidiophores, conidia, and classification. Can. J. Botany 31: 577-659

Kendrick, W.B., Cole, G.T., and Bhatt, G.C. 1968. Conidium ontogeny in hyphomycetes. *Gonatobotryum apiculatum* and its botryose blastospores. Can. J. Botany 46: 591-6

Nobles, M.K. 1935. Conidial formation, mutation and hybridization in *Peniophora allescheri*. Mycologia 27: 286-301

- 1942. Secondary spores in *Corticium effuscatum*. Can. J. Res., C, 20: 352

Robinow, C.F. 1961. Mitosis in the yeast *Lipomyces lipofer*. J. Biophys. Biochem. Cytol. 9: 879-92

Robinow, C.F., and Caten, C.E. 1969. Mitosis in *Aspergillus nidulans*. J. Cell Sci. 5: 403-31

# 19

# Scanning and Transmission Electron Microscopy and Freeze-Etching Techniques used in Ultrastructural Studies of Hyphomycetes

G.T. COLE* and H.C. ALDRICH

Attempts simply to air-dry thin-walled biological specimens in preparation for examining the material with a scanning electron microscope commonly result in collapsed and badly distorted material. Several successful techniques for preserving the structure of soft tissues have been developed recently (Boyde and Barber 1969; Small and Marszalek 1969; Horridge and Tamm 1969; Zachariah and Pasternak 1970) but, as yet, there have been few attempts to examine the ultrastructure of hyphomycetes. As a result of the many unanswered questions which have arisen from light microscope developmental studies of the Fungi Imperfecti, questions primarily concerning wall relations and cytological changes during conidium formation, we first began to explore the ultrastructure of conidiogenous cells with the transmission electron microscope. At an early stage in this work, we found it desirable to supplement our transmission studies with observations of external morphology at high magnifications to assist in our interpretations of the processes of conidium formation. The availability of a Balzers freeze-etching apparatus allowed us to develop a simple technique for preparing our material for the scanning electron microscope. The procedure involved no structural modifications to the Balzers machine, and its normal function could be resumed directly after completion of each preparation.

Fortunately, the fungi which we have examined with the scanning electron microscope sporulate profusely in plate culture when grown on either potato dextrose agar or malt extract agar. Using a clean razor blade, a 5-mm square of mycelium and agar is removed from the culture in the zone of densest sporulation. The agar layer, which should be relatively thin (1.0-1.5 mm thick), acts as a supporting substratum for the mycelium during the rest of the procedure. The material is

Department of Botany, University of Florida, Gainesville, Florida, U.S.A.
*Now Assistant Professor in the Department of Botany, University of Texas, Austin, Texas.

handled with the flattened tip of an applicator stick under the agar base, thus preventing any extensive damage to, or disorientation of, the conidiogenous cells and chains of conidia. The fungus is fixed with 2.5 per cent glutaraldehyde followed by 1 per cent osmium tetroxide, both buffered to pH 7.4 with cacodylate buffer. The fixation time is usually 8 hours in osmium tetroxide but varies from 1 to 2 hours in glutaraldehyde. The material is dehydrated in an ethanol series and left in 100 per cent ethanol for at least 45 minutes. We find that this preliminary fixation procedure promotes some rigidity in the fungal wall and is a necessary step for the subsequent sublimation procedure. The square of mycelium is gently removed from the alcohol and floated on liquid Freon-22 (-150°C) for approximately 2 minutes before it is quickly transferred to liquid nitrogen (-196°C) using a pair of fine forceps. The advantage of precooling the mycelium in liquid freon is that it eliminates the vigorous boiling which occurs when objects at higher temperatures are immersed in liquid nitrogen, and which might dislodge or damage the fungal cells. The material is left in liquid nitrogen for at least 5 minutes, but it can be stored at this temperature indefinitely. The freeze-etching apparatus has a thermostatically controlled stage which can be cooled with liquid nitrogen to -150°C or heated electrically to +50°C. A bell jar fits over the stage and can be readily evacuated to at least $1 \times 10^{-5}$ Torr. The frozen mycelium is removed from the liquid nitrogen, quickly placed on the precooled stage (-150°C), and the bell jar is sealed and evacuated. This process is carried out rapidly in order to prevent the formation of ice crystals. When the vacuum within the bell jar reaches $1 \times 10^{-2}$ Torr, the temperature of the stage is raised to -100°C and the material is allowed to sublime for at least one hour at $1 \times 10^{-5}$ Torr. The dried mycelium is then lifted from the stage and oriented under a dissecting microscope before being stuck to the surface of an aluminium stub with conductive silver paint. The fungus is now ready to be coated with a thin layer of gold/palladium (60:40). This final process is carried out under high vacuum ($2 \times 10^{-5}$ Torr or better) on a rotating stage (11 rpm) held at an angle of approximately 45°. We have outlined the above procedure in Figure 19.1. The micrographs in Figures 19.2 and 19.3 were taken with a Cambridge Stereoscan scanning electron microscope using Polaroid Type 55 P/N film (Figure 19.2 B, D) and Panatomic-X sheet film (Figure 19.2 A, C and Figure 19.3). We have found that the latter produces a sharper and more contrasty negative, probably because of its antihalation backing.

As already stated, specimen preparation involved fixation with glutaraldehyde followed by osmium tetroxide. Our initial experimentation with these fixatives was carried out during a transmission electron microscope study of *Scopulariopsis brevicaulis* (Sacc.) Bain., and we soon found that the same chemical concentrations and exposure periods were adequate for scanning electron microscope preparations as well. It was therefore convenient to fix material simultaneously for both electron microscopes following an identical process up to the 100 per cent ethanol stage, at which point the fungus is placed in either liquid freon or absolute acetone according to its ultimate destination (Figure 19.1). Figure 19.3 B, C and Figure 19.4 illustrate conidiogenous cells of *S. brevicaulis* prepared by this integrated technique. In Figure 19.3 B, C the annellated apices of two conidiogenous cells are

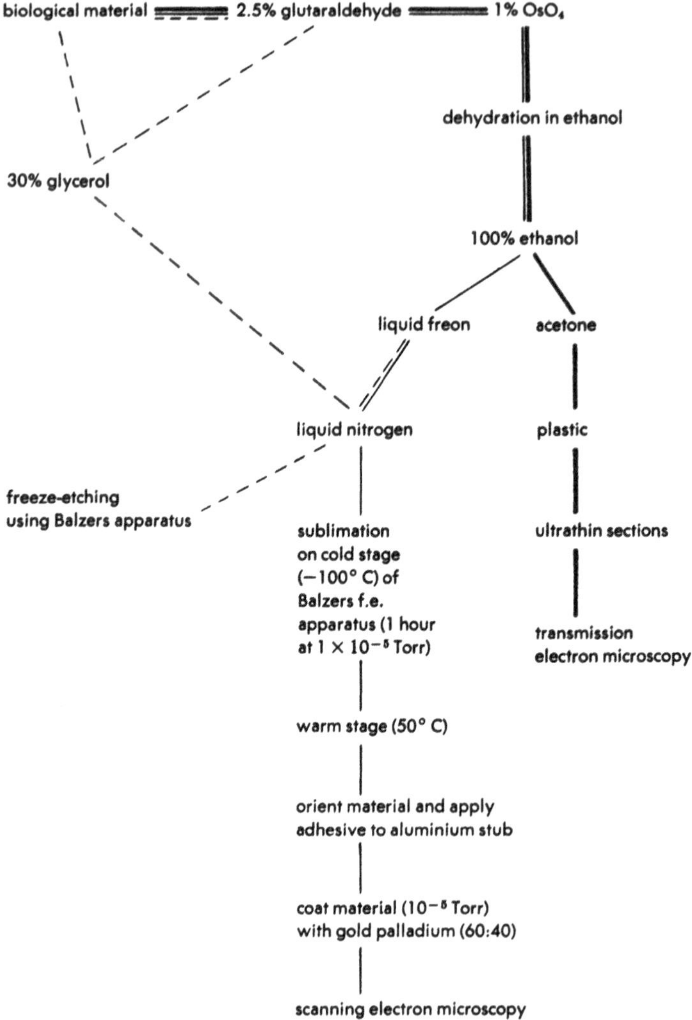


Figure 19.1 Diagrammatic explanation of the integrated techniques used in scanning and transmission electron microscopy and freeze-etching.

shown. Two annellations have been produced by the conidiogenous cell in B, and a third undehisced conidium remains at its apex. Successive wall layers are labelled a, b, and c and the conidium is labelled 3 to indicate its place in the sequence (see Figure 10.8). The annellated conidiogenous cell in Figure 19.3 C shows an apical zone of proliferation (zp) resulting from the formation of a new wall layer which caused the base of the terminal conidium (4) to separate partially from the previously formed wall layer (d). A comparable stage of development is illustrated in the transmission electron micrographs of *S. brevicaulis* (Figure 19.4). In these cases, a single annellation has formed when the outer wall of the conidium separated in a

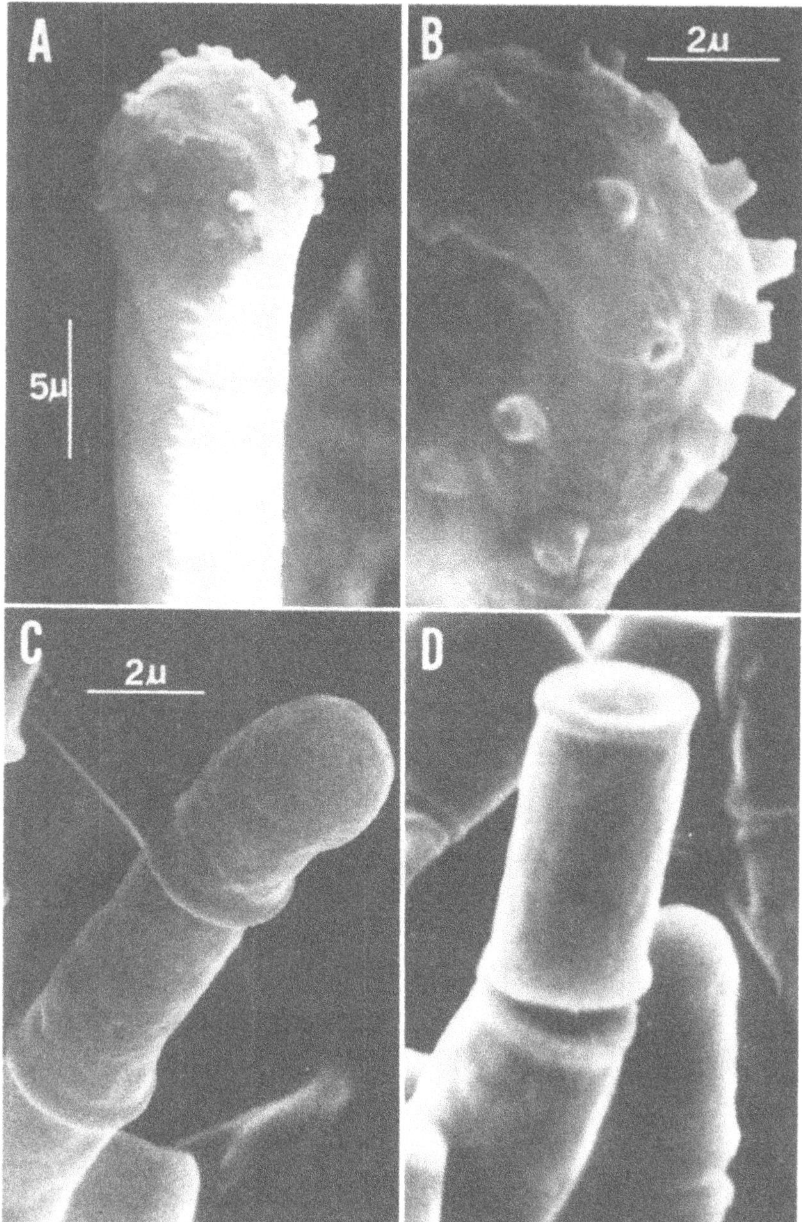


Figure 19.2 Scanning electron micrographs of several species of hyphomycetes. A,B: *Gonatobotryum apiculatum* (Peck) Hughes; A, conidiogenous cell showing apical ampulla and cluster of denticles; B, higher magnification of the same ampulla showing the cylindrical shape and apparently hollow nature of the denticles. C,D: *Geotrichum candidum* Link ex Persoon; C, septate, fertile hypha; D, arthroconidia formed by the disjunction of fertile hyphae.

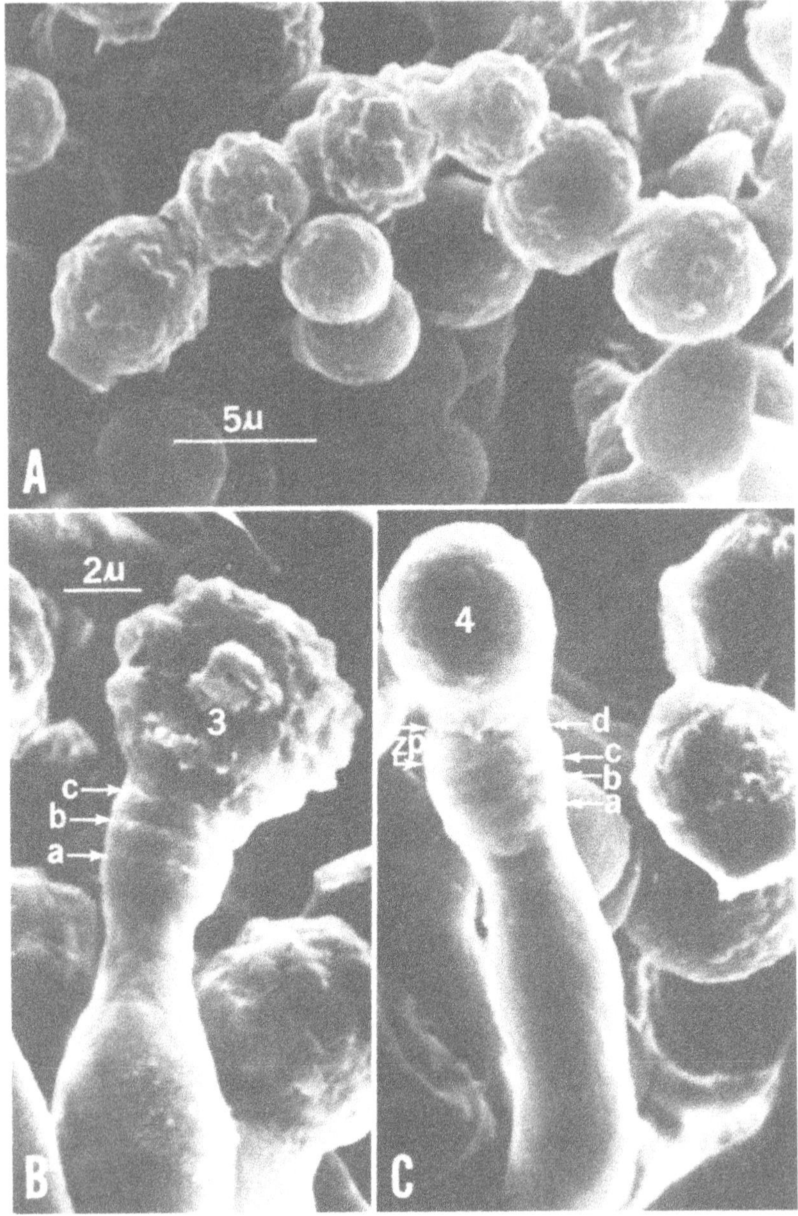


Figure 19.3 *Scopulariopsis brevicaulis* (Sacc.) Bain: A, chains of annelloconidia; B, C, annellate conidiogenous cells, each with an undehisced, terminal conidium. The wall layers and conidia are labelled to indicate their order of formation. A zone of proliferation (zp) is indicated at the apex of the conidiogenous cell in C.

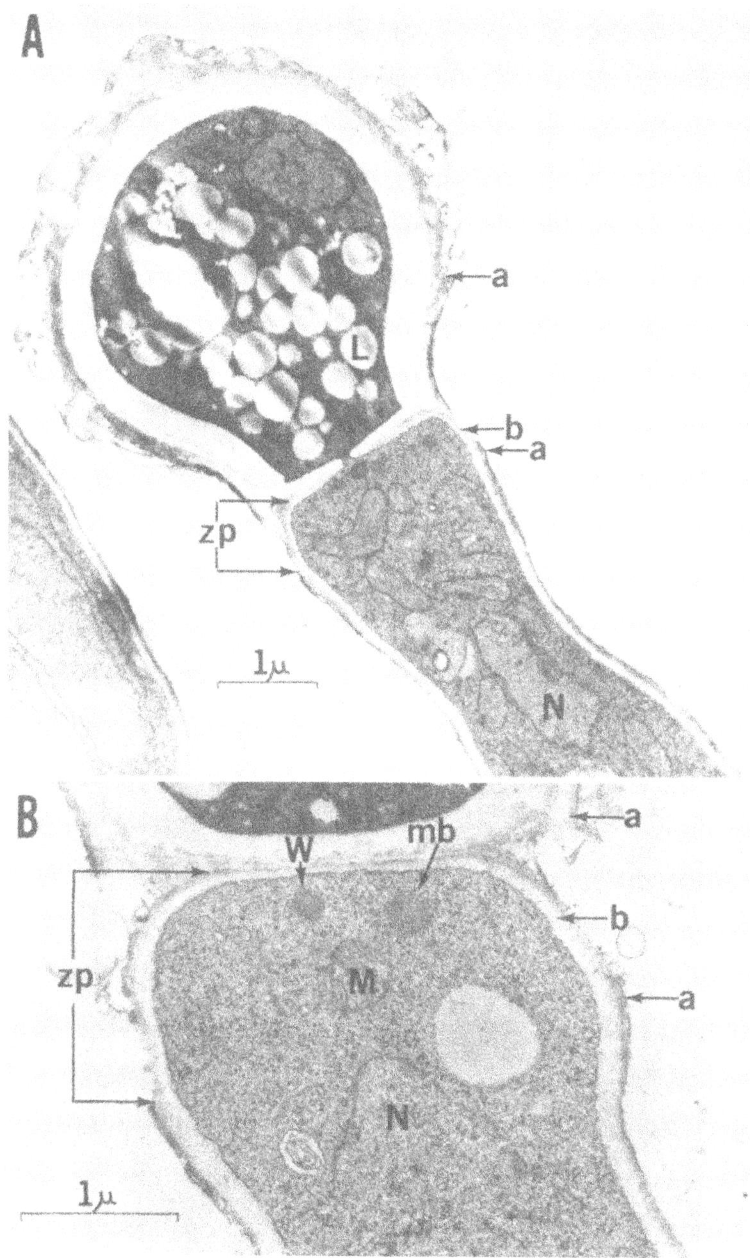


Figure 19.4 *Scopulariopsis brevicaulis*, transmission electron micrographs of proliferating conidiogenous cells and terminal conidia. Wall layers are labelled a and b to indicate their order of formation. N, nucleus; L, lipid droplet; W, woronin body; M, mitochondria; mb, microbody.

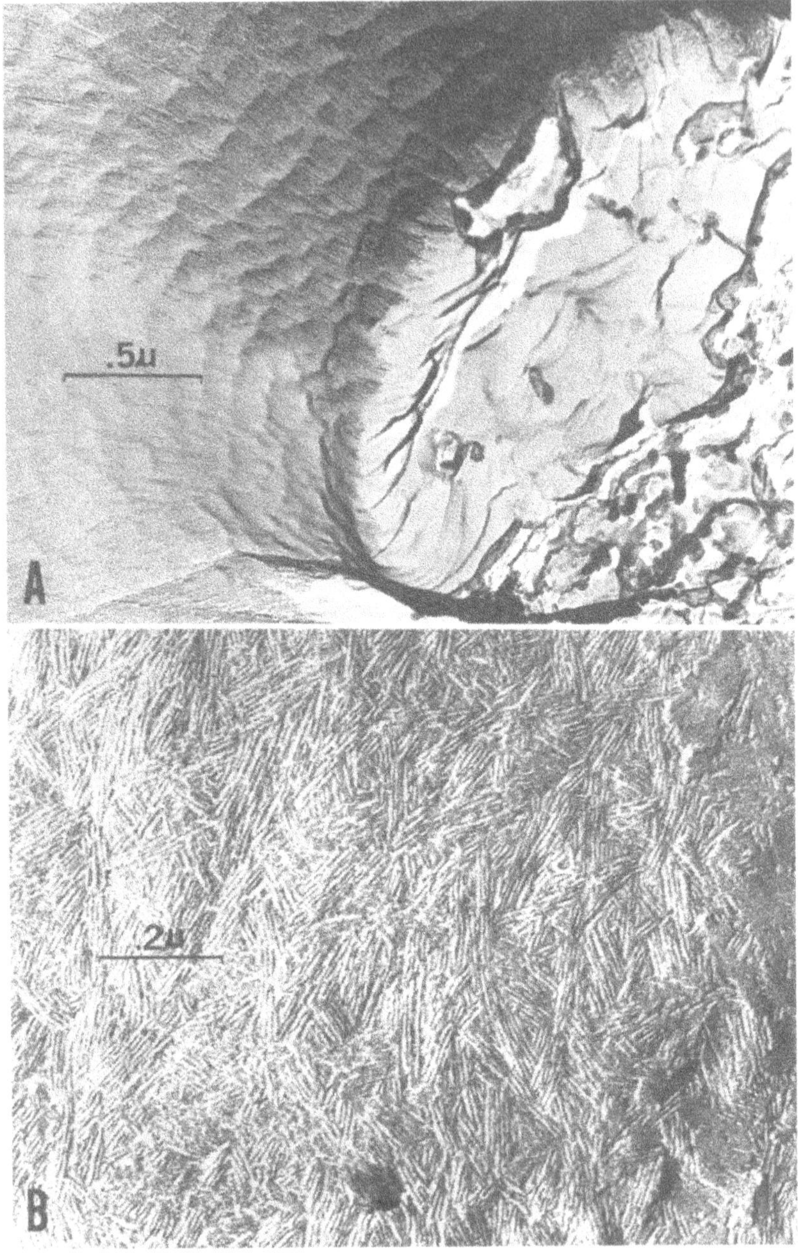


Figure 19.5 *Scopulariopsis brevicaulis*, frozen-etched conidia showing the nature of rodlet arrangement in the wall.

circumscissile manner from the outer wall of the conidiogenous cell. The prolifera- 
tion zones (zp) of the two conidiogenous cells are clearly visible.

We consider that the correlation of data from these two sources of ultrastructural information is both instructive and necessary. The stereoscan electron microscope, at least as it applies to our investigations of ultrastructure in hyphomycetes, is ineffective by itself in solving problems concerning the ontogeny of conidiogenous cells and conidia. However, the information obtained from both kinds of electron microscope studies enables us to build a three-dimensional, morphological-developmental concept.

More recently, our studies have included the use of the freeze-etching technique (Moor and Mühlethaler 1963) to examine fine structure of the wall layers of developing conidia. The fungal material is either placed initially in 3.5 per cent glutaraldehyde followed by 30 per cent glycerol, or immediately immersed in 30 per cent glycerol, before being frozen in liquid freon (Figure 19.1).

The conidia of *S. brevicaulis* are encased in two wall layers (Figure 19.4 A). The outer wall, which was initially continuous with the conidiogenous cell wall, appears to be degenerating as the conidium matures (Figures 19.3 and 19.4). The inner wall layer, which is laid down outside the plasma membrane of the differentiating conidium initial, becomes quite thick (Figure 19.4 A). In Figure 19.5 the "rodlets" (Hess, Sassen, and Ramsen 1966; Hess 1968; Hess and Stocks 1969; Sassen, Ramsen, and Hess 1967) comprising a part of the wall structure are illustrated using the freeze-etching technique. In order to determine the extent of the rodlet distribution in the two-layered conidium wall, we are now examining whole mount and deep-etched replicas.

We do not suggest that all the rather elaborate equipment mentioned here is always necessary for developmental-ultrastructural studies of microfungi; and we do wish to stress that one should be as cautious about interpreting development in terms of morphology at the ultrastructural level as one must be at the macroscopic level. Correlation of information from the electron microscopes and the light microscope in our work is mandatory; correlation studies using the three ultrastructural research tools listed above are certainly helpful.

## ACKNOWLEDGMENTS

We thank the Graduate School of the University of Florida and the National Science Foundation (GB 8748) for financial support.

## REFERENCES

Boyde, A., and Barber, V.C. 1969. Freeze-drying methods for the scanning electron-microscopical study of the protozoan *Spirostomum ambiguum* and statocyst of the cephalopod mollusc *Loligo vulgaris*. J. Cell Sci. 4: 223-39

Hess, W.M. 1968. Surface characteristics of *Penicillium* conidia. Mycologia 60: 290-303

Hess, W.M., Sassen, M.M.A., and Ramsen, C.C. 1966. Surface structures of frozen-etched *Penicillium* conidiospores. Naturwissenschaften 53: 708-9

Hess, W.M., and Stocks, D.L. 1969. Surface characteristics of *Aspergillus* conidia. Mycologia 61: 560-71

Horridge, G.A., and Tamm, S.L. 1969. Critical point drying for scanning electron microscopic study of ciliary motion. Science 163: 817-18

Moor, H., and Mühlethaler, K. 1963. Fine structure of frozen-etched yeast cells. J. Cell. Biol. 17: 609-28

Sassen, M.M.A., Ramsen, C.C., and Hess, W.M. 1967. Fine structure of *Penicillium megasporum* conidiospores. Protoplasma 64: 75-88

Small, E.G., and Marszalek, D.S. 1969. Scanning electron microscopy of fixed, frozen dried protozoa. Science 163: 1064-5

Zachariah, K., and Pasternak, J. 1970. Processing soft tissues for scanning electron microscopy: simplification of the freeze drying procedure. Stain Technol. 1: 43-5

## ADDENDUM: TECHNIQUES OF LIGHT MICROSCOPY WHICH MAY HELP TO INCREASE OUR UNDERSTANDING OF CONIDIUM ONTOGENY IN FUNGI IMPERFECTI

G.C. CARROLL

1 *Nomarski interference-phase optics.* This is a special optical system using polarized light and phase optics, capable of producing images of very thin optical sections of cells without optical interference from parts of the cell above or below the plane of focus. A shadowed three-dimensional effect is produced. This technique is particularly valuable for the elucidation of surface detail. The resolution is almost comparable with that obtained with the scanning electron microscope, and the Nomarski technique has the advantage that it can be used to examine living cells.

2 *Fluorescence microscopy.* Although of limited application, this should be useful for demonstrating such structures as annellations, because of its potential for differentiating old and new wall material. The material is soaked in a 0.5 per cent solution of primulin dye for 5 minutes, then washed several times in distilled water and observed on a microscope equipped with fluorescence equipment. Different colours are produced by different combinations of excitation and barrier filters. Not all material is stained by primulin, and a number of other dyes mentioned in the Zeiss manual on fluorescence microscopy are worthy of experimentation. The technique was initially, and very productively, applied to yeasts by Streiblová and Beran (1963) (see Figure 2.6).

# 20
# Conidium Shape*

BRYCE KENDRICK

At Kananaskis, I gave a very brief talk listing features, other than ontogeny, of conidia and conidiogenous cells which could be used to classify Fungi Imperfecti. Conidium shape was one.

The shape of an object is generally one of the first of its attributes to be used in describing it because, in our strongly visual approach to life, it is one of the first things we perceive. This has been equally true for microscopic objects like the conidia of hyphomycetes. Some groups of Saccardo's classification are distinguished by their production of "helico-," "stauro-," and "scoleco-spores." At the generic and specific levels much more limited or discrete shape concepts are used: spherical, ovoid, ellipsoid, biconic, curved, flame-shaped, etc. We have tended to use these characters without questioning, because they were convenient and accessible. But as the taxonomic spotlight shifts from mature morphology to ontogeny, we should re-examine shape from two angles: (i) What are the mechanisms which produce the myriad of shapes we know? (ii) How stable a feature is shape - how much is it genetically controlled, and therefore how taxonomically valid is it?

I have an answer of sorts for the first question. As to the genetic control of the factors involved, I know nothing. But the postulation of a possible mechanism must precede any answer to the second question. Let me say at once that I am concerned here only with the over-all shape of a spore, not with acellular appendages, outgrowths, or ornamentations. Further, I am concerned only with conidia

Department of Biology, University of Waterloo, Waterloo, Ontario, Canada.

*I wrote this brief chapter in hopes of stimulating a little rethinking and a little new thinking among students of the Fungi Imperfecti. I have subsequently discovered Professor N. F. Robertson's presidential address to the British Mycological Society (*Trans. Brit. Mycol. Soc.* 48 [1965]: 1-8) in which he makes some comparable suggestions. I am pleased to have arrived independently at somewhat similar views to those held by Professor Robertson, but I am disturbed that relatively little progress appears to have been made in solving the problems he raised in 1965.

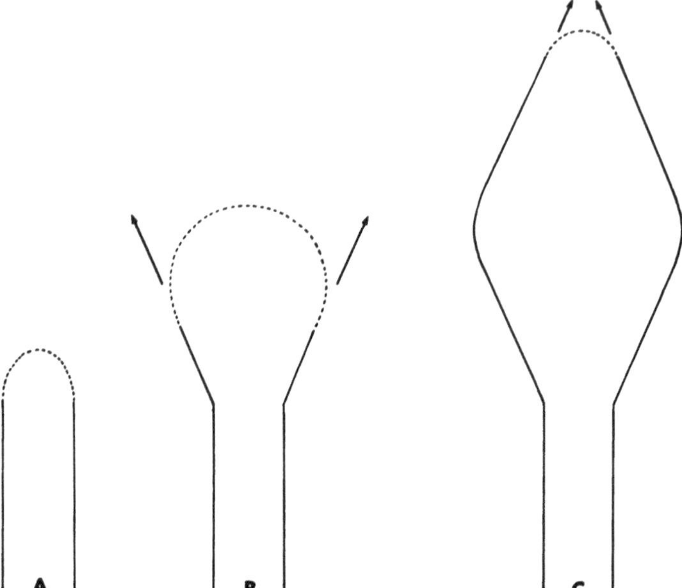


Figure 20.1 A: Diagram of hyphal tip; the dotted line indicates the hemisphere of "unset" wall. B: During spore widening, wall setting lags behind blowing out, and there is more than a hemisphere of unset wall. Wall setting is proceeding along the tangents shown by the arrows. C: During spore narrowing, wall setting overtakes blowing out, there is less than a hemisphere of unset wall, and wall setting is moving along the tangents shown by the arrows.

which are produced entirely by apical growth; this excludes those in which there is considerable secondary swelling or intussusception of new wall material, as in *Basipetospora, Cladobotryum,* and *Amblyosporium.*

A simple analysis of the situation with which I am concerned here will show that one fluid (protoplasm) is bounded by a membrane (the cell wall) beyond which is another fluid (air or water). As far as the cell wall is concerned, I assume for the sake of simplicity in model-building that it may exist in two conditions, which I will call "unset" and "set." For the purposes of this discussion, the "set" wall is inelastic and incapable of further deformation except by rupture due to trauma or to secondary enzymatic softening. The "unset" wall is elastic and can yield and stretch in response to pressure differentials. One further assumption is necessary for simplicity's sake: that there is at any given moment a sharp line of demarcation between set and unset wall (in reality there must usually be a zone rather than a line).

The first and simplest case is the growing hyphal tip (Figure 20.1 A), but remember that all other fungal manifestations at the apices of hyphae are derived from this case. The hyphal tip can be represented as a cylinder of set wall, with a hemisphere of unset wall (the shape dictated by surface tension) at its end. As hyphal growth proceeds, fluid pressure generated within the hypha finds expression at the unset tip, tending to blow it out like a balloon. If the diameter of the hypha is to remain relatively constant, then wall setting must move forward at exactly the same rate as the tip is blowing out.* Conidium production (and the production of many other basically hyphal structures) may be regarded as a result of a change or series of changes in the relationship between rate of blowing out and rate of wall setting. Let us now consider a number of specific examples. 

1. The spherical conidium. Here I assume that while blowing out continues, wall setting ceases temporarily. When the conidium has reached full size, rapid and uniform wall setting occurs all over its surface, and the internal growth pressure either ceases or is diverted elsewhere. The area of the hyphal tip which remains or becomes unset at the beginning of this process may vary from the full width of the hypha, as in the case just outlined, down to a very small circular spot (Figure 15.1), and all intermediate cases may be found, giving rise to conidia with basal scars of different widths - in fact, everything from Subramanian's "gangliospore" to Hughes's "blastospore."

When a conidium is not ultimately spherical (the shape dictated by surface tension, minimizing surface area for any given volume), I assume this deviation to be due either to differential growth or to differential wall setting. Since the latter is much easier to deal with in a model system, I shall expand on its various expressions.

2. The long, cylindrical conidium. This may originate from a narrow base, but essentially it repeats the phenomenon of hyphal growth: tip blowing out just keeping pace with wall setting.

3. The conidium widest in the middle, tapered towards both ends. This shape may be due to two shifts in the relationship between rate of wall setting and rate of blowing out. Initially, wall setting lags behind blowing out by a slight but relatively constant amount. There will thus be more than a hemisphere of unset wall (Figure 20.1 B), and setting will thus move along the tangent drawn in the figure. Later, the rate of wall setting overtakes the rate of blowing out, and exceeds the latter by a slight but relatively constant amount. Now there will be less than a hemisphere of unset wall (Figure 20.1 C), and setting will move along the tangent drawn in the figure, steadily narrowing the conidium to its apex.

4. The curved cylindrical conidium. The curvature may be the result of slightly advanced wall setting on one side of the growing apex (the outside of the curve), or slightly retarded wall setting on the other (the inside of the curve), or a combination of both processes going on simultaneously.

5. The "helicospore," coiled in two dimensions. When the differential in wall-setting rates on opposite sides of the apex gradually becomes more extreme, and is so maintained over an extended period of elongation growth, the conidium may become spirally coiled in two dimensions.

*Note how different this process is from the "allometric" growth of most higher plant cells, in which there is increase in both length and width, while the relative growth rates of the two dimensions remain constant.

6. The "helicospore," coiled in three dimensions. In this case, the asymmetry of wall setting must be in two directions: skewed "back/front" as well as "right/left."

It seems to me that all these examples, and more, could be analysed mathematically, using calculus and fluid mechanics; that the dynamic processes of growth and the changing equilibria between blowing out and wall setting could be simulated on an analogue computer; and that ultimately the techniques of computer graphics could be tapped to give visual simulations whose driving functions could be rapidly corrected as the simulations approached or diverged from the desired pattern. In this way we may arrive at some understanding of the forces involved in conidium formation. Work on such computer models is proceeding in my laboratory.

George Wald (*Scientific American*, September 1958) wrote: "Confronting any phenomenon in living organisms, the biologist has always to ask three kinds of questions, each independent of the others: the question of mechanism (how does it work?), the question of adaptation (what does it do for the organism?), the twin questions of embryogeny and evolution (how did it come about?) ... one really understands only when all three have been answered."

For the Hyphomycetes, the answers lie in the future, but we must at least begin to ask the questions.

# Postscript

The Kananaskis Conference has surely set a world-wide seal of approval on the ontogenetic approach to the systematics of the Fungi Imperfecti suggested long ago by Vuillemin, and carried forward by Mason, but pioneered for our generation by Dr Stan Hughes in his 1953 paper, "Conidiophores, Conidia, and Classification" (*Can. J. Botany,* 31:577-659). As should be apparent from the early chapter headings in this book, it was essentially along the lines Dr Hughes suggested in 1953 that the subject matter of the conference was initially divided. That we came out of the meeting with a considerably modified view of the system is no discredit to Dr Hughes's scheme, but rather a tribute to the intellectual stimulus his thought has provided for many of us. It was with great pleasure that I, as one of Dr Hughes's colleagues and collaborators, learned during the XIth International Botanical Congress, which immediately preceded our conference, that his work had been recognized by the award of the Jakob Eriksson Gold Medal for Mycology. His colleagues throughout the world will surely wish to join me in congratulating him on this well-deserved honour.

At Kananaskis many new and untried ideas were tentatively tossed into the ring, or dogmatically declaimed. These ideas - freely sprinkled throughout both the keynote addresses and the discussions - are often very thought-provoking. They may be debated, accepted or rejected, praised or scornfully spurned, by the coming generation of mycologists, but I do not think they can be ignored. There are theories waiting for enthusiastic graduate students to test - questions or suggestions that could easily become the subjects of M.Sc. or Ph.D. theses.

Among the achievements of the conference, I would list the following:

1. For the first time we have recognized and accepted the fact that there is often a clear separation between those ontogenetic features involved in the production of an individual conidium and those ontogenetic features which permit a conidio-

genous cell to produce a succession of conidia. No longer will we say dogmatically that a specific kind of conidiogenous cell called a "sympodula" produces a specific kind of conidium called a "sympodioconidium." We have now devised separate terminologies for these two aspects of conidiogenesis; and I am convinced that, if those who study Fungi Imperfecti will seek conscientiously to elucidate both mechanisms and to apply and improve the terminologies we have supplied, our knowledge and understanding of these fungi will make even more rapid progress than it has in the last few decades.

2, We have recognized basic underlying similarities in wall-layer involvement in methods of conidium ontogeny that at first sight appear quite different. We have used these similarities in an experimental grouping of ontogenetic types.

3. We have devised what we hope is a more logical and more easily understood terminology for describing the various aspects of conidium ontogeny than has hitherto been available. This does not mean, of course, that the new terminology will be easy to apply, or that it could not be improved.

Detailed and painstaking ontogenetic studies will be required before some fungi will be sufficiently well understood to place them in the scheme. The techniques of time-lapse photomicrography and transmission electron microscopy, especially if applied in parallel, show great promise of supplying much of the necessary information. An enormous amount of work remains to be done before the ideas propounded at the Conference can be fully vindicated (or rejected!). I hope this book will have stimulated some students of mycology to help in the attempts to grapple with that fascinating but enigmatic group of organisms, the Fungi Imperfecti.

B.K.

# Members of the Conference

Girish C. Bhatt, Department of Biology, University of Calgary, Calgary 44, Alberta, Canada

J.W. Carmichael, Mold Herbarium and Culture Collection, University of Alberta, Edmonton, Alberta, Canada

George C. Carroll, Department of Biology, University of Oregon, Eugene, Oregon, 97403, USA

Garry T. Cole, Department of Biology, University of Waterloo, Waterloo, Ontario, Canada

Wm. Bridge Cooke, 1135 Wiltshire Ct., Cincinnati, Ohio, 45230, USA

J. Leland Crane, Illinois Natural History Survey, Natural Resources Building, Urbana, Illinois, 61801, USA

Martin B. Ellis, Commonwealth Mycological Institute, Ferry Lane, Kew, Surrey, England

A. Funk, Forest Research Laboratory, 506 West Burnside Road, Victoria, British Columbia, Canada

Roger D. Goos, Department of Botany, University of Hawaii, Honolulu, Hawaii, 96822, USA

Gregoire L. Hennebert, Laboratoire de Mycologie systématique et appliqué, Université de Louvain, Parc d'Arenberg, Heverlee, Belgium

Gilbert C. Hughes, Department of Botany, University of British Columbia, Vancouver 8, British Columbia, Canada

Stanley J. Hughes

椿 啓介

J. Leland Crane

Gilbert C. Hughes

G. L. Hennebert

George Carroll

Roger D. Goos

Wm. Bridge Cooke

Flora G. Pollack

Garry T. Cole

Bryce Kendrick

Stanley J. Hughes, Mycology Unit, Plant Research Institute, Canada Department of Agriculture, Central Experimental Farm, Ottawa, Ontario, Canada

Bryce Kendrick, Department of Biology, University of Waterloo, Waterloo, Ontario, Canada

Emil Müller, Institut für Spezielle Botanik, Eidg. Technische Hochschule, Universitätstrasse Z, CH-8006 Zurich 6, Switzerland

T.R. Nag Raj, Department of Biology, University of Waterloo, Waterloo, Ontario, Canada

Mme Jacqueline Nicot, Laboratoire de Cryptogamie, Muséum National d'Histoire Naturelle, 12 rue de Buffon, Paris V$^{e}$, France

Kris A. Pirozynski, Mycology Unit, Plant Research Institute, Canada Department of Agriculture, Central Experimental Farm, Ottawa, Ontario, Canada

Mrs Flora G. Pollack, Mycology Investigations, Plant Industry Station, Beltsville, Maryland, 20705, USA

C.V. Subramanian, Centre for Advanced Studies, University Botany Laboratory, Chepauk, Madras-5, India

Keisuke Tubaki, Institute for Fermentation, Juso-Nishino-cho, Higashiyodogawa-ku, Osaka, Japan

## ADDITIONAL CONTRIBUTING AUTHORS

Henry Aldrich, Department of Botany, University of Florida, Gainesville, Florida, 32601, USA

Mrs Fanny E. Carroll, Graduate Student, Department of Biology, University of Oregon, Eugene, Oregon, 97403, USA

Miss Ming-en Grace Chang, Department of Biology, University of Waterloo, Waterloo, Ontario, Canada

Philip Metitiri, Department of Biology, University of Waterloo, Waterloo, Ontario, Canada

Brian C. Sutton, Commonwealth Mycological Institute, Ferry Lane, Kew, Surrey, England

K. Zachariah, Department of Biology, University of Waterloo, Waterloo, Ontario, Canada

www.ingramcontent.com/pod-product-compliance
Lightning Source LLC
Chambersburg PA
CBHW080618030426
42336CB00018B/3011

* 9 7 8 1 4 8 7 5 9 2 2 3 3 *